Systema Naturae 250
The Linnaean Ark

Systema Naturae 250
The Linnaean Ark

Edited by Andrew Polaszek

CRC Press
Taylor & Francis Group
Boca Raton   London   New York

CRC Press is an imprint of the
Taylor & Francis Group, an **informa** business

Cover image of "La fontaine d'os" by surrealist Wolfgang Paalen (1907–1959) used with permission of Paalen Archiv Berlin.

CRC Press
Taylor & Francis Group
6000 Broken Sound Parkway NW, Suite 300
Boca Raton, FL 33487-2742

© 2010 by Taylor and Francis Group, LLC
CRC Press is an imprint of Taylor & Francis Group, an Informa business

Library of Congress Cataloging-in-Publication Data

Systema naturae 250 : the Linnaean ark / editor, Andrew Polaszek.
    p. cm.
  Includes bibliographical references and index.
    ISBN 978-1-4200-9501-2 (hardcover : alk. paper)
  1. Linné, Carl von, 1707-1778. Systema naturae. 2. Animals--Nomenclature. 3. Biology--Classification. I. Polaszek, Andrew. II. Title: System naturae two hundred fifty.

QH44.S97 2010
578.01'2--dc22
                                                             2009047098

Visit the Taylor & Francis Web site at
http://www.taylorandfrancis.com

and the CRC Press Web site at
http://www.crcpress.com

Contents

Preface

The opportunity to prepare to celebrate a quarter millennium of the scientific naming of animals coincided with my position as Executive Secretary of the International Commission on Zoological Nomenclature (ICZN) from January 2003 to December 2007. For those entirely unfamiliar with the role, or even the existence of ICZN, it may be helpful to briefly put in context the *Systema Naturae* 250 event that took place in Paris in August 2008, and this resulting collection of essays.

ICZN, the *Commission*, was formally established in 1895, having evolved from several previous organizations or associations that had long-term stability of organismal nomenclature as their mission. ICZN's greatest achievement is the adoption, by the majority of zoological taxonomists, of its series of codes of zoological nomenclature from 1905 to 1999 and beyond. Regulating the way the scientific names of animals are formed, published, and used directly influences diverse areas of human activity, but primarily taxonomy, and consequently those fields in which taxonomy is important. Zoological nomenclature and its rules therefore provide either direct or less direct support for such fields as conservation, biological control, biodiversity studies, quarantine regulations, and the importation of animal products from agriculture and fisheries, to animal products controlled or banned under the Convention on International Trade in Endangered Species (CITES). Despite this pivotal role, the Code of Zoological Nomenclature should not overly influence activity in these fields, any more than the highway code should dictate progress in vehicle design or advances in transportation.

Binominal nomenclature for the scientific names of animals is deemed to have started on January 1, 1758, with the 10th edition of Linnaeus's *Systema Naturae*. Other central principles of animal nomenclature such as the principle of priority and that of typification came later, and owe little to Linnaeus. The continuum in the history of natural history, and organismal nomenclature in particular, extends before Linnaeus to John Ray and Joseph Pitton de Tournefort, among others, but the fact that modern zoological nomenclature begins officially with the 1758 edition of *Systema Naturae* ties animal nomenclature inseparably to Linnaeus. For this reason, we in ICZN saw 2008—250 years later—as a perfect opportunity to celebrate the scientific naming of animals.

Organismal nomenclature and its associated rules are viewed by many as dull, dry, outside the realm of science, and less and less relevant to modern methods in zootaxonomy. This idea seems to be based largely on a widespread misconception of the purposes of both the Code and ICZN as somehow aiming to restrict activity in zoological taxonomy by imposing authoritarian and arbitrary rules and decisions. In fact, the Code has been virtually universally adopted and followed by zoologists since 1905, testimony to the willingness of most zoological taxonomists to adopt, share, and conform to this universal set of regulations. Far from being authoritarian and imposing, the zoological Code, as with the highway code mentioned above, is an excellent example of a set of regulations, formulated after more than a century of open debate, that provides essential guidance for its users. If there are problems with the Code—and there are many—these stem from it being too permissive. With each zoological code, and the debates and discussions that led to them, there is an inherent conflict between a willingness to establish and improve the rules (by definition restrictive) of nomenclature, and a desire to maintain freedom of nomenclatural (and by extension taxonomic) expression. This conflict is evident in the longstanding and ongoing discussions regarding the necessity for type specimens and, more recently, the acceptable criteria for, and definitions of, hardcopy and electronic publication.

The *Commission* is an international body of experts, elected by other zoologists, whose current role is to lead the writing and editing of successive editions of the Code and to rule on the 40 or so yearly "cases" that require its ruling. Much of the work of the Commission, both in preparing successive editions of the Code and in dealing with the annual applications for ICZN decisions, results from elements of the Code that essentially allow work that is of a standard that would be unacceptable in any other scientific discipline to manage to conform to its rules (i.e., to be Code-compliant). These substandard nomenclatural acts, albeit a tiny minority, thus enter the *corpus* of zoological nomenclature, and therefore have to be dealt with sooner or later. Commissioners are not rewarded in any material way for their painstaking work in attempting (mostly successfully) to resolve this plethora of nomenclatural problems. Their reward is the knowledge that they are contributing to nomenclatural stability and removing obstacles to progress in zootaxonomy. My experience dealing with the Commission as ICZN Executive Secretary led me to conclude that their vast knowledge, experience, and ability would be far better employed in addressing the major issues facing nomenclature and zootaxonomy, such as online-only journals, copyright of taxonomic works, registration of names and nomenclatural acts, and the requirement for type specimens.

It was therefore with a view to addressing these major issues that the program for the *Systema Naturae 250* symposium was developed. The 20th International Congress of Zoology took place in Paris, August 26–29, 2008. *Systema Naturae 250* was one of many symposia contributing to the congress, and was spread over the first two days. While the overarching theme of the symposium was the celebration of 250 years of the scientific naming of animals, within this broad theme several more specific areas were addressed. The historical aspect, central to this celebration, was visited in a series of presentations beginning with Professor Edward O. Wilson's introductory presentation on the major historical trends in biodiversity studies, followed by David Quammen's essay on Linnaeus's life, and Jim Dobreff's highly original account of Daniel Rolander, one of Linnaeus's apostles, dubbed by Jim "the invisible naturalist." Ellinor Michel (ICZN's current executive secretary) and I focused on Charles Davies Sherborn's contribution to the emergence of biodiversity informatics. Hans Sues addressed fossils in the context of Linnaean classification, and Gordon Reid looked at animal nomenclature in the context of conservation. The future of taxonomy was inevitably a major theme of the symposium, addressed by several invited speakers: Charles Godfray, Norm Johnson, David Patterson, Rich Pyle, David Remsen, David Schindel, Simon Tillier, Alfried Vogler, and Quentin Wheeler. These authors highlighted current advances in various technologies from large-scale databasing through imaging technology to DNA sequencing that will both enhance and accelerate taxonomy. Examples of progress in particular taxonomic areas or ecosystems included molluscs (Philippe Bouchet), flies (Thomas Pape), and freshwater biodiversity (Koen Martens).

Electronic publication of taxon descriptions and other nomenclatural acts, and the related issue of copyright of those acts, were addressed by Sandra Knapp and Donat Agosti, respectively. Zhi-Qiang Zhang, founding editor of *Zootaxa*, contributed a brief history of that very successful young journal. The current amount of debate about these issues on various discussion sites would fill several volumes such as the present one, as both subjects are absolutely critical to progress in nomenclature and taxonomy. Two global information cataloguing projects, one already well-established (Species2000) and one in its infancy (Encyclopedia of Life) were covered by Frank Bisby and Jim Hanken, respectively.

As editor, in asking for contributions both to the Paris symposium and to this resulting volume, I tried to give authors maximum freedom regarding their chapters, especially in terms of page length. The consequences of this approach are immediately obvious. Reviewers will, with justification, point to the unevenness of the length of the chapters, from Ed Wilson's brief overview of biodiversity studies to Benoît Dayrat's epic history of zoological nomenclature. Finally, Fred Ronquist's account of the monumental Swedish Malaise Trap Project brings matters full circle, following precisely in the Linnaean tradition in a way that other countries might consider emulating.

To all of the authors and co-authors who contributed to the Paris meeting and to this volume I extend my very grateful thanks. I also thank the following for peer-review of many of the chapters published here: Philippe Bouchet, Mark Costello, Gathorne Cranbrook, Benoît Dayrat, Sandy Knapp, Shai Meiri, Sandro Minelli, Malcolm Scoble, and Vincent Smith.

The meeting also provided a unique opportunity to honor two major figures in biodiversity informatics. The presentation of the Sherborn Award (Figure 0.1) was aimed to demonstrate appreciation

FIGURE 0.1 The Sherborn Award, a silver sculpture of the perciform fish *Howella sherborni* (Norman), named for Sherborn by his biographer ichthyologist J. R. Norman. The sculpture was commissioned from British sculptress Juliet Simpson, who examined the holotype and other specimens of *H. sherborni* in the Natural History Museum.

both for its namesake, Charles Davies Sherborn, and for its first-time recipient Professor Alessandro Minelli. The award was presented by Professor E.O. Wilson to Professor Minelli in the presence of Hans Odöo (in his persona of Linnaeus himself, see Figure 0.2) and the Director of the Paris Museum M. Bertrand-Pierre Galey.

Alessandro Minelli, born in Treviso, Italy, is professor of zoology at the University of Padova. Until the mid-1990s his main scientific interests were myriapod phylogeny and taxonomy, as well as the principles and methods of biological systematics. He is a past president of ICZN and has served as a Commissioner for 18 years, through some of the Commission's most difficult periods, including the drafting of the 4th Code. He was head of the organizing committee and chief editor of the *Fauna d'Italia* project, coordinating the work of 273 specialists. The work was completed within four years, in 1995, with a list of some 57,000 species. He then mobilized colleagues from many different European countries to extend the project to the whole of Europe, and to what became the very successful *Fauna Europaea* project. During the last decade, his main research interests have turned toward evolutionary developmental biology ("Evo-Devo"), with emphasis on the origin and evolution of appendages and segmentation. His most recent contribution to the diffusion of taxonomic information is ChiloBase, the Web-based catalogue of world centipedes, developed with the aid of a team of colleagues from all continents under the aegis of the Global Biodiversity Information Facility (GBIF).

The Sherborn Award itself is a silver sculpture of the perciform fish *Howella sherborni* (Norman), named for Sherborn by his biographer, ichthyologist J.R. Norman. The sculpture was commissioned by the International Trust for Zoological Nomenclature from British sculptress Juliet Simpson. She examined the holotype and other specimens of *H. sherborni* in the Natural History Museum,

FIGURE 0.2   The Sherborn Award presented by Professor E.O. Wilson (right) to Professor Minelli (left) in the presence of Hans Odöo (as Linnaeus himself) and the Director of the Paris Museum, M. Bertrand-Pierre Galey.

London, to perfect the final details of the sculpture, and Juliet and I are both grateful to fish expert Oliver Crimmen of the Department of Zoology for facilitating this work.

The *Systema Naturae* symposium took place in the Grand Amphithéâtre du Muséum, situated a short distance from the main Museum National d'Histoire Naturelle (MNHN) building within the Jardin des Plantes. The streets in this part of Paris, within the 5th Arrondissement, are named Rue Buffon, Rue Cuvier, Rue Geoffroy Saint-Hilaire, Rue Lacépède, Rue Linné, Rue Tournefort—the list honoring great biologists goes on. For the congress participants, it was sheer pleasure to visit the cafés and restaurants in an area commemorating these figures while discussing their ideas in a modern context. The great Lamarck, whose statue stands at the west end of the Jardin des Plantes (Figure 5.3; see color insert following page 16), is celebrated with Rue Lamarck in Montmartre, running next to the Basilique du Sacre Coeur and parallel to Rue Darwin. Great scientists are honored in Paris, as they are in capital cities of the world from Mexico City to Warsaw. To find Darwin Street in London, one has to seek out Southwark, SE17 (in the vicinity of Elephant & Castle); Darwin Road is in Haringey, N22.

While celebrating a quarter millennium of the scientific naming of animals, the *Systema Naturae* 250 symposium brought attention to most of the issues currently facing animal nomenclature, taxonomy, and the publication of their products. The zoological community is clearly united in recognizing a need to both facilitate and accelerate the description of biological taxa, and the zoological Code is of central importance in this facilitation process. Modification of the Code to accommodate publication of electronic-only zootaxonomic and nomenclatural acts is an urgent step in this process, but also appears to be divisive. The central issue is the adoption of a registration system for these electronic-only zootaxonomic acts. Even in a pre-Internet world, the usefulness of a registration service for such acts is apparent, but for our cybertaxonomic future, registration is essential.

Most concerns regarding electronic-only publication of taxonomy appear to be based on doubts concerning permanence and archiving of such publications. I am convinced that electronic archiving is potentially as permanent, and often probably safer than hardcopy archiving. In 2005 a group of 29 scientists published a commentary in *Nature* advocating a universal open-access register for the scientific names of animals, to be called *ZooBank*. The original proposal was for mandatory registration of all nomenclatural acts in zoology. Regarded as draconian by some, it is self-evident that only *mandatory* registration achieves completeness of the register, and an incomplete *ZooBank* would be of far less value to the zootaxonomic community. Currently, *ZooBank* is still under development, but it is hoped that as we move further toward Web-based taxonomy, it will assume a role there as central as Genbank currently is to the fields of molecular evolution and molecular phylogenetics. To achieve this, registration must be a straightforward, user-friendly process that maximizes benefits for those registering taxa and nomenclatural acts, and those using *ZooBank* to study those acts, many of whom also belong to the former category.

Advances in taxonomy and systematics are currently occurring on many connected fronts. DNA sequencing technology is improving alongside specimen imaging capabilities, and the resulting data are being fed into matrix-based applications such as the MX system developed by Matt Yoder, or within the scratchpads developed by Vince Smith. The U.S. National Science Foundation is supporting progress via its Planetary Biodiversity Inventories (PBIs) and Partnerships for Enhancing Expertise in Taxonomy (PEET) program, while the European Union's 6th Framework Programme has funded the European Distributed Institute of Taxonomy (EDIT) project, within which the scratchpads evolved. These advances are bringing true Web-based taxonomy closer to Charles Godfray's original vision. Many of the problems associated with the advent of Web-taxonomy that have been predicted by its detractors can be solved by mandatory registration in *ZooBank*. Other problems are less easily soluble, for example, the description of taxa that are fully Code-compliant while being otherwise highly questionable—in extreme circumstances, fakes. There is a continuum stretching from zootaxonomic work of the highest standards possible published in the best-quality journals or monographs, down to dubiously delineated taxa often published in poor-quality, non-peer-reviewed journals. The latter kind of output has been described as "taxonomic vandalism." Unfortunately, there are several incentives for unscrupulous people to publish work of this kind, from boosting their publication list to profiting financially by selling names or "naming rights" of taxa purportedly new to science. It is unlikely that registration in *ZooBank* alone could solve this problem, and some have advocated a system of certification that would be recognized under the Code, which, together with registration in *ZooBank*, would have the potential to eliminate many of the opportunities for taxonomic vandalism, either deliberate or due to ignorance. As with many issues in modern taxonomy, opinions are deeply divided, due partly to the natural conservatism of many of those involved in taxonomy.

One of the most controversial of such issues is copyright of taxonomic work, in particular, descriptions of new taxa. Donat Agosti has long championed the idea that all taxonomic work should be free from any form of copyright. His position that all taxonomy, past, present, and future, should be completely and freely open-access is again one that has generated divided opinions. In my experience, many enlightened commercial publishers agree with his view in principle and are eager to develop ways to achieve this goal.

In summary, *Systema Naturae* 250 gave us a rich Linnaean smörgåsbord of presentations, resulting in the chapters herein. The unifying theme—a celebration of a quarter millennium of the scientific naming of animals—provided us with a unique opportunity to focus on where zoological taxonomy should be at the start of the 21st century. We are now in the midst of a taxonomic revolution through which our science can advance proportionately much further than much of its quarter-millennium history would otherwise suggest. But for this to happen requires an unprecedented level of unified, concerted action by us, the practitioners and publishers of zoological taxonomy. We must work together to get as close as possible to universal agreement within our community on the critical issues of electronic publication, copyright, registration, and type specimens. By achieving this, we will guarantee that the next quarter millennium produces zoological taxonomy to a standard far surpassing even the greatest achievements of the last 250 years.

I thank the following for their institutions' generous financial support of the Paris meeting, without which neither the meeting nor this volume would have been possible: Jim Edwards (EOL); Hugo von Linstow and David Remsen (GBIF); Richard Greatrex (Syngenta Bioline), Richard Lane (NHM), David Cutler, Gren Lucas, Malcolm Scoble, and Vaughan Southgate (Linnaean Society of London); John Sulzycki (Taylor & Francis, CRC Press); Gordon MacGregor Reid (North of England Zoological Society); Dale Seaton (Elsevier); and Martin Smith (Syncroscopy). Also Jean-Marc Jallon (Université Paris-Sud) of the organizing committee of the 20th International Congress of Entomology.

The front cover image is "La fontaine d'os" by the surrealist painter Wolfgang Paalen (1907–1959). I thank the following for helping me through the process of obtaining the copyright from the Paalen Archiv Berlin, and for producing the image from the original painting in the Kahlo Museum, Mexico City: Alejandra López, Alejandro Zaldívar Riverón, Andreas Neufert, Hilda Trujillo, and Catherine and Tessa Farmer.

Finally, I thank my wife Maud and sons Tim and Philip for their patience during my organizing the Paris meeting and editing this volume.

Andrew Polaszek
London

Editor

Andrew Polaszek is a researcher in entomology at the Natural History Museum, London. He specializes in parasitoid wasps, especially chalcidoids, but also has an active interest in bees, ants, and other Hymenoptera. He has participated in several classical biological control programs with parasitoid wasps, including the successful introduction of two species into Trinidad against the citrus blackfly, a major pest.

Dr. Polaszek completed his PhD at Imperial College, London, before joining the Commonwealth Institute of Entomology in 1985. From 1990–1994 he was Principal Investigator in the Department of Entomology at Wageningen Agricultural University (the Netherlands) studying cereal pests and their natural enemies in sub-Saharan Africa. In 2004 he was appointed Executive Secretary of the International Commission on Zoological Nomenclature (ICZN). At ICZN he was involved in the establishment of the ZooBank register of animal names and nomenclatural acts, and as a consequence, maintains a strong interest in the development of the Code, especially in relation to electronic publication of systematic work, including Web-based taxonomy. He has authored more than 100 scientific publications, edited the book *Cereal Stem Borers in Africa* (CABI Bioscience, 1998) and its French translation (CIRAD, 2000) and has co-edited *Journal of Natural History* since 1995.

Contributors

Philippe Bouchet
Muséum National d'Histoire Naturelle
Paris, France

Benoît Dayrat
School of Natural Sciences
University of California
Merced, California

James Dobreff
The Daniel Rolander Project
Lund University
Lund, Sweden

Neal L. Evenhuis
Bishop Museum
Honololu, Hawaii

James Hanken
Department of Organismic and Evolutionary
  Biology
and
Museum of Comparative Zoology
Harvard University
Cambridge, Massachusetts

Norman F. Johnson
Department of Entomology
The Ohio State University
Columbus, Ohio

Sandra Knapp
Department of Botany
Natural History Museum
London, United Kingdom

Ellinor Michel
International Commission on Zoological
  Nomenclature
Natural History Museum
London, United Kingdom

Scott E. Miller
Consortium for the Barcode of Life
Smithsonian Institution
Washington, D.C.

Thomas Pape
Natural History Museum of Denmark
Copenhagen, Denmark

David J. Patterson
Marine Biological Laboratory
Woods Hole, Massachusetts

Andrew Polaszek
Department of Entomology
Natural History Museum
London, United Kingdom

Adrian C. Pont
Oxford University Museum of Natural History
Oxford, United Kingdom

Richard L. Pyle
Department of Natural Sciences
Bishop Museum
Honolulu, Hawaii

David Quammen
Contributing Writer
National Geographic Magazine
Bozeman, Montana

Gordon McGregor Reid
North of England Zoological Society
  Zoological Gardens
Chester, United Kingdom

David Remsen
Global Biodiversity Information Facility
  Secretariat
Copenhagen, Denmark

Fredrik Ronquist
Department of Entomology
Swedish Museum of Natural History
Stockholm, Sweden

David E. Schindel
Consortium for the Barcode of Life
Smithsonian Institution
Washington, D.C.

Ellen E. Strong
National Museum of Natural History
Smithsonian Institution
Washington, D.C.

F. Christian Thompson
Systematic Entomology Laboratory
United States Department of Agriculture
Washington, D.C.

Quentin D. Wheeler
International Institute for Species Exploration
Arizona State University
Tempe, Arizona

Edward O. Wilson
Museum of Comparative Zoology
Harvard University
Cambridge, Massachusetts

Debbie Wright
Wiley-Blackwell
Oxford, United Kingdom

Zhi-Qiang Zhang
New Zealand Arthropod Collection
Landcare Research
Auckland, New Zealand

1 The Major Historical Trends of Biodiversity Studies

Edward O. Wilson

During the past 2,300 years, systematics, the science of classification, evolved in Western culture through four stages. The first was the hierarchical system introduced by Aristotle. Although this first systematics of recorded history muddled the picture somewhat by strict formal criteria based on Platonic essentials, he did establish the concept of taxonomic hierarchy—in this case the *eidos* of a particular form, such as horse, dog, or lion, and the *genos,* a class of such forms that can be regarded as part of Earth's natural order. Aristotle recognized some 520 animal species, mostly from Greece, that were consistent with his definition of *eidos.*

During the Middle Ages and into the Enlightenment, in an effort to perfect a system of classification for all known plants and animals, much of the research of early life scientists consisted of systematics in the Aristotelian mode. Microorganisms and the smaller multicellular organisms, of course, remained largely unknown until the invention of the microscope in the 1600s. The work of these authors, including Andrea Cesalpino, Caspar Bauhin, Joseph Pitton de Tournefort, and John Ray, culminated with the system devised during the mid-1700s by Carolus Linnaeus.

The great Swedish biologist, whose name is virtually synonymous with the modern era of systematics, made three decisively influential contributions. The first, presented in the Leiden *Systema Naturae* of 1735, formalized the hierarchical system of classification used today. A direct philosophical descendant of Aristotle's first scheme, it grouped all known organisms into two kingdoms (a third included minerals), which were then divided successively downward into classes, orders, genera, and species. The basic unit Linnaeus recognized is the species, and he aggregated the higher taxonomic categories into successively larger clusters of species according to their anatomical similarity. Although Linnaeus believed in special creation, he nevertheless spent his entire career striving to define the diversity of life as a natural comprehensible system, as opposed to an arbitrary chaotic system.

Linnaeus's second major contribution was the binomial nomenclatural system, introduced in the *Species Plantarum* in 1753 for plants and *Systema Naturae* Vol. 10, 1758 for animals. These publications were made the standard as the starting point of biological classification. The early procedure he had used was that of the very capable Joseph Pitton de Tournefort, who in 1700 characterized each genus by a single term and the species within it by a brief diagnostic description. Linnaeus took the next step by simplifying the procedure with a single Latinized name for the genus coupled with a single Latinized name assigned to the species, followed by a diagnostic description. Thus, we have today our own species *Homo sapiens* and our faithful companion species *Canis familiaris.* Subsequently came the development of the modern rule of nomenclatural priority observed by all taxonomists: in order to ensure stability, the first binomen applied to a species with a formally published description must thereafter prevail and exclude the use of other binomens. Another stabilizing rule is the designation of types, the specimens used in the original description, and as a result designated as the final reference for the species and its binomen.

Linnaeus's binomial system facilitated his third great contribution, the initiative to find and diagnose the entirety of biodiversity, from the local Swedish biota to all those around the world. Such an effort became technically possible in Linnaeus's scheme because large numbers of species, including novelties, could be diagnosed and labeled efficiently. Linnaeus himself limited his field trips to Sweden, going as far as Lapland and the Baltic Island of Öland. But, ever productive in his Uppsala professorial headquarters, Linnaeus inspired students, some of whom traveled widely, to collect and describe as many new species as they could find. Among these apostles, as Linnaeus modestly called them, were Peter Forsskål and Karl Peter Thunberg, the first field taxonomists to visit Asia. Another pioneering insect taxonomist was the prolific Johan Christian Fabricius, though he is not usually included among Linnaeus's apostles. I am currently in the process, with José María Gómez Durán of Madrid, of bringing another apostle to the pantheon, José Celestino Mutis, who in the late 1700s founded the Botanical Garden at Bogotá, Colombia. Most of Mutis's prodigious work was lost in shipping and war, but now Gómez Durán and I have been able to reconstruct another large piece of his newly discovered diaries and letters.

The systematic exploration of the biosphere had begun, in what today can legitimately be called the Linnaean enterprise. Where the launching of global biodiversity exploration was an eighteenth-century achievement, the great advance of the nineteenth century, the third landmark in the series of four I would like to recognize, was the introduction of evolutionary theory as the leitmotif of biodiversity studies. The first to promote this idea was Jean Baptiste Pierre Antoine de Monet de Lamarck, often called Chevalier de Lamarck. His *Philosophie zoologique,* published in 1809, argued that the world's multitudinous life forms can be organized into a phylogeny, a history of ancestors and descendants.

But Lamarck's reasoning convinced few scholars of the value of phylogenetic classification or even of the fact of evolution. In fact, it ultimately turned out that his proposed mechanism of evolution was wrong. This is the theory of inheritance of acquired characters, in which changes in one generation induced by interaction with the environment tend to be passed on to the next generation.

It remained for Charles Darwin, in his masterwork *On the Origin of Species,* fifty years later, to provide massive and compelling evidence for the ongoing process of evolution. He also put forward the correct explanation for it: natural selection, whereby spontaneous mutations create hereditary variants that compete for survival and reproduction, resulting in the gradual replacement of some variants by others over many generations. Darwin's theory of evolution by natural selection, although at first opposed on both religious and philosophical grounds, spread in influence steadily thereafter. In time it not only succeeded but became fully transformative throughout biology. Applied to systematics, evolutionary theory cemented the concept of phylogeny and validated classification above the species level, based on phylogenetic reconstructions.

What, then, is the fourth and current advance in systematics? It is nothing less than the attempted completion of the Linnaean enterprise by a full mapping of Earth's biodiversity, pole to pole, bacteria to whales, at every level of biological organization from the genome to the ecosystem. It aims to yield as complete as possible a cause-and-effect explanation of the biosphere and a correct and verifiable family tree for all the millions of species—in short, a unified biology.

This is a task which, in spite of centuries of effort already devoted to it, has today scarcely begun. At the present time, almost 250 years later, we still have discovered only as few as 10 percent of the species of organisms living on Earth. Most kinds of flowering plants and birds have been discovered, but our knowledge of insects and other small invertebrates, of fungi, and bacteria and other microorganisms is shockingly incomplete. For example, about 60,000 species of molds, mushrooms, and other kinds of fungi are known to science, but the true number has been estimated to exceed 1.5 million. The number of known species of nematode roundworms, the most abundant animals on Earth, with four out of every five animals being nematode roundworms, is about 16,000, but the number could easily be in the millions. In the order of 10,000 species of bacteria are known to science, but 5,000 to 6,000 are found in a handful of fertile soil, almost all unknown to science, and some *4 million* are estimated to live in a ton of soil. As far as our knowledge of them goes, they might as well be on Mars.

Each of these and millions of other species are exquisitely well adapted and interlocked in intricate ecological webs of interaction we have scarcely begun to understand. Yet they are a large part of the foundation of the world's ecosystems. Our lives depend utterly on this largely unknown living world.

We live, in short, on a little known planet. When dealing with the living world, we are flying mostly blind. When we try to diagnose the health of an ecosystem such as a lake or forest, for example, in order to save and stabilize it, we're in the position of a doctor trying to treat a patient knowing only 10 percent of the organs.

Now, new advances in technology, including genomics (reading the genetic codes of each species), high-resolution digital photography, and Internet publication, allow us to speed the exploration of the living world by as much as ten times, and further, to organize the information and make it immediately accessible as an open source everywhere in the world.

This dream brought to reality is called the Encyclopedia of Life. Launched on May 9, 2007, it represents the convergence of efforts by scientists principally working in museums and herbaria and other centers of the major collections of species diversity, as well as planners in the large libraries that contain the totality of already published information on biological diversity.

There are compelling reasons to build such an encyclopedia, with an indefinitely expansible page for each species containing everything known about it. New phenomena, and new connections among phenomena, will come to light. Only with such encyclopedic knowledge can biology as a whole fully mature as a science and acquire predictive power species by species and ecosystem by ecosystem.

The Encyclopedia of Life will serve human welfare in immediate practical ways. The discovery of wild plant species adaptable for agriculture, new genes for enhancement of crop productivity, and new classes of pharmaceuticals, all will be accelerated. The outbreak of pathogens and harmful plant and animal invasives will be better anticipated. Never again, with knowledge of adequate extent, need we overlook so many golden opportunities in the living world around us, or be so often surprised by the sudden appearance of destructive aliens that spring from that world.

Those of us who planned the Encyclopedia of Life are grateful for the support now promised by the MacArthur and Sloan Foundations for this initiative, and by the favorable attention we hope the wider public will give it. We are also grateful to the memory of Carl Linnaeus, who led the way in the systematic exploration of life on Earth.

ACKNOWLEDGMENTS

This article is based on a lecture given by the author on the occasion of the tercentenary of Linnaeus's birth and 250th anniversary of the publication of his *Systema Naturae*, 10th Edition (1758); and in part on an earlier article by the author, "The Linnaean enterprise: Past, present, and future," *Proceedings of the American Philosophical Society*, 149: 344–348 (2005).

2 Linnaeus
*A Passion for Order**

David Quammen

Springtime comes late in Sweden. So it was still springtime, on May 23, 1707, when a son was born to the wife of the curate of a small Swedish village called Stenbrohult. The season was raw, the ground was wet, the trees were in leaf but not yet flowering as the baby arrived, raw and wet himself. The child's father, Nils Linnaeus, was an amateur botanist and avid gardener as well as a Lutheran priest, who had concocted his own surname (a bureaucratic necessity for university enrollment, replacing his traditional patronymic, son-of-Ingemar) from the Swedish word *lind*, meaning linden tree. Nils Linnaeus loved plants. The child's mother, a rector's daughter, was only eighteen. They christened the boy Carl and, as the story comes down, filtered through mythic retrospection flattering a man who became the world's preeminent botanist, they decorated his cradle with flowers.

When he was cranky as a toddler, they put a flower in his hand, which calmed him. Or anyway, again, that's what later testimony claims. Flowers were his point of entrance to appreciating beauty and diversity in nature. He seems even to have sensed, at an early age, that they were more than just beautiful and diverse—that they also encoded some sort of meaning. He grew quickly into a boy fascinated not just by flowers, and by the plants that produce them, but by the *names* of those plants. He pestered his father for identifications of wildflowers brought home from the meadows around Stenbrohult. "But he was still only a child," according to one account, "and often forgot them." His father, reaching a point of impatience, scolded little Carl, saying that he would not tell him any more names if he continued to forget them. After that, the boy gave his whole mind to remembering them, so that he might not be deprived of his greatest pleasure. This is the sort of detail, like Rosebud the sled, that seems too perfectly portentous for real history as opposed to screen drama or hagiography; still, it might just be true. Names and their storage in memory, along with the packets of information they reference, are abiding themes of his scientific maturity. But to understand the huge renown he enjoyed during his lifetime, and his lasting significance, you need to recognize that Carl Linnaeus wasn't simply a great botanist and a prolific deviser and memorizer of names. He was something more modern: an information architect. If you read a thumbnail biography in an encyclopedia or on a Website, you're liable to be told that Carl Linnaeus was "the father of taxonomy"—that is, of biological classification—or that he created the Latin binomial system of naming species still used today. Those statements are roughly accurate, but they don't convey what made the man so important to biology during his era and afterward. You might read that he coined the name *Homo sapiens* for our own species and placed us, daringly, within a category of mammals that included also monkeys and apes. That's true too, but somewhat misleading. Linnaeus was no evolutionist. On the contrary, he heartily embraced the prevailing creationist view of biological origins, which stipulated that studying nature reveals evidence for the creative powers and mysterious orderliness of God.

* "Linnaeus. A Passion for Order" is reprinted from *National Geographic* and *Natural Acts: A Sidelong View of Science and Nature*, published by W. W. Norton & Co. Copyright © (2009) David Quammen. The author is indebted to Wilfrid Blunt's biography *Linnaeus: The Compleat Naturalist*, among other sources; to the Linnaean Society of London for access to their holdings; and to officials and curators at Uppsala University.

He wasn't such a pious man, though, that he sought *nothing but* godliness in the material world. Here's what makes him a hero for our time: he treasured the diversity of nature for its own sake, not just for its theological edification, and he hungered to embrace every possible bit of it within his own mind. He believed that mankind should discover, name, count, understand, and appreciate every kind of creature on Earth. In order to assemble all that knowledge, two things were required—tireless and acute observation, and a system.

In the spring of 1732, at age twenty-five, Linnaeus set off on an expedition through Lapland, the wild northern region of the Swedish kingdom inhabited by a sparse population of the Sami people, who lived as herders of reindeer. Over the next five months he traveled 3,000 miles by horseback, on foot, and by boat, making collections and taking notes as he went. He was interested in everything—birds, insects, fishes, geology, the customs and technology of the Sami—but especially in plants. He made drawings in his journal, some of which were crude sketches, some of which (again, those of plants) were delicate and lovingly precise. Eventually he produced a book, *Flora Lapponica*, describing the botanical data he had gathered.

He went abroad in 1735 to advance his career prospects. He spent three years on the continent, mostly in Holland, taking a medical doctor's degree quickly, then turning back to plants. It wasn't a stretch to combine both activities, since botany in that era was closely related to medicine through the pharmaceutical uses of vegetation. He found temporary work with a rich man named George Clifford, a director of the Dutch East India Company, as botanical curator and house physician at Clifford's country estate near Leiden. Linnaeus's work there led to another book, a descriptive catalog of Clifford's botanical holdings titled *Hortus Cliffortianus* and gorgeously illustrated by a young artist named Georg Dionysius Ehret. Although they became lifelong friends, Ehret later recalled the Linnaeus of these years as a self-aggrandizing opportunist. By any account, he was full of energy and plans, full of ideas and opinions, and hungry for success as well as for deeper knowledge. Confident to the point of arrogance but charming enough to compensate, he proved good at making friends, finding sponsors, and cultivating powerful contacts. During the three years abroad he published eight books—an amazing spurt of productivity, partly explained by the fact that he had left Sweden carrying some manuscripts written earlier. One of those manuscripts became *Systema Naturae*, now considered the founding text of modern taxonomy.

Linnaeus wasn't the first naturalist to try to roster and systematize nature. His predecessors included Aristotle (who had classified animals as "bloodless" and "blooded"), Leonhart Fuchs in the sixteenth century (who described 500 genera of plants, listing them in alphabetical order), the Englishman John Ray (whose *Historia Plantarum*, published in 1686, helped define the species concept), and the French botanist Joseph Pitton de Tournefort, contemporary with Ray, who sorted the plant world into roughly 700 genera, based on the appearance of their flowers, their fruit, and their other anatomical parts.

Linnaeus emerged from this tradition and went beyond it. His *Systema Naturae*, as published in 1735, was a unique and peculiar thing: a folio volume of barely more than a dozen pages, in which he outlined a classification system for all members of what he considered the three kingdoms of nature—plants, animals, and minerals. Notwithstanding the inclusion of minerals, what really mattered were his views on the kingdoms of *life*. His treatment of animals, presented on one double-page spread, was organized into six major columns, each topped with a name for one of his classes: Quadrupedia, Aves, Amphibia, Pisces, Insecta, Vermes. Quadrupedia was divided into several four-limbed orders, including Anthropomorpha (mainly primates), Ferae (such as canids, felids, bears), and others. His Amphibia encompassed reptiles as well as amphibians, and his Vermes was a catch-all group, containing not just worms and leeches and flukes but also slugs, sea cucumbers, starfish, barnacles, and other sea animals. He divided each order further into genera (some with recognizable names such as Leo, Ursus, Hippopotamus, and Homo), and each genus into species.

Apart from the six classes, Linnaeus also gave half a column to what he called Paradoxa, a wild-card group of chimerical or simply befuddling creatures such as the unicorn, the phoenix, the dragon, the satyr, and a certain giant tadpole (now known as *Pseudis paradoxa*) that, weirdly, shrinks during metamorphosis into a much smaller frog. Across the top of the chart ran large letters: CAROLI LINNAEI REGNUM ANIMALE. It was a provisional effort, grand in scope, integrated, but not especially original, to make sense of faunal diversity based on what was known and believed at the time. Then again, animals weren't his specialty. Plants were. His classification of the vegetable kingdom was more innovative, more informed, and more orderly. It became known as the "sexual system" because he recognized that flowers are sexual structures and he used their male and female organs—their stamens and pistils—to characterize his groups. He defined 23 classes, into which he placed all the flowering plants (with a 24th class for cryptogams, those that don't flower), based on the number, size, and arrangement of their stamens. Then he broke each class into orders, based on their pistils. To the classes he gave names such as Monandria, Diandria, Triandria (meaning: one husband, two husbands, three husbands) and, within each, ordinal names such as Monogynia, Digynia, Trigynia, thereby evoking all sorts of scandalous menage (a plant of the Monogynia order within the Tetrandria class: one wife with four husbands) that caused lewd smirks and disapproving scowls among some of his contemporaries. Linnaeus himself seems to have enjoyed the sexy subtext. And it didn't prevent his botanical schema from becoming the accepted system of plant classification throughout Europe.

The artist Georg Ehret again helped popularize his ideas by producing a handsome *tabella*, a poster, illustrating the diagnostic features for Linnaeus's twenty-four classes. The tabella sold well and earned Ehret some gulden. Linnaeus himself, always stingy about sharing credit, included Ehret's drawing without acknowledgment in one of his later books. But he wouldn't forget his old pal, and evidence left after his death—we'll come to it—suggests that he valued Ehret's botanical vision as he valued few aside from his own. After returning to Sweden, becoming a husband and a father and a professor at Uppsala University, Linnaeus continued to churn out books. He published revised and expanded editions of *Systema Naturae*, as well as strictly botanical volumes such as *Flora Suecica* ("Swedish Flora") in 1745, *Philosophia Botanica* (1751), and *Species Plantarum* (1753).

Philosophia Botanica is a compendium of terse, numbered postulates in which he lays out his botanical philosophy. For instance: "The foundation of botany is two-fold, arrangement and nomenclature." Arrangement of plants into rational categories and subcategories is crucial for three reasons: because there are so many kinds (and more every year, during the great age of discovery in which Linnaeus lived), because much is known about many of those kinds, and because classification makes that knowledge accessible. Alphabetical listing may have worked well enough with 500 plant species, but as the count rose into thousands, it didn't serve. There was also a deeper purpose, for Linnaeus, to this enterprise. Find the "natural method" of arranging plants into groups, and you would have discovered God's own secret logic of biological creation, just as Isaac Newton had discovered God's physical mathematics. Linnaeus knew that he hadn't achieved that, not even with his twenty-four-class sexual system, which was convenient but artificial. He couldn't see, couldn't imagine, that the most natural classification of species reflects their degree of relatedness based on evolutionary descent. But his passion for order—for seeking a *natural* order—did move taxonomy toward the insights later delivered by Charles Darwin. As for nomenclature, it contributes to the same purpose. "If you do not know the names of things, the knowledge of them is lost too," he wrote in *Philosophia Botanica*. Naming species, like arranging them, became increasingly problematic as more and more were discovered; the old-fashioned method, linking long chains of adjectives and references into fully descriptive labels, grew unwieldy. In *Species Plantarum* he established the Latin binomial system for naming plants, and then in the tenth edition of *System Naturae*, published in 1758–59 as two fat volumes, he extended it to all species, both plant and animal. A pondweed clumsily known as *Potamogeton caule compresso, folio Graminis canini ...* etc., became *Potamogeton compressum*. We became *Homo sapiens*.

Linnaeus's life back in Uppsala entailed more than authorship. He was a wonderful teacher with a vivid speaking style, clear and witty, and a terrific memory for facts. His lectures often packed the hall, his private tutoring earned him extra money, and he made botany both empirical and fun by leading big festive fieldtrips into the countryside on summer Saturdays, complete with picnic lunches, banners and kettledrums, and a bugle sounding whenever someone found a rare plant. He had the instincts of an impresario. But he was also quietly effective in mentoring the most talented and serious of his students, of whom a dozen went off on adventuresome natural-history explorations around the world, sending data and specimens faithfully back to the old man. With his typically sublime absence of modesty, he called those travelers "the apostles."

In 1761, the government ennobled him, whereupon he upgraded his linden-tree name to von Linné. By then he was the most famous naturalist in Europe. His wife sternly guarded their privacy, and his son became only a middling botanist, but his teaching life delivered rich satisfactions, and he had an abundance of brilliant intellectual offspring. Despite the limitations of his language skills (he may have known some Dutch and German but did all his writing in Swedish and Latin), and of his geographical experience (he never left Sweden again), he became a global encyclopedist of flora and fauna through his written correspondence with naturalists all over the world, and through the information he received from the apostles, such as Daniel Solander (who sailed on Cook's first voyage), Pehr Kalm (in North America), and Anders Sparrman (China, South Africa, then Cook's second voyage). Linnaeus himself had no appetite for the rigors and climate of the tropics, though he was voraciously curious about tropical plant diversity. Let the young men gather the information; he would systematize it.

On a recent afternoon in Uppsala, I discussed this manipulative homebody aspect with Professor Carl-Olof Jacobson, a retired zoologist who serves as chairman of the Swedish Linnaeus Society. No, Linnaeus didn't want to travel, Professor Jacobson told me. "What he wanted to be was a spider in the net." The center point of that net, that vast web of scientific silk, was in and around Uppsala— including the university, its splendid botanical garden, and a small farm known as Hammarby about five miles outside the city. Linnaeus bought Hammarby and built a large, simple house there to be his summer retreat. It might have served also as his retirement home, though he never retired. Each autumn, having savored his time in this getaway, he moved back into town, where the living was less austere. He grew feeble and ill, then suffered a seizure after one last escape to the countryside strictly against doctor's orders, and died on January 10, 1778.

They buried him beneath the stone floor of Uppsala's cathedral, the Westminster Abbey of Sweden. Six years later, following Linnaeus's posthumous instructions, his widow sold his library, his manuscripts, and most of his collections to a buyer who would care for them well. That buyer, a young Englishman named James Edward Smith, founded a scientific society to receive those treasures and called it the Linnean Society (its spelling derived not from "Linnaeus" but from the noble version, "von Linné") of London, where they lie protected today in a basement vault but available in physical (and, soon, digitized) form to scholars. Linnaeus himself would approve; knowledge, he believed, is meant to be communicated and used. But his country home, Hammarby, remained in the family for a century and then was bought by the Swedish state to be made a museum. Although his house near the university in Uppsala has also been saved and lately restored, Hammarby conveys a more vivid sense of his character, his foibles, his loneliest joys. Inside the old farmhouse, overlooking muddy crop fields, his collection of walking sticks is on display. So is the red skull cap he often wore over his short-cropped hair in lieu of a formal wig. There are oil portraits of his four daughters, his son, and his pet monkey, in no particular order of fondness. His wife and he kept separate bedrooms at opposite ends of the second floor. His is tucked away, accessible only through another room that functioned as his study.

The bedroom, preserved much as he left it, contains a small curtained bed of the sort known in Sweden as a *himmelssäng*, a bed of heaven. Against the west wall is a wooden desk and, above it, a window. The walls are covered with flowers. That is, they are wallpapered wildly from floor to ceiling with large floral images cut from books. The plants are robust, exuberant, some of them garish, some elegant, all suggesting fecundity and fruition: pineapple, banana, magnolia, lily, cactus, papaya, frangipani, and others. Many of these hand-colored engravings came from paintings by his old friend Georg Dionysius Ehret. Rare and magnificent, they would be collectibles in their own right, even absent the association with Linnaeus. But, once bright and crisp, they are now faded, smeary, streaked with the punishments of moisture and time. On the day I visited, accompanied by a botanical curator named Karin Martinsson, still another damp January chill hung in the air. Linnaeus was warned that such damage would occur, but evidently he didn't care. He wanted the pictures around him. Never mind if they decayed. So what? His own body was doing that too.

Even now these antique prints could be peeled carefully off, Martinsson told me, and preserved under better conditions. But that's not going to happen. "Taking them down from the walls," she said, "would be like ripping the heart out of Hammarby." Left as is, the heart of the house reflects the heart of its original owner: full of plants. The pilgrims who visit this room during the tercentenary year—presumably there will be many, from around the world—can look at that improvised wallpaper and sense an important truth about the life's work of Carl Linnaeus. It wasn't just about knowledge. It was about knowledge and love.

3 Daniel Rolander
The Invisible Naturalist

James Dobreff

CONTENTS

Words define the human condition. Their interplay with our thoughts enables reasoning. Yet many speak with great dexterity without knowing the elements of individual words. The adjective *invisible*, in general terms, consists of the prefix *in-*, the root or core *-vis-*, and the suffix *-ible*. This root signifies the idea of *seeing*. The suffix marks it as an adjective, while the prefix adds a negative sense to the word, shifting the meaning from that which can be seen to that which cannot be seen—even when present. People understand the meaning of the word and use it correctly without being conscious of the elements or genesis of *invisible*. Bad luck, and a number of mistakes by scholars, have made the Swedish naturalist Daniel Rolander (c.1722–1795) into an invisible naturalist. We cannot see him, although his work survives in the publications and collections of several famous men of science.

The following pages will cast some light on Rolander the person, his expedition to Surinam, his work as an entomologist in particular, and how his Surinam insects crept into European scientific literature without garnering any fame for their collector and his detailed descriptions.

A SHORT BIOGRAPHY OF A LINNAEAN APOSTLE[1]

Daniel Rolander was born and raised in the Swedish province of Småland, as was Carl Linnaeus.[2] Little is known about his parents, who were said to have been farmers. He attended gymnasium [elementary school] in the town of Växjö (1736–1741) and university at Uppsala (1741–1754). He defended no dissertations, though he published five entomological articles in the proceedings of the

Royal Swedish Academy of Sciences (*Kungliga Vetenskapsakademien*).[3] His writings caught the eye of Linnaeus, who urged him to submit them to the Academy. Linnaeus wrote to the Academy:

> "Daniel Rolander, who studied insects with me some years ago, has allowed me to read a couple of his observations. They greatly impressed me, so I asked that he submit them to the Academy of Sciences— they were marvellous and the sort I wouldn't have expected from anyone but a Réaumur or De Geer. When he comes to the Academy, I advise that this creative spirit be separated from simple collectors. As you know, for every creative spirit we have a thousand collectors."[4]

Linnaeus took a liking to the young Rolander, whose enthusiasm, diligence, and humble origins perhaps reminded him of himself. While flowers and botany had fascinated Linnaeus as a boy, insects took hold of Rolander in the pastoral Swedish countryside. En route to southern Sweden from Uppsala in 1754, Rolander revisited his home in the rural parish of Hälleberga. He didn't tell his readers about a first kiss or his family and friends; he reminisced about insects:[5]

> "On the tenth day of this month I arrived at Stibbetorp, a country house in the parish of Helleberga. It is situated seven Swedish miles from Växjö, some 404 miles from Stockholm. This is where I was born and passed my childhood. Insects were my particular delight. I now found myself seized by no small desire to revisit those pleasant areas that had so very often separated me as a boy, intent upon the business of insects, from my comrades. The areas I had in mind were those where I had seen the wasp *Vespa rupestris* fix its tubular nests to rocks by building them on the southern-facing sides of the rocks. I even found one or two of their nests occupied by the wasps in those same tracts despite the time of year, when the great majority of such nests have been despoiled of their inhabitants. Neither I, nor anyone else as far as I know, have observed these insects, or wasps, in any other localities. Seeing that no one has ever attended to the transmission of their description and history, I can here touch on each aspect, though my itinerary forces me to be brief."

Rolander fills the next manuscript page with his observations of this wasp. It is a fantastic nest builder. He found the threatening attacks of this wasp to be rather empty, since it either cannot sting or lacks the strength to inflict a sting. The dedicated observer used to tear the nests from the rocks and carry them to his room, where the wasps went about their construction business paying no heed to his presence. The passage is typically Rolandrian in its deceptively casual blend of personal reflections and scientific observations.

These images of a country boy obsessed with the operations of insects call to mind many of the stories of the young Linnaeus, whose fascination with plants knew no bounds.[6] Linnaeus is over-joyed to finally have a dedicated and skilled entomologist among his disciples.[7] It also reinforces Linnaeus's own picture of Rolander as a uniquely skilled observer of insects. When Per Löfling departed on his journey to document the natural flora and fauna of Spain in early 1750, Linnaeus selected Rolander to take his place as resident tutor to Carl Jr. Rolander lived with Linnaeus in Uppsala, and at his country residence Hammarby (Figure 3.1) from 1750 to October 1754, when he departed for his own expedition to Surinam. Rolander, then, had daily contact with Linnaeus and lived at the center of Linnaeus's myriad of activities. Linnaeus and many of his colleagues conducted private colleges or instructional sessions at their private residences, where discussions and investigations could be conducted at a much more detailed and personal level. Besides these sessions, Linnaeus was perpetually busy composing dissertations for his students to defend pub-licly, and he also had an extensive correspondence to maintain with scientists and former students in Europe and around the globe. His homes also included extensive collections of preserved plants and animals, as well as minerals and, of course, books and manuscripts.[8] There Rolander spent four years imbibing the teachings of one of the great figures in modern science and reading the letters and reports of Linnaeus's apostles who were in, or had been to, the most exotic corners of the world.

FIGURE 3.1   Hammarby, summer residence of Linnaeus, 1864. (E. Schenson. 1864. *In memoriam Caroli A Linnè*. Uppsala. Reprinted with the permission of Hagströmer Library.)

It was hardly an accident that Linnaeus selected Rolander to go to Surinam when the opportunity arose. Carl Gustaf Dahlberg, a Swedish mercenary officer in Dutch service, had married a wealthy widow while in Surinam. The marriage made him the owner of extensive plantation holdings in Surinam. While revisiting Sweden in 1754, he offered to take one of Linnaeus's students back to Surinam to serve as tutor to his daughters, with the understanding that the student would be allowed to study the flora and fauna of the region.[9]

LEAVING UPPSALA FOR SURINAM

Rolander departed from Uppsala on October 21, 1754 by land for southern Sweden. There he met up with Dahlberg and his family, including a number of African slaves from Surinam. They set sail on the 17th of December from Ystad in southern Sweden and landed at a village not far from Rostock in northern Germany on the 19th. They traveled by coach to Amsterdam, where they arrived on January 5, 1755. Rolander fell seriously ill with a severe fever in early February. Abandoned for lost by a Dutch doctor named Koelmann, Rolander made a final plea for help to Dr. Rudolf Fortstèn, who had an excellent reputation among Amsterdam's wealthier citizens. Despite warnings from others that the busy Dr. Fortstèn would never have time to see him, Rolander was warmly welcomed by the doctor, whose skill exceeded his reputation.[10] Thanks to Fortstèn's intervention and care, Rolander was healthy enough to leave Amsterdam on April 12, 1755, when he sailed to the Dutch port of Texel to await favorable winds. His voyage to Surinam started on the 25th. He disembarked at Paramaribo, Surinam, on June 21, 1755 in the earliest hours of the morning. He remained in Surinam for seven months, departing by ship on January 20, 1756. Adverse winds kept his ship around the West Indies for three weeks. On February 12 it dropped anchor in the port of St. Eustatius to load supplies and wait for favorable winds. Rolander explored the island for several days prior to his ship's departure on the 23rd. He returned to Texel on April 14, 1756. Several bouts of poor health and a lack of funds kept him in Hamburg until late August. He arrived by ship in Stockholm on October 2.

Back in Stockholm: October 1756

Rolander's return should have been a great victory; he had survived, unlike many other Linnaeus apostles. Christopher Ternström had died in 1746, while sailing with the Swedish East India Company, not even a year after leaving Sweden. Carl Fredrik Adler had also sailed with the East India Company in 1748, and he was buried thirteen years later off the coast of Java. Fredrik Hasselquist died in Smyrna in 1752, barely three years after leaving Linnaeus. Rolander had seen his colleague Per Löfling depart for Spain in 1750. Löfling, Linnaeus's favourite apostle, died in Spanish Guyana, not far from Rolander's Surinam, on February 22, 1756. In 1753, Olof Torén died in Sweden, just a year after returning from a voyage with the East India Company. Of the apostles who had journeyed abroad prior to 1754, only Per Kalm (North America 1747–1751), Pehr Osbeck (East India Company 1750–1752), and Martin Kähler survived their expeditions. Rolander had seen them all depart Uppsala, where Linnaeus was professor.

Linnaeus welcomed him home, offering him room and board in his own house.[11] Rolander did not take up the offer, though he promised Linnaeus a specimen of the plant *Sauvagesia*. Linnaeus eventually understood that he would not be allowed to use or even see Rolander's specimens— not even *Sauvagesia*. In a fit of anger while in Stockholm, Linnaeus barged into Rolander's apartment and took the promised specimen.[12] That was in late 1756 or early 1757 and marks the definite end of Linnaeus's patronage. Rolander never again allowed Linnaeus to see anything.

Rolander's apparent ingratitude has no definite explanation, but a few possible ones are apparent. Rolander may have considered his expedition diary and specimens the only means he would have eventually to obtain a professorship. If he let Linnaeus use them, he knew Linnaeus would publish substantial portions of this material in the coming tenth edition of *Systema Naturae* (1758). Rather than seeing the honor of appearing in such a publication, Rolander may have feared that it would leave him with nothing to offer a future employer.

Another probability is that Rolander was deeply disappointed and blamed Linnaeus for the physical suffering caused by the journey. He had suffered two dire bouts with fevers in Amsterdam in 1755 and Hamburg in 1756. In both instances he received little or no aid from Linnaeus's extensive network of friends and colleagues. His stay in Hamburg was apparently lengthened considerably by a lack of funds; here, again, Rolander was forced to wait more than a few weeks for sufficient funds from Sweden to pay debts in Hamburg and purchase passage home to Stockholm. When Rolander saw Linnaeus again in 1756, he may have seen not the man who had enabled him to undertake a remarkable expedition to Surinam, but rather the man who had nearly cost him his life and had left him to flounder in Hamburg in debt and poor health. In such circumstances, Linnaeus's keen interest in Rolander and his specimens in 1756 may have been interpreted by Rolander as an intention to exploit his hard-won observations and specimens from Surinam. Despite encouragement from friends such as P. G. Bergius, Rolander never published the results of his Surinam expedition.[13] In fact, apart from an article on *Doliocarpus* published in the proceedings of the Swedish Academy of Sciences in 1756, Rolander never published again.

Despite the new enmity between Rolander and Linnaeus, Abraham Bäck continued to support Rolander, first with an appointment to head a new botanical garden at the Serafimer Hospital in Stockholm in January 1757, and then with a commission to give several lectures on medicine to Stockholm students the following year.[14] Linnaeus opposed the lectures, considering Rolander completely unqualified.[15] Rolander either abandoned or lost his position at the hospital's garden in 1761, and after being passed over for a new professorship of natural history at Stockholm, he left Stockholm for Copenhagen to organize the curiosity cabinet of a Danish minister.[16]

Copenhagen: 1761–1765

Rolander's luck was no better in Copenhagen. After the position with the minister ended, he found no academic positions and was out of money.[17] At one point in 1763 he had to live for two weeks on just apples, until Danish-German professor Christian Gottlieb Kratzenstein met him and took him into his home. In exchange for room and board, he demanded that Rolander dedicate himself to preparing his travel journal from Surinam for publication. Rolander finished the two-volume work by early 1765 and entitled it *Diarium Surinamicum, quod sub itinere exotico conscripsit Daniel Rolander.*[18] Rolander left the manuscript with Kratzenstein, who found it impressive and well written. He apparently tried to get the Swedes (perhaps Linnaeus's publisher Lars Salvius) to publish it. The only stipulation Kratzenstein made was that profits from the sale of the manuscript and a portion of eventual income from its publication would go to Rolander.[19] As for the herbarium, it is not clear whether Rolander sold or gave some of it to Rottbøll, botany professor in Copenhagen and former Linnaeus student. Claims that Rottbøll got the entire herbarium are pure fiction.[20]

From 1775 to 1786, Rottbøll published three works based on Rolander's accounts in *Diarium Surinamicum* and the herbarium specimens from Surinam.[21] Rottbøll followed Rolander closely, openly crediting him as the source of the plants and their documentation, to the extent that one can say that many of Rolander's plants have actually been published officially. This is no secret to a number of plant taxonomists who work closely with plants from Surinam.[22] Rottbøll's *Descriptiones rariorum plantarum, nec non materiae medicae atque oeconomicae e terra Surinamnsi fragmentum* (1776) treated some 160 plants from the *Diarium*, including several new genera and many new species. The book's numerous discussions of the medicinal uses of Surinam plants were taken directly from Rolander's manuscript.

Rolander's Surinam herbarium was apparently immense. According to Rottbøll, Rolander had enough samples of the new genera to supply every botanist in Europe with a specimen.[23] Rottbøll honored Rolander with the genus *Rolandria.* Nevertheless, some modern scholars maintain that Rolander was mentally ill even while in Surinam and accomplished little or nothing.[24] Rolander the botanist is then invisible, present but not visible in the shadow of Rottbøll's name and Latin publications.

Rolander left Copenhagen in 1765, for Landskrona, Sweden, where he enjoyed the patronage of Arvid Schauw (1711–1788), a powerful politician and businessman in southern Sweden.[25] Rolander completed an inventory of plants of the island of Ven for Schauw, the manuscript of which has yet to be located.[26] He eventually (1770 or 1778?) moved to Lund, where he died on August 9 or 10, 1793. He had lived his last years poor and in bad health.

ROLANDER'S INSECTS

In 1758 Linnaeus published the tenth edition of his famous *Systema Naturae*, the first edition of which had left the press in Leiden in December 1735. Each subsequent edition was augmented with the names and brief descriptions of whatever plants, animals, and minerals Linnaeus had obtained since the previous edition. As a custom, in the introduction Linnaeus lists the authors and works cited in his books. The tenth edition includes Daniel Rolander and his journey to Surinam and St. Eustatius as one of Linnaeus's sources. A careful review of the extensive section on insects reveals that Linnaeus cites Rolander as the unique source for eighty-five insects. Moreover, Charles De Geer, the most famous Swedish entomologist of Linnaeus's age, used Rolander the same way in his own publications.[27] All but a few of those insects came from Surinam (Figure 3.2). However, I have mentioned above that Rolander had never allowed Linnaeus to examine his collections. In 1758, just two years after Rolander's return from Surinam, Linnaeus had managed to publish his own names for more than eighty Surinam insects collected by Rolander. Here again, Rolander is cited, although for two and a half centuries he remained invisible to all but those taxonomists in entomology whose research has treated those citations in Linnaeus's *Systema Naturae.*[28] (See Figure 3.3)

A considerable number of Linnaeus scholars, intellectual historians, and bibliographers have not noticed Rolander's presence in the most important edition of Linnaeus's *Systema Naturae.* Many of them have severely criticized Rolander for being a coward who did not dare to enter the forests of Surinam,[29] and a recent publication from Uppsala University went so far as to claim that

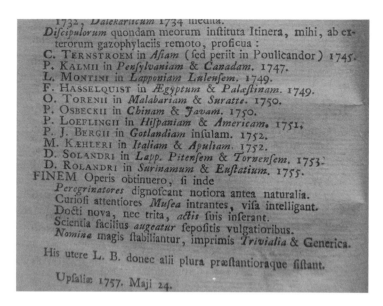

FIGURE 3.2   Linnaeus cites "D. Rolandri in *Surinamum & Eustatium*. 1755." *Systema Naturae* (Stockholm: Lars Salvius, 1758), introduction.

Rolander suffered from a fear of insects.[30] There can be no doubt now, however, that Linnaeus and Rottbøll must be considered the best witnesses of the success of Rolander's expedition to Surinam. The appearance of so many, if not all, of Rolander's Surinam insects in *Systema Naturae* certifies Rolander's ability, skill, and diligence as a collector of insects. Rottbøll's extensive use of Rolander's plant specimens and descriptions from *Diarium Surinamicum* likewise shows that Rolander's botanical work in Surinam was first-rate.[31]

SOLVING A MYSTERY

If we turn back to the mystery of how Linnaeus got hold of Rolander's insects, we only have to travel a short distance from Linnaeus's residence in Uppsala to Charles De Geer's Leufsta manor, a country estate. That is where specimens, probably duplicates, of Rolander's insects spent several decades in the collections of De Geer (1720–1778), one of the wealthiest citizens of Sweden in the 18th century, and a professional entomologist. After arriving back in Amsterdam in mid-April 1756, Rolander sent a crate of insects to De Geer.[32] The shipment of insects most probably fulfilled a stipulation that De Geer must have made back in 1754, when he contributed 600 Swedish dollars to Rolander's travel funds.[33] Because the selection of the individual who would travel to Surinam was not made until the last minute, Linnaeus canvassed desperately for financial contributions to cover Rolander's travel costs. De Geer probably demanded a crate of Surinam insects in return for his support. Rolander's refusal to give Linnaeus access to the Surinam specimens and journal created a dilemma for Linnaeus, who was in the midst of preparing a major revision of *Systema Naturae*. That revision would be the tenth edition. Linnaeus already had the collections of Löfling and Per Kalm, but he needed Rolander's Surinam insects to substantially augment the insect section. A letter from former Linnaeus student Daniel Solander (1733–1782) to his fellow Linnean Eric Gustaf Lidbeck (professor of natural history at Lund University) solves everything.[34] Linnaeus, according to Solander, had gone to Leufsta in either December 1756 or January 1757 to study Rolander's insects. With Linnaeus's new names, brief Latin descriptions and a listing of Rolander's insect collections, the Surinam insects of Daniel Rolander enter the annals of scientific history in the tenth edition of Linnaeus's *Systema Naturae*.

FIGURE 0.2   See page x.

FIGURE 3.5   See page 19.

FIGURE 5.3 See page 60.

FIGURE 6.1 See page 66.

FIGURE 6.2 See page 67.

FIGURE 7.5   See page 79.

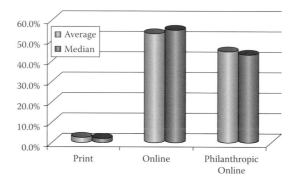

FIGURE 8.1   See page 88.

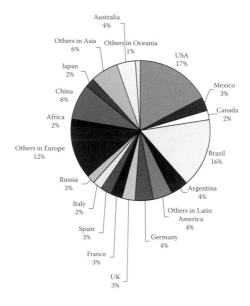

FIGURE 9.8   See page 103.

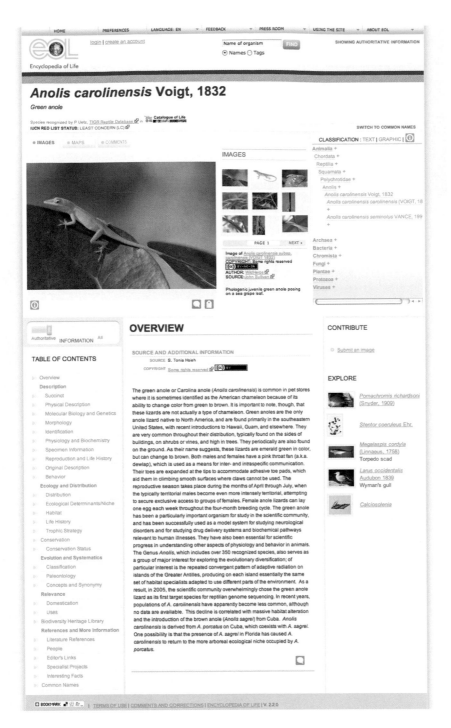

FIGURE 12.1   See page 129.

FIGURE 13.2   See page 139.

FIGURE 15.2   See page 168.

FIGURE 18.2 See page 244.

FIGURE 18.3 See page 245.

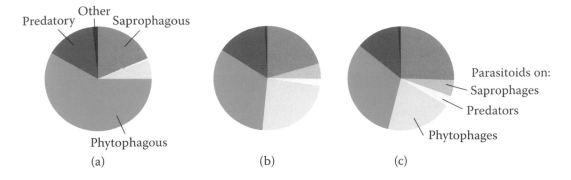

Predatory Other Saprophagous

Phytophagous

(a)

(b)

Parasitoids on:
Saprophages
Predators
Phytophages

(c)

FIGURE 18.4 See page 246.

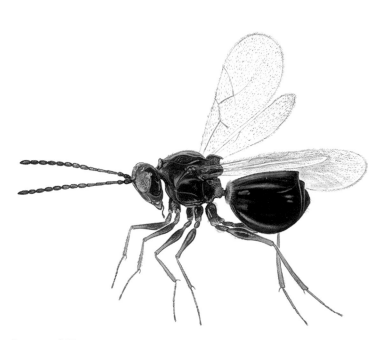

FIGURE 18.5 See page 247.

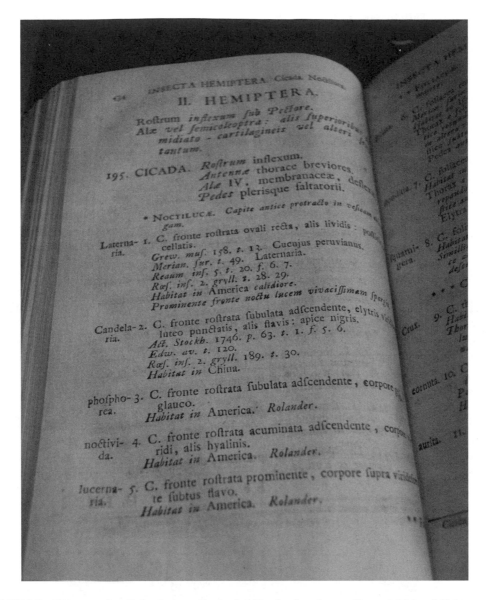

FIGURE 3.3 Linnaeus cites Rolander as collector for *Cicada phosphorea, C. noctivida*, and *C. lucernaria*, specifying only "America." *Systema Naturae* (Stockholm: Lars Salvius, 1758), vol. 1, p. 434.

INSECTS IN THE ARCHIVES

My first attempts to locate Rolander's insect specimens in January 2007 were unsuccessful. I conjectured that they were in De Geer's collections, which had been traced to the Natural History Museum of Sweden. But the museum's collections did not list or index the field collector, nor were the insects listed by Rolander's names. It would be necessary to have Linnaeus's or De Geer's names to identify Rolander's specimens definitely. While preparing the first edition of Rolander's *Diarium Surinamicum*, I noticed that Rolander had either made additions to, or had altered completely, his original names for many of the insects in the *Diarium*. The new names often contained elements from Linnean names first found in the tenth edition of *Systema Naturae* from 1758. Their presence proved that Rolander had revised the insect names in his *Diarium* to reflect the names published by his former patron, now his nemesis. Chronologically, it all made sense. The final revision had taken place in Copenhagen from 1763 to 1765. *Systema Naturae* had doubtlessly been on Rolander's desk during the final revision. Those alterations to Rolander's insect names, which will all be documented and published in the first edition of the Latin manuscript of *Diarium Surinamicum*, led me to the tenth edition of *Systema Naturae*. It contained more than eighty references to Rolander as the collector of specific insects from Surinam. Using my new list of Linnean names, curator Bert Gustafsson at the Natural History Museum of Sweden required hardly an hour to confirm that many of the insects survived in De Geer's collection (Figure 3.4 and Figure 3.5).[35] My subsequent review of the De Geer collections located a number of the specimens. Some are missing, although their tags and needles are still present. Most of the Rolander insects seem to be present. The verification of the insects will be left to an interested entomologist. The concordance of Rolander's insects with Linnaeus's names, located in Appendix 1 of this volume, includes detailed cross-references to De Geer's well known eight-volume series *Mémoires pour Servir à L'Histoire des Insectes* (1752–1778). It turns out that De Geer also described and published Rolander's insects. The invisible Rolander has again been found to be hiding in the publications of another great author.

During my preparation of the first edition of the *Diarium*, Ove Hagelin, the founding father of Hagströmer Library, and Gertie Johansson, the library's sole librarian, generously allowed me to research and write at the library for almost a year. Their generosity greatly facilitated my work by providing access to many of the books Rolander and Linnaeus would have had on their desks. It also pointed me in the direction of the rare books and manuscript collections of Uppsala University. Hagelin had been an important factor in the sale and transfer of the extensive library of De Geer's Leufsta manor to Uppsala University's library. He assured me that Uppsala held De Geer's original drawings that had served as the models for the hundreds of insects depicted in the copperplates of De Geer's *Mémoires*. On a recent visit to Uppsala, I confirmed Hagelin's information. Two bound volumes contain De Geer's drawings. The final section of the larger of the two volumes contains drawings of many of Rolander's insects. Rolander's insects survive, then, in written form in the works of Linnaeus and De Geer. Extremely accurate depictions of how these insects looked shortly after their arrival in Sweden in 1756 survive in copperplates in many of De Geer's eight volumes of the *Mémoires* as well as in De Geer's own original drawings. The cross-references to De Geer in the appended concordance should greatly facilitate the work of any entomologists who decide to put the finishing touches on the findings documented in this essay. It will also lead them to De Geer's illustrations, which provide detailed drawings of these insects as they appeared just after arriving from Surinam.

FIGURE 3.4 Rolander's specimen of *Carabus americanus* L, in Charles De Geer's collections, KVA, Swedish Museum of Natural History.

FIGURE 3.5   Rolander's specimen of *Cerambyx festivus* L = *C. spinosus* De Geer, in Charles De Geer's collections, KVA, Swedish Museum of Natural History.

WITNESSES FOR THE DEFENSE

Linnaeus, almost rabidly, had repeatedly insisted that Rolander's expertise, albeit profound, did not extend beyond entomology.[36] But the hundreds of detailed descriptions of plants and animals in *Diarium Surinamicum* strikingly contradict Linnaeus's inveterate contention that Rolander was strictly an entomologist. Nevertheless, the particularly elegant style and architecture of the insect passages in the *Diarium* bear witness to Rolander's special fascination with insects. These passages Rolander crafted into essays and folded them carefully into the longer, multi-topic daily entries. The Swedish entomologist N.A. Kemner (1887–1948) was so taken with these descriptions that he appended the Latin text and his own German translation of the manuscript's pages 462–467 (December 16, 1755) and 552–567 (January 17, 1756).[37] Among the numerous praises Kemner heaps upon Rolander, the following is perhaps the most telling: "On the whole Rolander's observations on the termites of Surinam are so interesting that they are deserving of their own separate publication. Had they been published immediately upon Rolander's return, they would have doubtlessly been prized as the first detailed report on the habits of the termites that had even been studied in small experiments. In fact, König and Smeathmann had established their reputations as pioneers in termite biology by publishing important observations of these same animals twenty years after Rolander had made the same observations in his South American studies of these animals. Moreover, the bounding or jumping of the termite, which is first described here [in Rolander], was not rediscovered until our own time [i.e., the 1930s]. Hagen considered, as I mentioned earlier, that Linnaeus's statements on this point were nonsense."[38] Kemner then, a fellow entomologist, found Rolander's observations to be of astounding quality. His opinion resonates true all the more since Linnaeus repeatedly had bragged of Rolander's entomological talents in much the same tones.

REHABILITATION

This chapter has not provided detailed lists of the many publications that have unfoundedly criti-cized Rolander because none of their authors bothered to examine his *Diarium Surinamicum*. In that respect, they have no place in a discussion of Rolander's merits as an entomologist or biologist. Those who had worked with the text or read portions of it were impressed, including Kratzenstein and Rottbøll, as well as Linnaeus's student and Danish professor Martin Vahl (1749–1804), Vahl's successor Professor Jens Wilkens Horneman (1770–1841), Kemner in the 20th century and Brazilian plant taxonomist Dr. Pedro Luís Rodrigues de Moraes in the 21st century.[39] Ironically, in light of the severe criticisms that have been aimed at Rolander in the twentieth century, his Surinam inves-tigations garnered just the sort of praise that Rolander's Swedish entomological studies had earned from Linnaeus. Linnaeus's foresight and perspicacity identified Rolander as a uniquely talented entomologist well before the Surinam expedition. The testimonies of Rottbøll, Kemner, and others strongly suggest that Linnaeus was wrong on just one point—Rolander was no longer strictly an entomologist; the Surinam expedition had forged him into a full-fledged naturalist.

The rehabilitation of Rolander's reputation is well under way. Rolander will speak for himself soon. My first edition of the Latin *Diarium*, which is due out in 2010, will provide scholars with a definitive text of the myriad of descriptions and essays that are Rolander's *Diarium*. A volume of scientific commentary and essays in English is being composed under my editorship to accompany the Latin text. Independent of the Latin edition, IK Foundation & Company of London published an English translation of the *Diarium* in December 2008, which will provide a considerably easier means of accessing Rolander for many readers.

The coming Latin edition and the English translation present two distinct versions of Rolander's *Diarium*. Many of the manuscript pages contain numerous changes made by Rolander, some numbering more than forty per page. The Latin edition will present all the changes, while the English translation is based on the final results of the changes. The changes are of two distinct types. The first, and most common, type includes grammatical, syntactical, and stylistic changes. These changes improved the linguistic clothing of the *Diarium* without augmenting or diminish-ing the information conveyed. Researchers interested in how authors prepare their manuscripts for publication will find such changes of great importance. Many other readers will find them of no interest at all.

The second type of changes consists of additions, deletions, or corrections that affect the meaning and content of the text. Most importantly, he changed many of the names he had originally given to those insects that had never been formally described prior to his visit to Surinam. These changes were probably implemented in Copenhagen from 1763–1765, when Kratzenstein's financial and moral support enabled Rolander to complete the final draft. Most such changes either add the species names that Linnaeus had applied to Rolander's insects in De Geer's collection or they are deletions of Rolander's names (either both genus and species, or only the species).

Such changes provide a fascinating testimony to Rolander's difficult life. He risked his life by undertaking the expedition to Surinam. He carried out his duties with diligence and care, despite poor health and trying circumstances. Yet Linnaeus won the race to publish the names of Rolander's beloved insects. Rolander's dedication to the advancement of natural history required that he update the nomenclature of the *Diarium*. He added Linnaeus's names, even when it meant that his own would have to be deleted—the bitterness of having to make such changes must have been nearly unbearable. The Latin edition of the *Diarium* will provide readers with a detailed picture of all the changes Rolander made. It also will clearly indicate with separate markings Rolander's changes to nomenclature, providing, whenever possible, the actual texts and names deleted by Rolander. In these respects, the Latin text will serve as the definitive source for all future research on this important European natural historian and explorer of the New World.

The rehabilitation of Rolander got under way in the late 1990s, when Lund University professor Arne Jönsson acquired a black-and-white microfilm copy of Rolander's manuscript *Diarium Surinamicum* from the library of the Botanical Museum in Copenhagen, Denmark.[40] Two of his Master's students later wrote their theses on Rolander and his *Diarium*.[41] Both works are well documented and effectively move the scholarly focus toward the text itself, while providing useful revisions of the biographical material.

Around the same time as Jönsson acquired his copy of the manuscript, Lars Hansen, a Swede now residing in England, initiated an ambitious project to produce and publish English translations of the diaries or expedition reports of seventeen Linnaeus apostles. He initially contracted Claes Dahlman, Jönsson's former student, to translate Rolander. Hansen, with Dahlman's consent, transferred the translation to me to complete the remaining 600 pages of the 699-page manuscript. Working together with Professor David Morgan of Furman University and Latinist Joseph Tipton, I translated 350 pages and Morgan and Tipton the remainder.[42] The translation was published in December 2008. Professor Jönsson received a two-year grant from the Bank of Sweden Tercentenary Foundation in 2007 that included funding for me to prepare a first edition of the Latin text along with extensive historical and scientific commentary. My work with the manuscript revealed the relationship to the tenth edition of *Systema Naturae*, which led to the identification of Rolander's insect specimens in the Swedish Museum of Natural History and their linking with the originals of De Geer's drawings. While reviewing manuscripts of Rolander's correspondence, I also discovered a 1,041-page manuscript containing Rolander's completely unknown translation of portions of René Antoine Ferchault Réaumur's *Mémoires pour Servir à L'Histoire des Insectes*.[43] Linnaeus's references in the early 1750s to Rolander as the next *Réaumur* suggests that Rolander made the long translation while living with Linnaeus (1750–1754).[44]

The evidence presented in this essay clearly shows that Linnaeus and De Geer, Rolander's two most influential contemporaries, greatly valued and fully exploited Rolander's Surinam insect collections, just as Rottbøll did with *Diarium Surinamicum* and Rolander's herbarium. From the statements of Kemner and others, it is apparent that had Rolander's *Diarium* been published, Rolander would have been considered a pioneer in entomology.[45] As it is, his dominating presence in the influential works of Linnaeus and De Geer placed him in the center of a great deal of entomological research; though to see him there, one had to know that the majority of Linnaeus's American insects from the tenth edition of *Systema Naturae* came from Rolander's collections. While it may be too late to credit Rolander with naming these eighty-some insects (he named them all in the *Diarium*, providing detailed descriptions of their habitats and habits), scientists and the reading public will soon be able to read for themselves just how profoundly entertaining Rolander could be while teaching the world about the minerals, plants, and animals (including man) of Surinam in 1755. The detail, knowledge, and craftsmanship of the entomological essays in the *Diarium* attest to Rolander's obsession with insects—an obsession that readers will enjoy for generations once Rolander's works are published.[46]

While Linnaeus seemed to have been obsessed with framing Rolander's expertise as strictly entomological, Rolander's own writings show us a natural historian who was far more than only a Linnean apostle. He was a talented observer with a sharp wit and a diligent explorer of the New World, a worthy successor to the great Dutch natural historian Willem Piso. Where other men of science in the 18th century had abandoned literary journals for long catalogs of plants and animals, Rolander retains the older tradition of creating a text that engages readers and educates as it entertains. In this and many other respects, Rolander remains an author who has a great deal to offer readers in the 21st century.

EPILOGUE

After this chapter had been reviewed and accepted, I obtained a copy of the German-Swedish historian Felix Bryk's *Linné als praktischer Entomologe* (Stockholm 1924). Bryk can be added to the list of Rolander's enthusiastic admirers. In four separate passages, he highlights the importance of Rolander's early publications. He even argues that Rolander was the first to "discover" that insects contributed to the fertilization of plants (p. 7).

ACKNOWLEDGMENTS

Access to the collections of the Hagströmer Library at the Karolinska Institute have greatly facilitated much of the research behind this article. I am particularly grateful for the insightful discussions enjoyed with Ove Hagelin, representative of the Hagströmer Library, and the skillful aid of the library's librarian Gertie Johansson. I thank Dr. Diane Warne Anderson whose criticism proved very helpful. The staff of the Swedish Museum of Natural History and the Rare Books and Manuscript Room at Uppsala University Library have my thanks for their prompt and professional aid.

APPENDIX

PRIMARY SOURCES

1. 20 Nov. 1750 old style; 1 Dec. 1750 new style
   En studiosus Dan. Rolander, som för ett par åhr sedan lärde hos mig något litet uti insecterne, har communicerat med mig ett par observationer, wid hwilka jag just baxnade och bad honom ingifwa dem till Wettenskaps Academien, ty de woro merveillieuse och sådane jag ej wäntat utan af en Réaumur äller en De Geer. ... När [Rolander] kommer till wettenskaps academien, skulle jag råda at en inventor distinguerades ifrån compilatores, ty då wij hafwom 1 inventor, hafwom wij 1,000 compilatores.[47]

2. Carl Linnaeus to Abraham Bäck: 13 January 1751 new style; 2 Jan. 1751, old style[48]
   Hvad tycker Min Bror om mine disciplar Kalm, Hasselquist, äller om skiutflugan och sichtflugan [this is a reference to two articles by Rolander], som äro af min schola?

   What does my Brother think of my disciples Kalm and Hasselquist or [the articles about] *skiutflugan* [Carabus crepitans L] and *sichtflugan* [Vespa cribraria L] that come [from Rolander] from my school?

3. Carl Linnaeus to Kungliga Svenska Vetenskapsakademien (Royal Swedish Academy of Sciences) :12 March 1752, new style; 1 March 1752, old style[49]
   Det är mer än et år sedan student Rolander ingaf til Kongel. Vetenskaps Academien et par rön om två mer än undransvärde Insecter, sikt-Flugan och skiut-Flugan, som voro i mitt tycke så stora, at man ej kunnat lofva sig sådane af Franssosernes mirakel Reaumur. Jag märkte straxt at den, som desse giort, vore originel ock icke copia, såsom vi mäste Autorer äro i verlden. iag giorde nogare bekantskap med denna ungkarlen ock fant honom vara af rätta sporrhundslaget ock at hafva det bästa väderkornet. En Professor, som skall upodla ingenia, kan aldrig mer distinguera sig vid sin sysla än igenom käcka Elevers anskaffande och upmuntrande, hvarvid största konsten består uti *selectu ingeniorum*, ty de rätte originalerne eller observatores äro ibland den andra hopen sälsynte, som Cometer ibland stiernorna.

   More than a year has passed since the student Rolander submitted to the Royal Academy of Sciences his two findings on those curious insects *sikt-Flugan* [Vespa cribraria L] and *skiut-Flugan* [Carabus crepitans L], which in my opinion were of such quality and import that one could hardly expect them from the Frenchmen's great Réaumur. I quickly realized that the author of those was unique, and not just another imitation—as are most of us authors in this world. I made a closer acquaintance with the young man and found him to be the right sort of natural investigator with a genuine knack for this work. There is no better way for a professor, whose duty it is to cultivate intellects, to distinguish himself than by attracting and encouraging sharp students. The key to all this is the art of selection, since the truly unique intellects and observers are rare among the others—indeed, as rare as comets among the stars.

4. Carl Linnaeus to Abraham Bäck: 12 March 1752, new style; 1 March 1752, old style[50]

Jag har icke haft någon rätt Eleve uti Insecterne, sedan iag kom till Academien, förän [Rolander], som är den endaste; får han [stipendium Thunianum], så är han bärgad ock Academien får i honom en ärlig skattdragare ock Fäderneslandet med tiden en Reaumur: det är intet fel på karlen, mer än det, at han icke är Medicus, utan totus quantus Theologus och Insectologus; hans makalöse rön om skiut-flugan, sikt-Flugan och pediculus pulsatorius har Min Bror set, ock nu har han med sig det som giör hvitaxen.

I have not had a suitable student of insects since I came to the Academy [Uppsala], until Rolander. He is the only one. If he gets the Thun stipendium, he'll be all set and the Academy (Swedish Academy of Sciences) will have in him a conscientious contributor and in time our Fatherland its own Réaumur. There is nothing wrong with him more than that he is not interested in the field of medicine. He is through and through a theologian and entomologist. You, my Brother, have already seen the peerless contributions he submitted on *skiut-flugan* [Carabus crepitans L], *sikt-Flugan* [Vespa cribraria L], and *pediculus pulsatorius*.[51] Now he has just finished one on *hvitaxen* [Phalaena (noctua) secalis L].[52]

5. Carl Linnaeus to Abraham Bäck: 12 March 1752, new style; 27 March 1752, old style[53]

Gud straffe mig, om Rolander någonsin tänkt, tänker äller blifwer Medicus. owännen har warit framme; liksom ingen bör studera naturen mer än endast Medici äfter gambla monke principiis.

I'll be damned, if Rolander considered or is considering or becomes a medical doctor. Our enemy has been busy! As if only medical doctors should study nature, just as the old monks recommended!

6. Carl Linnaeus to Domkapitlet i Växjö: 25 December 1753[54]

Herr Daniel Rolander har, utom wackra Philosophiska och Theologiska wettenskaper, rätt stor insigt uti scientia Naturali, och besynnerl. uti Historia Insectorum, hwaruti Han näppeligen lärer cedera någon uti Europa utom endast Fransmännernas Reaumur, som med denna wettenskapen uplyst sitt fädernes land.

Mr. Daniel Rolander possesses considerable knowledge of the natural sciences and particularly the natural history of insects, to say nothing of his fine grasp of philosophy and theology. In the history of insects he cedes to no one out in Europe, with the exception of the Frenchmen's Réaumur who has enlightened his fatherland with that science.

7. Rolander, *Diarium Surinamicum*: 10 November 1754, ms p. 4

Stibbetorp, villam ruralem, in paroecia Helleberga sitam, leucis septem Wexionia, et circiter quadraginta quatuor Holmia, distantem, perveni die 10 hujus mensis. Haec mihi natales dedit: annosque aetatis puerilis hic transigens, Insecta scrutari, in deliciis habui. Jam loca revisendi amoena, quae me puerum, negotiis ejusmodi occupatum, a sodalibus saepissime disjunxerant, haud parum me tenuit desiderii, et ea inprimis, ubi *Vespas rupestres*, nidos tubulosos lapidum lateribus, austrum spectantibus, construendo affigere vidi. Etiam, per eosdem telluris tractus, nidos earum, qua maximam partem, pro anni tempore, suis spoliatos habitatoribus, unum tantum, alterumve, illis occupatum, reperi. Alibi locorum nec ego, nec, quod sciam, haec Insecta, seu Vespas,[55] observavit quisquam, unde descriptio et historia earum a nemine tradita est: verbo tantum utramque hic tangere, per imperatam brevitatem, liceat. (10 Nov. 1754, ms p. 4)

8. Linnaeus to Abraham Bäck: 17 Feb. 1758[56]

Rolander borde hafwa någon pension, men aldrig läsa för medicis Mat. Medicam, det essentiellaste af Medicinen, utan wähl insecterne. Huru kan man undgå att höra opprobria?

Rolander should be given some sort of pension, but he ought not be lecturing medical students in *Materia Medica*, which is the very foundation of medicine. He should stick to insects. How can one not hear the insults?

9. Linnaeus to Kungliga Svenska Vetenskapsakademien: 26 April 1757[57]

Om min intercession kan något giälla, beder jag för Hr Osbeck; Rolander är braf i Insecterne; den andre i Botaniquen och hela naturkunnogheten.

If my intercession is of any importance, I am all for Mr. Osbeck [as a member of the Swedish Academy of Sciences]. Rolander is good with insects, while the other is so in Botany and all of natural history.

12. L2143 Linnaeus to Carl Gustaf Tessin 25 Feb. 1757 (n.s.)

Att Rolander fått understöd, tackar jag aldra ödmjukast; han är uti europeiska Insecter den habilaste, vi egt; och som denna vetenskapen nu blifvit så vidlyftig, tyckes vara värdt att hafva en enda i riket, som det förstår. Han är här uti så skicklig, som han skulle vara osk-icklig till Professor uti Botaniquen eller Mineralogien, som aldrig varit hans studium, fast han något lärt att samla uti hvarje; ty jag uppammade honom endast till Insecterne, under all den tid, han var i mitt hus. Jag har genom vänner rådt honom ej rodomontera, att han ej må en gång lida derföre.

I send my humble thanks that Rolander has got his support. Among Europeans he is the leading man we have had in insects. As that science has grown so much, it seems worth-while to the one [entomologist] in the kingdom who knows his stuff. He is so competent in this field as to be incompetent as a professor in Botany or Mineralogy, which have never been his topics of study. He has however learned to collect in both these fields. You see, I encouraged him only to study insects for the entire time he resided in my home. Through friends, I have encouraged him not to change this focus, lest he suffer for it at some time.

11. N. A. Kemner (1934), p. 28.

Als Ganzes sind diese Beobachtungen über die Termiten Surinams aber so interessant, dass eine Publikation derselben als erwünscht betrachtet werden kann. Wären sie nach Rolanders Rückkehr sofort publiziert worden, so würden sie zweifelsohne als die erste eingehendere Schilderung der Lebensweise der Termiten, die sogar durch kleine Experimente studiert wurde, sicher geschätzt worden sein. Die bedeutenden Beobachtungen über diese Tiere, die König und Smeathmann zu den Pionieren der Termitenbiologie gemacht haben, sind doch beide über 20 Jahre nach Rolanders Südamerikanischen Studium über diese Tiere gemacht worden. Und das Springen der Termiten, das hier zuerst beschrieben wird, ist erst in unsren Tagen wieder ent-deckt worden. Hagen betrachtete, wie gesagt, Linnés Worte darüber als einen Irrtum.[58]

MANUSCRIPTS AND LETTERS

Linnaean correspondence. Svenska Linnésällskapet, & Centre international d'étude du XVIIIᵉ siècle. (2005). *The Linnaean correspondence.* Ferney-Voltaire, France: Centre International d'Étude du XVIIIe Siècle. http://www.linnaeus.c18.net/.

Hagströmer Library. The Hagströmer Medico-Historical Library, Karolinska Institute, Stockholm, Sweden.

    Rolander to Abraham Bäck

    Rolander's translation of Réaumur's *Mémoires pour Servir à L'Histoire des Insectes*

    Bäck to Dr. Rudolf Fortstén

KVA. Kungliga Vetenskapsakademien, Stockholm, Sweden. See index for Bergianska brevsamlingen (The letters of P. J. Bergius).

    Rolander to Bergius

    Rolander to Wargentin

Lund University Library.

    Rolander to Eric Gustaf Lidbeck

    C. G. Kratzenstein to E. G. Lidbeck

ENDNOTES

1. The terms disciple and apostles are now and were in the age of Linnaeus applied to his students. The best of the best were selected to undertake natural-history expeditions in Sweden and abroad. These select few were his apostles.
2. Bibliographic and biographic information is provided by Anfält (1998), Krok (1925), Wilstadius (1978), Dobreff (2008), Ericsson (1969), and Sandermann Olson (1997). Krok is the most useful source for an overview of Rolander's influence in the publications of other authors.
3. Rolander (1750), (1751), (1752), (1754), and (1755).
4. See text 1 in the appendix Primary Sources at the end of this chapter.
5. Text 7 in appendix Primary Sources.
6. Blunt (2001), 12–18.
7. Text 4 in appendix Primary Sources.
8. The majority of his library is now housed at The Linnean Society of London.
9. Grape (1949), pp. 634–635, describes how complicated the selection of Rolander became. Queen Lovisa Ulrika had proposed another candidate. Only the last-minute intercession of Linnaeus's friend and Rolander's second patron Abraham Bäck secured the spot for Rolander.
10. Rolander's time in Amsterdam is *Diarium Surinamicum*, ms pp. 47–56. His illness is at pp. 55–56. The episode of his illness is described in detail in his letter to Linnaeus from Paramaribo, Surinam, dated 11 July 1755; L1932 in Linnaean Correspondence.
11. Linnaeus to Rolander 8 Oct. 1756; copy in KVA.
12. Beckmann (1911), p. 110.
13. Peter Jonas Bergius (1730–1790) had studied with Linnaeus when Rolander was also his student. Bergius became a famous Stockholm medical doctor. He was appointed the first professor of natural history and pharmacology in Stockholm in 1761. Like Linnaeus and Rolander, he came from the Swedish province of Småland. His herbarium is at the Bergianska botaniska trädgården in Stockholm. It includes the largest single collection (to date, over three hundred specimens) of Rolander's impressive herbarium. His success and growing reputation apparently pained Linnaeus, who wrote that since Bergius's appointment he was tired of being a professor at Uppsala and was packing his bags (*colligere sarcinas meas*), Linnaeus to Bäck, 2 Oct. 1762, L3148. He was perhaps joking, since he also jokes in reference to an impending visit of the king to Uppsala: *jag skall sättia mig in obscuro, unum coelum non fert duos soles*, i.e., I'm going to hide out, since one sky cannot endure two suns.
14. Dr. Abraham Bäck (1713–1795) was perhaps Linnaeus closest friend from the early 1740s to Linnaeus's death in 1778. He was a leading figure in Swedish medical circles. His titles included president of Swedish college of medical doctors (*Collegium medicum*), doctor to the royal court, and doctor to the royal crown. A considerable correspondence between Bäck and Linnaeus has survived, providing an intimate background to Linnaeus's scientific achievements.
15. Grape (1949), p. 638 and Linnaeus to Bäck 17 Feb. 1758, L 2303; *ibid*, no date in 1758, L2298; Rolander's *Diarium Surinamicum* depicts Rolander as skilled in the medicinal use of plants.
16. Kratzenstein to Lidbeck: Lund University Library.
17. To date, the only two sources for Rolander's stay in Denmark are a handful of letters from German-Danish professor Christian Gottlieb Kratzenstein (1723–1795) to Eric Gustaf Lidbeck (Linnacus student and professor of natural history at Lund University) and a single letter from Rolander to Lidbeck dated 6 Aug. 1763. Kratzenstein's letters date from 1763 to 1765.
18. The *Diarium* was left by Rolander with Kratzenstein and has been in the library collections of the Danish Botanical Museum in Copenhagen since 1811. I am preparing the first edition of the original Latin with modern scientific commentary. It is due out in early 2010.
19. No letter to Lars Salvius (Linnaeus's publisher), P. J. Bergius, or Linnaeus has been located, which might have discussed a proposal to purchase or sell the *Diarium* manuscript.
20. Wilstadius (1978).
21. Rottbøll (1775), (1776) and (1786).
22. De Moraes (under submission) and (under submission).
23. Rottbøll (1776), p. 7: *Tantam speciminum copiam secum advexit Dominus Rolander, ut cunctis Europae Botanicis sufficerent.*
24. For example, Rausing (2003), pp. 194–195, for claims of mental illness in Surinam; Sandermann Olsen (1997), p. 324, for claims that Rolander accomplished little or nothing in Surinam; Fries (1903), pp. 51–55 for both claims; and Lindroth (1967) follows Fries.
25. Wilstadius (1978).

26. Gertz (1936), p. 123.
27. Appendix 1 provides a full index of the Rolandrian insects cited by Linnaeus. The list also cross-references the same insects in De Geer's volumes; Linnaeus (1758) and De Geer (1752–1778).
28. Progress in taxonomy has meant that fewer and fewer of Linnaeus's names from 1758 for insects remain in use. That process has slowly pushed the little that was credited to Rolander further into the shadows of the history of science. Kemner (1934), Papavero (1971), pp. 8–9, and Ball (1996), p. 10, all express very positive opinions of Rolander's entomological work.
29. Fries (1903), pp. 51–55, is the main source of the modern misconceptions about Rolander. His very influential biography of Linnaeus (*Linné: lefnadsteckning*) remains a standard source for scholars and amateurs writing about Rolander and anyone else related to Linnaeus.
30. National resurscentrum för biologi och bioteknik (2007), p. 85. This publication was almost assuredly based on Fries (1903) cited above.
31. Dobreff (2008) provides more details on Rolander as a botanist.
32. Rolander to Abraham Bäck: 23 July 1756, Hagströmer Library.
33. Grape (1949) mentions De Geer's contribution, p. 636. The sources are Linnaeus to Bäck: 1 Nov. 1754 (old style), L1839; and Linnaeus to Bäck 19 Nov. 1754 (old style), L1834.
34. Anfält (1998) knew that De Geer had gotten a considerable collection of insects from Rolander and that Linnaeus had eventually studied them. He never pursued the issue further. While noting in some detail Rottbøll's use of Rolander plants, Anfält does not mention the presence of Rolander's insects in Linnaeus's or De Geer's publications or drawings.
35. Appendix 1 lists these insects with Linnaeus's names.
36. Appendix Primary Sources 5, 8 and 9; he continued to present Rolander as only an entomologist even after Rolander's return from Surinam.
37. Kemner (1934), pp. 31–55.
38. Kemner (1934), pp. 23–24, Primary Sources 11 (translation by Dobreff). Kemner had gotten wind of Rolander's *Diarium* in Danish entomologist J. C. Schiödte's article *Corotoca og Spirachtha: Staphyliner, som föde levende Unger, og ere Huusdyr hos en Termit* from 1854. Schiödte also quotes two long passages on Rolander's *Termes saltatorium, s. fatale*, which is Linnaeus's *Termes fatale*. He notes that Rolander had correctly observed that this insect's large mandibles were not weapons but part of a mechanism of locomotion. Generations after Rolander, ignorant of his observations, had considered this insect as a soldier termite.
39. The library of the Danish Botanical Museum in Copenhagen preserves a manuscript from Vahl, which shows that Vahl was planning to publish an extensive work on the animals in Rolander's *Diarium*. Horneman (1811) includes translations of several sections of the *Diarium* into Danish. He provides insightful observations on Rolander's personality and why he had such a poor reputation among the Swedes. De Moraes, as part of the Daniel Rolander Project, is currently reconstituting Rolander's herbarium, which now numbers over one hundred specimens; De Moraes (under submission).
40. I describe Rolander's life and the modern period leading up to the English translation and the preparations of the Latin edition, see Dobreff (2008b).
41. Ohlsson (1999) and Dahlman (2000).
42. I was the leading translator, establishing guidelines for the translators, proofing their translations against the manuscript and performing the final three proofs for the publisher IK Foundation & Company.
43. Dobreff (2008), where the manuscript is discussed and described. I also describe my other advancements in Rolandrian scholarship as well as those of Dr. Pedro De Moraes.
44. Appendix Primary Sources 1 and 3. This topic is dealt with in detail in Dobreff (2008), though I now consider a composition date of the early 1750s as more likely.
45. Kemner (1934), pp. 23–24.
46. Rolander (2008).
47. Linnaean Correspondence: L1193 - Carl Linnaeus to Kungliga Svenska Vetenskapsakademien (20 Nov. 1750, old style; 1 Dec. 1750 new style).
48. Linnaean Correspondence: L1224.
49. Linnaean Correspondence: L1397.
50. Linnaean Correspondence: L1396.
51. P. pulsatorius, see Rolander (1754).
52. Rolander (1752).
53. Linnaean Correspondence. This comment from Linnaeus to his old and trusted friend Abraham Bäck, himself a leading medical doctor, cannot be taken completely at face value. Linnaeus had to present Rolander as not being a student of medicine, since medical students could not apply for the stipendium Thunianum for which Linnaeus had been promoting Rolander.

54. Linnaean Correspondence: L5797. *Domkapitlet* is the cathedral chapter in Växjö. One of Linnaeus's responsibilities at Uppsala University was to report on the progress of students from his native province of Småland. This quote comes from one such report.

55. *Rolander added* seu Vespas.

56. Linnaean Correspondence: L2302.

57. Linnaean Correspondence: L2175.

58. English translation given in the last section of the essay.

BIBLIOGRAPHY

Anfält, T. 1998. Daniel Rolander, in *Svenskt Biografiskt Lexikon.* Stockholm.

Beckmann, J. 1911. *Johann Beckmanns Schwedische Reise in dem Jahren 1765–1766. Tagebuch, mit Einleitung und Anmerkungen* (Fries, Th. M. Ed.). Uppsala: Almqvist & Wiksell.

Blunt, W. and W. T. Stearn. 2001. *The compleat naturalist: A life of Linnaeus.* London: Lincoln.

Browne, P. 1756. *The civil and natural history of Jamaica.* London.

Cockerell, T. D. A. 1899. Some notes on Coccidae. *Proceedings of the Academy of Natural Sciences of Philadelphia* 51: 259–75.

Dahlman, C. 2000. *Diarium Surinamicum: om Linnéapostln Daniel Rolander och hans latinska resedagbok.* Master's thesis, Lund University, Sweden.

De Geer, C. 1752–1778. *Mémoires pour servir à l'histoire des Insectes* (vols 1–8). Stockholm: Pierre Hesselberg.

de Moraes, P. L. R., J. Dobreff, and L.-G. Reinhammar. 2010. Lectotypification of *Nectandra sanguinea* Rolander, *typ. gen.* (Lauraceae). *Harvard Papers in Botany.*

de Moraes, P. L. R., J. Dobreff, L.-G. and Reinhammar. (submitted). Current taxonomic status of Daniel Rolander's species published by Rottbøll in 1776. *Taxon.* Dobreff, J. 2008. Redux Rolander: Advancements in Rolander research as well as the identification of an unknown Rolandrian manuscript. *Svenska Linnésällskapets Årsskrift (Yearbook of the Swedish Linnaeus Society).*

Dobreff, J. 2008b. Translator's and Transcriber's Notes. In Rolander, D. (2008). *The Surinam Journal: Composed during an exotic journey.* (Dobreff, J. et al. trans.; Hansen, L. Ed.). London.

Eriksson, G. 1969. *Botanikens historia i Sverige intill år 1800.* In series *Naturvetenskapernas historia i sverige intill år 1800,* 17: 3 (J. Lindroth and Johan Nordström, Eds.). Uppsala: Almqvist & Wiksell.

Faber, B., A. Buchnerus, C. Cellarius, J. M. Gesnerus, J. G. Graevius, and A. Stübelius. (1735). *Basilii Fabri Sorani Thesaurus Eruditionis Scholasticæ omnium usui et disciplinis omnibus accommodatus / post cele-berrimorum virorum Bvchneri, Cellarii, Graevii, operas et adnotationes et multiplices Andreae Stübelii curas iterum recensitus, emendatus, locupletatus a Jo. Matthia Gesnero. Accessit primum hac editione verborvm et formularum interpretatio Gallica.* Lipsiae: Saalbach.

Fries, Th. M. 1903. *Linné : lefnadsteckning* (2 vols.). Stockholm: Fahlcrantz.

Gertz, O. 1936. Linné och cecidologien. III. *Svenska Linnésällskapets Årsskrift.*

Horneman, J. W. 1811. Om den svenske Naturforsker Daniel Rolander og Manuscript af hans Reise til Surinam. *Det skandinaviske Literaturselskabs Skrifter 7.*

Horneman, J. W. 1812. *Om de svenske naturforsker Daniel Rolander og manuscript af hans reise til Surinam.* København: Andreas Seidelin.

Jonston, J. 1653. *Historiae Naturalis de Insectis Libri III. De Serpentibus et Draconibus Libri II.* Frankfurt [Main]: Merian.

Kemner, N. V. A. 1934. Über Linnés Termes fatale 1758 und Rolanders springende Termite aus Surinam. *Entomologisk Tidskrift* 55: 10-53.

Lindroth, S. 1967. *Kungl. Svenska vetenskapsakademiens historia 1739–1818.* Stockholm: Almqvist & Wiksell.

Linnaeus, C. 1758. *Systema Naturae* (10th ed.). Stockholm: Lars Salvius.

Linnaeus, C. 1759. Von der schwedischen Cochenille. *Kongliga Vetenskapsakademiens handlingar* 21: 28–30.

Linnaeus, C. 1957. *Vita Caroli Linnaei: Carl von Linnés självbiografier* (E. Malmeström and A. Hj. Uggla, Eds.). Stockholm: Almqvist & Wiksell.

National resurscentrum för biologi och bioteknik. 2007. *Med livet som insats i fjärran länder.* Uppsala: National resurscentrum för biologi och biotekni. Also on line at www.bioresurs.uu.se (accessed September 23, 2007).

Ohlsson, J. 1990. *Daniel Rolander, Diarium Surinamicum. En Linné-lärjunges resedagbok.* Master's thesis, Lund University, Sweden.

Papavero, N. 1971. *Essays on the history of Neotropical Dipterology, with special reference to collectors (1750–1905).* (vol. 1). São Paulo: Universidade de São Paulo.

Plinius, C. S. 1603. *The Historie of the World. Commonly called, The Naturall Historie of C. Plinius Secundus* (P. Holland, Trans.). London: Adam Fslip. http://penelope.uchicago.edu/holland/index.html (accessed October 26, 2008).

Plinius Secundus, G., and L. Ianus and K. Mayhoff. 1967. *Naturalis historiae libri XXXVII.* Stutgardiae: Teubner.

Piso, W., J. D. Laet, C. Marggraf. 1648. *Historia naturalis Brasiliae ... in qua non tantum plantae et animalia, sed et indigenarum morbi, ingenia et mores describuntur et iconibus supra quingentis illustrantur.* Lugduni Batavorum & Amstelodami: Franciscus Hackius & Ludovicus Elzevirius.

Piso, W., C. Marggraf, J. D. Bondt. 1658. *Gulielmi Pisonis medici Amstelaedamensis De Indiae utriusque re naturali et medica: libri quatuordecim, quorum contenta pagina sequens exhibit.* Amstelaedami: Ludovicus et Danielis Elzevirs.

Rausing, L. 2003. Underwriting the Oeconomy: Linnaeus on Nature and Mind. *History of Political Economy* 35: 173–203.

Réaumur, R.-A. F. d. 1734–42. *Mémoires pour servir à l'histoire des insectes.* Paris: De l'imprimerie royale.

Retzius, A. J. 1783. *Caroli. De Geer. Genera et Species Insectorum.* Lipsiae: S. L. Crusius.

Rolander, D. 1750. Skjutflugan upptäckt och beskrifven in *Kongliga Vetenskapsakademiens handlingar.* XI, 290–295. Also published in German in *Der Königl. schwedischen Akademie der Wissenschaften Abhandlungen, aus der Naturlehre, Haushaltung und Mechanik,* 1754, and in Latin in *Analecta transalpina,* vol. 2, 1762.

Rolander, D. 1751. Beskrifning på Sikt-Biet in *Kongliga Vetenskapsakademiens handlingar.* XII, 56–60. Also published in German in *Der Königl. schwedischen Akademie der Wissenschaften Abhandlungen, aus der Naturlehre, Haushaltung und Mechanik,* 1755, and in Latin in *Analecta transalpina,* vol. 2, 1762.

Rolander, D. 1752. Hvitax-masken beskrifven in *Kongliga Vetenskapsakademiens handlingar.* XIII, 62–66. Also published in German in *Der Königl. schwedischen Akademie der Wissenschaften Abhandlungen, aus der Naturlehre, Haushaltung und Mechanik,* 1755, and in Latin in *Analecta transalpina,* vol. 2, 1762.

Rolander, D. 1754. Beskrifning på Vägg-smeden in *Kongliga Vetenskapsakademiens handlingar.* XV, 152–156. Also published in German in *Der Königl. schwedischen Akademie der Wissenschaften Abhandlungen, aus der Naturlehre, Haushaltung und Mechanik,* 1756.

Rolander, D. 1755. Anmärkningar öfver en bar Larve, med 16 fötter och tvådelta leder, som lefver af Såfvel-mat in *Kongliga Vetenskapsakademiens handlingar.* XVI, 51–55. Also published in German in *Der Königl. schwedischen Akademie der Wissenschaften Abhandlungen, aus der Naturlehre, Haushaltung und Mechanik,* 1757.

Rolander, D. 1756. Doliocarpus. En ört af nytt Genus från America, in *Kongliga Vetenskapsakademiens handlingar.* XVII, 256–261. Also published in German in *Der Königl. schwedischen Akademie der Wissenschaften Abhandlungen, aus der Naturlehre, Haushaltung und Mechanik,* 1757.

Rottbøll, C. F. 1775. Observationes ad genera quaedam rariora exoticarum plantarum, cum genere novo Rolandrae. *Societatis Medicae Havniensis Collectanea* 2: 245.

Rolander, D. 2008. *The Surinam Journal: Composed during an exotic journey.* (Dobreff, J. et al., Trans.; Hansen, L. Ed.). London.

Rolander, D. (forthcoming). *Diarium Surinamicum, quod sub itinere exotico conscripsit Daniel Rolander* (Dobreff, J. Ed.).

Rottbøll, C. F. 1776. *Descriptiones rariorum plantarum, nec non materiae medicae atque oeconomicae e terra Surinamnsi fragmentum placido ampliss. professorum examini, pro loco in consistorio rite tenendeo disputaturus, subjicit Christianus Friis Rottbøll.* ... Havniae: university printer N. C. Höpffner.

Rottbøll, C. F. 1786. *Descriptiones et icones rariorum et pro maxima parte novarum plantarum: cum Tab. XXI aeneis.* Havniae: Gyldendal.

Sandermann Olsen, S.-E. 1997. *Bibliographia discipuli Linnaei: Bibliographies of the 331 pupils of Linnaeus.* Copenhagen: Copenhagen Bibliotheca Linnaeana Danica.

Schenson, Emma. 1864. *In memoriam Caroli A Linné.* Uppsala.

Schiödte, J. C. 1854. Corotoca og Spirachtha: Staphyliner, som föde levende Unger, og ere Huusdyr hos en Termit. *Det kongelige danske videnskabernes selskabs skrifter: Naturvidenskabelig og Mathematisk Afdeling,* 5.4: 42–59.

Solander, D. 1995. *Daniel Solander: Collected correspondence, 1753–1782* (E. Duyker & P. Tingbrand, Eds.). Melbourne: Miegunyah Press.

Sörlin, S., and O. Fagerstedt. 2004. *Linné och hans apostlar.* Stockholm: Natur och kultur/Fakta etc.

Wilstadius, P. 1978. Daniel Rolander. In *Smolandi Upsalienses: Smålandstudenter i Uppsala: Biografier med genealogiska notiser: 1700–1744.* (vol. 5) Karlskrona: Uppsala Smålands Nation.

4 Taxonomy and the Survival of Threatened Animal Species
A Matter of Life and Death

Gordon McGregor Reid

CONTENTS

INTRODUCTION: NAMES AS DESCRIPTORS, DESIGNATORS, AND HYPOTHESES

Following the *Systema Naturae* (1758) of Carolus Linnaeus (1707–1778) the recognition and naming of species and other taxa became and still remains the fundamental activity of systematists working in biology, medicine, palaeontology, and even in mineralogy. In *The Meaning of Culture* (1929) Welsh Philosopher John Cowper Powys highlights the strong emotional and intellectual reassurance provided by names given to physical and biological entities:

> Each planet, each plant, each butterfly, each moth, each beetle becomes doubly real to you when you know its name. Lucky indeed are those who from their earliest childhood have all things named. This is no superficial pedantry. Deep in the oldest traditions of the human race dwells the secret of the magical power of names.

The philosophy and science of Linnaeus, particularly in his youth, was a curious blend of the rational and the irrational, drawing on traditional knowledge, folklore, and magic (Broberg, 2006; Reid, 2009). This reflected a culture, belief system and informal nomenclature widespread and deeply entrenched in Europe since at least the mediaeval period (Thomas, 1983). Named entities have always been the fundamental basis for Linnaean and other taxonomies, some numerical systems excepted. As Fortey (2008: 316) attests: "Every species on earth has its story to tell. But the first stage will always be the naming of names."

As well as providing a general emotional and intellectual reassurance in natural philosophy, names act at different levels of generality as descriptors, designators, and scientific hypotheses in systematic and general biology and as legal entities in conservation and environmental law. Modern challenges to the science and utility of the Linnaean system are emerging strongly, particularly from the bioinformatic pressures of conservation science. Substitutes, in full or in part, such as the Phyllocode, DNA bar-coding, and Web-based systems are often advocated (Pickett, 2005; Savolainen et al. 2005; Alves et al., 2007; Web of Life electronic resources at http://www.nhm. ac.uk). However, no satisfactory generally accepted alternative to Linnaean taxonomy has so far been found.

Of course, philosophers long before and since Linnaeus have debated the "nominal fallacy" perhaps implicit in his system. That is to say by naming physical or biological phenomena we are not necessarily gaining a greater understanding of them. For example, the "lumbago" diagnosis of physicians is simply another way of telling patients that they have a sore back—which fact they already knew and which condition may have multiple causes. Similarly, in taxonomic or conservational diagnoses one cannot necessarily be sure that the name corresponds with an objective reality and has universal application. Names in themselves do not provide certainty in the real world that we see and otherwise sense and that we know or believe to be under increasing environmental threat. It has yet to be determined (if such is possible) whether animal, plant, and other species are the same general biological phenomenon across all taxa, and whether they exist as objective entities independent of the minds of the philosophers and scientists who perceive that they exist and who name and classify them. According to Thomas (1983: 60):

> The work of many anthropologists suggests that it is an enduring tendency of human thought to project upon the natural world (and particularly the animal kingdom) categories and values derived from human society and then to serve them back as a critique or reinforcement of the human order.... .

The question remains of whether named taxa are simply artificial constructs—perhaps, through circular arguments, reflecting human hierarchies, relationships or concerns. For example, the relationship to military and also theological ranks is discernable in the Linnaean hierarchy (Broberg, 2006, Reid 2009). The issue of rationality and objectivity is magnified when considering higher-level Linnaean categories such as genera, families, and orders.

The French naturalist Buffon (1707–88) considered that only individuals were objective entities (Reid, 2009). Today, many would argue that species are objective entities, while genera, families, orders, and other such categories are not. John Ray (1627—1705), whom Linnaeus admired, first suggested that species may be interbreeding units (Reid, 2009: 10). According to Ray in a letter to Edward Lhwyd (Lloyd) at the Ashmolean Museum, Oxford, July 18th 1692; p. 229 in correspondence edited by Gunther, 1928: "For Animals not pamper'd by man, and at their liberty, if they can find any of their own species will never couple with a strange one."

This prefigures the traditional "biological species" concept (Mayr, 1942) utilized more by conservationists than taxonomists, particularly in the identification, breeding, and genetic management of small threatened populations (IUCN-CBSG, 2009—http://www.cbsg.org/cbsg/ Pellier et al., 2009). For Mayr (1969: 29) a taxonomist:

> … uses the amount and kind of morphological difference only as an indication of reproductive isolation, only as evidence to draw an inference. … The experienced taxonomist knows what variation to expect within a biological species.

This raises fundamental, so far unresolved, questions over biological process (such as reproduction and genetic relationships) being used in whole or in part to define species versus traditional Linnaean "morphospecies" diagnoses based on form and pattern. Whether species are fixed and immutable, as many creationists believe and Linnaeus initially attested, or constantly evolving in space and geological time à la Darwin and Wallace (1858) and Darwin (1859) is still a matter of public (if not scientific) debate, with lingering theological aspects. Linnaeus gradually moved to the view that species might change to some degree through, perhaps, hybridization (Reid 2009). The emergence of the cladistic method in interpreting evolutionary or phylogenetic relationships (Hennig, 1966) later encouraged authors to reevaluate various species concepts and to first propose the alternative idea of a cladistic or phylogenetic species concept (Reid,1978: 9–22). This cladistic perspective, while still valid, does not take into account contemporary concerns over interspecific hybridization. In any event, some authors consider that classical Linnaean taxonomy lacks objectivity and is practically obsolete (Alves et al., 2007) while Groves (2004: 1105) argues that:

> Taxonomy has a well-defined role, which is much more than simply stamp-collecting and pigeon-holing. Species are the units of classification, biogeography and conservation. … The phylogenetic species concept—a pattern-based concept—is as nearly objective as we are likely to get.

Be that as it may, more immediately practical complications arise from a broad examination of the historical and recent taxonomic literature. This is replete with synonyms (one or more names applied to the same species, other than the name by which it should be "correctly" known) and homonyms (an identical name applied to what are deemed to be separate taxonomic or biological concepts). Even the "characters" by which species and other taxa are recognized regularly engender debate over anatomical and other correspondences that may imply a close phylogenetic relationship, versus those that may not, having perhaps "evolved in parallel." Today, phylogenetic (cladistic) hypotheses (sensu Henning, 1969) involving homologous synaptypic characters often determine levels of generality in taxonomic relationships. This increasingly informs contemporary biological classification, which in turn helps to determine conservation status and priorities.

Modern scientists, including taxonomists, are concerned to move beyond the notion of "magical properties" and the theological or psychological benefits provided by a name—albeit topics interesting in themselves and worth investigating. Today's systematic biologists would rather remove confusion, doubt, and uncertainty in naming. They focus instead on the sometimes complex and intertwined biological, ecological, evolutionary, biogeographical, genetic, and conservational concepts that may be implied by a name, and they seek to develop worthwhile and testable hypotheses to be associated with that name. From this point onward we can develop a body of knowledge removed from irrationality or theology and properly recognized as science, where names are descriptors, designators, and meaningful hypotheses that can be refuted (*sensu* Popper, 2002). But how do we reach this important and pragmatic position and what might we be progressing toward in terms of the vital significance of names in natural science, especially where the survival of geographical populations, species, or entire phyletic groups is at stake?

To tackle these issues we first need to reflect on the role of "folk taxonomy" and traditional knowledge in the survival and evolution of humans and other species. This leads to a consideration of the origins of the Linnaean system and its use in conservation; the actual and potential conservation role of preserved and living taxonomic collections; the origins, nature, and scale of conservation challenges; the taxonomic basis of threat assessment and conservation prioritization; the role of taxonomy in the statutory protection of species; the moral or legal ownership of names and, finally, their financial value or cost in conservational terms.

FOLK TAXONOMY, ECOSYSTEM SERVICES, AND SPECIES SURVIVAL

Linnaeus himself has a strong interest in what would now be termed ethnography and anthropology, including that of the Saami Lapps of northern Sweden and tribal peoples overseas (Reid, 2009). He was also greatly interested in the economics of exploiting the natural environment and the associated folk taxonomy. Humans have long been able to recognize, informally describe, illustrate, name, celebrate, and utilize many different "kinds" or "varieties" of animals, plants, or allied organisms and their products. Paleo-archaeology reveals that prehistoric humans gradually evolved the power of self-expression and organized themselves into simple communities and then more complex social or ethnographic groups. Palaeolithic cave art is a powerful medium of early tribal communities that illustrates favored or revered animals sometimes recognizable to species (Ucko and Rosenfeld, 1967). In one sense, cave art is an early example of biological illustration, nowadays a vital technique in support of taxonomic and other science. This artistic and cultural phenomenon probably developed in parallel with the human ability to verbalize effectively and chorus in a quasi-musical context (Merker, 2000), or perhaps just ahead of it. This early skill (present to some degree in other species, notably primates) to rapidly sort, categorize, discriminate, visualize, and communicate effectively has served and continues to serve a vital purpose, both in industrial and pre-industrial civilizations and rural communities.

The immediate pre-industrial or rural benefits lie in quickly and effectively promulgating the use, value, availability, or seasonality of any one kind of animal or plant, its associated products, or effects. This is often in terms of food, clothing, shelter, or culture, or as something to trade; or in relation to health hazards or medicinal benefits that might exist—as with an edible but dangerous prey animal, or useful herb versus a poisonous berry. A propensity to seek spiritual values through a kind of "biophilia" (Wilson, 1992) has often led to the worship of totemic animal items and the creation of societies associated with these, as in the "leopard spirit cult" of the Cross River region of Cameroon (personal observation). Nature worship through a deep cultural belief in animism is still a complex and globally widespread phenomenon (Bird-David, 2002). Evolutionary psychologists suggest that such religiosity may be generally adaptive and "hard-wired" in the brain (Wilson, 2003; Edmonds, 2009).

In one context or another, the important modern conservation concept of "rarity" (Hartley and Kunin, 2003) must have been developed early on in human evolution, with the value of animals and plants being in part predicated and enhanced by how unusual they are and (from contemporary research in zoological gardens) how far humans are attracted to them (Angulo et al., 2009). This enjoins the economic law of supply and demand concerning, for example, an uncommon and prized food item, a fine and difficult-to-acquire feather for a headdress, or rare shells and seeds for a necklace. From such rudimentary bio-economic, social, cultural, and spiritual beginnings we can reasonably visualize the emergence of a widespread and sometimes sophisticated oral folk taxonomy of the kind that persists around the globe today in many pre-industrial civilizations, and that thrives in parallel with the more formal scientific system and sometimes contributes to it. Traditional knowledge, including the concept of rarity and the need to somehow utilize and preserve it in economic terms was especially important in the time of Linnaeus (Koerner, 1999) and remains very relevant to contemporary systematic biology and to conservation and environmental biology as a whole. By the process of empiricism, such traditional knowledge may have been reliably gained over millennia to be passed on orally down to the present day.

Some knowledge would have been acquired in difficult circumstances amounting to an evolutionary and psychosocial selection pressure critical to human survival, e.g., the utilization of insulating furs and dried food supplies to enable tribes to live through the winter or a dry season, when food and clothing become scarce or difficult to access; or the added benefits of domesticating and protecting animals and plants in supporting existence. Today, these immediate and obvious material, social, and cultural values devolve to a far wider range of "ecosystem services," including air to breathe, water to drink, food to eat, provision of a stable climate, and natural beauty to appreciate (MA, 2003; Watson and Smith, 2009). Surprisingly, these life-sustaining aspects are only now, in the 21st century, beginning to be fully understood and appreciated. Species survival today has in the human mind and popular culture seemingly become one step removed from the stark prehistoric situation of environmental dependency, even if ultimately the life or death reality is still actually the same in terms of risks associated with undermining the functionality of nature and its components. Species are generally regarded as vital units or "building blocks" in ecosystems (Groves, 2004) and so the accurate recognition, naming, and description of these become progressively more important for conservationists and environmental scientists.

Impressive ethnographic or anthropological traditions notwithstanding (Leakey, 1981: 108–109), studies of pre-industrial societies indicate that there are definite limits to the numbers of named kinds or species of animals and plants recognized in folk taxonomies (Berlin, 1973). This is usually only broadly specific or generic and numbering in the hundreds, with 900 (rather than thousands) as a characteristic maximum. Hence, any move to capitalize on widespread knowledge (beyond the limitations of individual human or tribal memory and oral communications) instantly creates a need for a more permanent, robust, and progressively more rapid taxonomic documentation and information handling system. This leads ultimately to modern museum-based reference collections, electronic computerization and bioinformatics supported by scientific illustration, photography, and other techniques for visualization. Following on from Linnaeus' *Systema Naturae* (1758) the original, relatively crude, folk-assortment of "kinds" can nowadays be referred to as populations, subspecies, species, genera, families, or other taxa (Winston, 1999).

Domestication or cultivation of animals and plants, beginning in prehistory, and their proliferation as named entities in space and time (particularly post-1500) is a central aspect of human evolution (Clutton-Brock, 1981). So, too, is a strong historical and recent propensity to translocate wild animals and plants. Linnaeus himself was a well-known enthusiast for domestications and translocations for agricultural and economic purposes (Koerner, 1999). Today we know that such animals and plants may become resident in areas well outside of their natural home range, frequently becoming invasive and environmentally problematic in the new location (Lowe et al. 2000; Global Invasive Species Database, 2009—http://www.issg.org/database/species/reference-files/). Sometimes this can pose theoretical and applied problems in modern taxonomy and conservation. For example, the Australian dingo or wild dog (*Canis familiaris dingo*) has probably existed in its present semi-domesticated form for some 40,000 years in association with Australian aboriginals and has become feral and naturalized. Although sometimes hybridizing with modern domestic dogs, it has been accorded a subspecific rank, implying some degree of natural wildness and biological purity. Not originally Australian, should the dingo now be regarded as a "wild" or a "domesticated" Australian dog and, in the former case, should we include it in a list of species to be protected under conservation legislation? If so, what are the taxon threshold limits (Elledge et al., 2006) in morphology, genetics, space, and time? Alternatively, should we treat the dingo as an alien invader to be destroyed in the quest to conserve the previous "more natural" ecosystem and original species? In native species lists adopted for any one country, time seems to lend enchantment to the view, with the baseline in accepting species as being indigenous sometimes being taken as prior to the years 1500 or 1600 because of the better availability of material or documentary evidence by that time (Fisher et al., 1972: 11). In any event, the earliest named domestications, naturalizations, or extinctions often acquire a separate, formal status in biological and regulatory documents.

At a lower level of generality, humans express a very strong compulsion to give individual, domesticated companion animals, working animals, and large, "charismatic" zoo and farm animals personal names, to speak to them and to otherwise anthropomorphize (Thomas, 1983: 95–96; Anglin, 1995; Angulo et al., 2009). This is, for instance, the case for the Saami Reindeer-herding dog in Sweden (Anderson, 1986) that would have been familiar to Linnaeus, who kept dogs and other companion animals. Returning to the example above, were Australian aboriginals among the first ever to give humanized pet names to their dogs, albeit from a specialized lexicon? This is, perhaps, more a matter for psycho-anthropologists than taxonomists, but it presages a large 21st-century contradiction or conflict between a personal identification and human sympathy with individual animals, versus a much more limited commitment to wildlife management for entire species and habitats. This is reflected in the disparate allocation of scarce economic resources to meet these often separate ends. In short, people are generally far more sentimental about, and financially well disposed to, supporting the welfare of individual animals than they are about the more abstract notions of saving species and the environment and preserving ecosystem services.

ORIGINS OF THE LINNAEAN SYSTEM AND TAXONOMIC COLLECTIONS OF USE IN CONSERVATION

Thomas (1983: 80, 81) considers that the move by late 17th- and 18th-century taxonomists to Latinize vernacular names for animals and plants, especially under the influence of Ray and Linnaeus, was a device to create a "… gulf between popular and learned ways of looking at the natural world." It was also an attempt to avoid "vulgar errors," separate fact from folklore and credible from incredible hypotheses concerning the existence of particular species or properties. For Thomas, a regrettable result of this was that "… it made serious eighteenth century naturalists contemptuous of popular lore." Nevertheless, building on the enterprises of successive generations of systematic scholars and museum curators, Linnaeus established his "universal" taxonomic system of binomial nomenclature in the 18th century, with *Species Plantarum* (1753) as the recognized formal starting point for names available in systematic botany, and *Systema Naturae* (1758) as the recognized formal starting point for names available in systematic zoology (Jeffrey, 1973, Winston, 1999). Long prior to this, the first written scientific approach to knowledge began in Babylon, now the site of contemporary Iraq. From here, the *Codex Hammurabi* (ca. 3,768 BP; 1,760 BC) developed as an important early attempt to organize and present knowledge in a systematic way, often in a quasi-legal context (Tamera, 2005). Nonetheless, the *Codex* is permeated with supernatural traditions and (to the scientific mind) irrational thought, although worthwhile biomedical details on anatomy, ailments, and medical remedies are contained within. This represents valuable wisdom passed down through the generations but for the first time communicated in a fixed, written form. The development of writing, record-keeping, and illustrative artwork hugely advanced the cause of science and society because there was now a permanent, testable, and challengeable record or reference point. This also formed the basis for a legal system that would eventually become important for conservation as well as many other purposes. Written communication and illustration remain fundamental components in good science, notably including taxonomy (Winston, 1999); and in interpreting and stabilizing the associated legalistic codes of biological nomenclature (Melville, 1995).

With Aristotle (ca. 2392 BP; 384–322 BC) we have the "founding father" of systematic zoology and medicine. Building on the written work of other Greek philosophers including Homer, many kinds of animals and their organs were named, categorized, ordered in series, and illustrated in a scientific manner (Lloyd, 1962). This includes, for the first time through reasoned processes, the recognition of dolphins as belonging to the class of warm-blooded mammals, rather than fishes—a disposition not at first accepted by Linnaeus and other 18th-century scholars (Reid, 2007). Names became not simply labels. In Aristotle's original view they signify forms or substances with "essences." "Human" has the significance it does because it is seen to represent the same generalized kind or species on all occasions when it is used. The "Aristotelean essence" is the fundamental feature (or set of features) that makes the species, form, or substance what it is and explains other intrinsic properties. In modern biological terms this may translate as the taxonomic characters, homologies, and synapomorphies that are used descriptively when defining taxa as unique, and setting levels of generality in phylogenetic hypotheses (Reid, 2009). However, there evidently were limits to Aristotelian discrimination (Lloyd, 1962) and cataloguing and the number of valid individual species recognized in his time might be limited to about 1,000. Aristotle did refer to living and preserved specimens, but these existed only in a transitory Alexandrian Muséon, a loose amalgam and equivalent of the modern university, zoo, and museum. There was then evidently no perceived need for an established collection to serve as a fixed reference point.

The much later establishment of permanent preserved museum collections in the 14th century, including by Conrad Gesner and Ulisse Aldrovandii, meant that named biological concepts could be readily associated with specimens and so ultimately be tested and challenged using material evidence (Reid, 1994a). Linnaeus adopted and promulgated this museological approach, establishing extensive reference collections of plants, animals, fossils, and minerals, and also curating the Swedish royal collections (Reid, 2007). Utilizing Linnaean taxonomy, this also represents the beginnings of bulk data handling or bioinformatic systems, capable of dealing with the growing avalanche of specimens representing exotic species being registered in European, North American, and other museum collections. The Musée d'Histoire Naturelle, Paris, alone houses some 60 million specimens, the great majority acquired in the 18th, 19th and 20th centuries. Today the total global resource may amount to 2.5 billion preserved museum specimens (Graham et al., 2004). The Linnaean foundation for what was eventually to become a typological system meant that specimens became in some way representative of species and the name bearers for them, their biology, and conservation.

Alongside museum collections of preserved animals, Linnaeus was interested in living collections and had a personal menagerie with exotic creatures such as an orang-utan (Reid, 2009: 10). He would have known about the first "modern" zoo in Schonbrunn, Vienna (established 1752) and certainly corresponded with Austrian zoological colleagues. The early zoological gardens, including the Menagerie du Jardin des Plantes, Paris (established 1794) often supplied material for or were otherwise closely associated with preserved collections and became the first "living museums" for named animals. Through repositories of mainly exotic specimens, they supported museum- or university-based studies in taxonomy and comparative anatomy (Reid, 1994b). They also brought home to taxonomists and the general public the wonder, educational value, and potential use of natural and domesticated plant and animal diversity, particularly from the tropics. However, it was not until the late 20th century that collections-based institutions began to develop a proper, organized role in conservation (WZO, 1993; Reid, 1994b; Davis, 1996; Kitchener 1997; Miller et al., 2004; WAZA, 2005; Dollinger et al., 2006; Reid et al., 2008). Indeed, the last individuals of some species have gone extinct in zoos and become individually named, anthropomorphized museum specimens, e.g. "Martha" a passenger pigeon (*Ectopistes migratorius*) who died at the Cincinnati Zoo on September 1st, 1914. This was formerly the commonest bird in North America, with flocks evidently numbering in the billions. Another tragic example is "Benjamin" a Tasmanian wolf (*Thylacinus cynocephalus*—the only large marsupial predator to exist in modern times) who died at the Hobart, formerly Beaumaris, Zoo on September 7th, 1936.

Public aquariums (or aquavivariums, as they were originally known) followed in the wake of the establishment of zoos and became living underwater museum collections of fishes and aquatic invertebrates (Reid, 1994b). They were first established following the publication of *The Aquarium* (Gosse, 1854). Zoos and aquariums today are still often exhibited on taxonomic/zoogeographical lines, but they are developing a new agenda in conservation and sustainability (WAZA, 2005; Penning et al., 2009). These separate institutions differ mainly in the relative proportions of terrestrial and aquatic taxa held. Zoos and aquariums also often include terrestrial and aquatic plants, held increasingly for conservation purposes. A first survey (WZO, 1993) by the World Zoo Organisation, now the World Association of Zoos and Aquariums, indicates a taxonomic library or "goldmine" of more than one million individual animals held ex situ across all major vertebrate taxa. These, alongside preserved or living specimen collections of museums, botanical gardens, universities, genetic resource banks, and wildlife conservation agencies, can potentially be used to support diverse activities in conservation, education and science (Reid, 1994b; Kitchener 1997; Miller et al., 2004; WAZA, 2005; Dollinger et al., 2006; Reid et al., 2008; Kouba and Vance, 2009; Lerman et al., 2009). Nevertheless, from George Rabb's "Evolution of Zoos" diagram in the original World Zoo Conservation Strategy (WZO, 1993: 5) one might reasonably assume that taxonomy is a somewhat historical and largely complete exercise, very much associated with the natural history "cabinet of curiosities" of the 19th century and not following through to the environmental resource center of the 21st century. For some conservation biologists the role of systematic biology and contemporary relevance of preserved and living taxonomic collections is not entirely clear. For them, the mills of taxonomic institutions seem to grind slowly, and obtaining coherent and consensual views from taxonomists on names and concepts as applied to conservation issues is sometimes seen as an attempt to herd cats. Indeed, a contrary and paradoxical view has been expressed that scientific description can actually imperil species (Stuart et al., 2006).

ORIGINS, NATURE, AND SCALE OF CONSERVATION CHALLENGES IN RELATION TO TAXONOMY

While not a conservationist in the contemporary sense, Linnaeus was concerned about and corresponded with others on issues in deforestation and other habitat destruction in Europe and elsewhere (Reid, 2009: 7). Definitions of conservation vary, but a useful one is "actions that substantially enhance the survival of named species and habitats, whether conducted in nature (in situ) or outside of the natural habitat (ex situ)." This means that collections-based institutions can develop a strong role alongside the more widely recognized field-based wildlife conservationists. For our purposes, "sustainability" can be defined as "meeting the resource needs of the present, including exploitation of the biota, without compromising the needs of future generations."

Clearly, without names and a full taxonomic, ecological, geographical, and phylogenetic vocabulary, the language of conservation and sustainability cannot properly develop. Curiously, the major questions of taxonomy are not included by McNeely et al. (1998) in an otherwise full account of the major conservation issues of the 1990s, developed from workshops held at the World Conservation Congress in Montreal in 1996. Remarkably, resolving species taxonomy does not feature explicitly among the "hundred most important questions" framed by leading conservation biologists for the 21st century (Sutherland et al., 2009). Beneficial progress in systematic biology has usually lagged well behind issues in conservation and sustainability that were becoming manifest in Europe as early as the 16th century. Linnaeus, during his academic sojourn in Holland in 1735, would probably have been made familiar with a woodcut by Willem van Zanen 1598, published in 1602 (Reid, 2007). This artwork was perhaps the very first to unwittingly illustrate the potentially severe and negative impact of humans on animals, plants, and the natural environment—in this case, Dutch seafarers of the Admiral Van Warwyck expedition enthusiastically exploiting the flora and fauna and other resources of Mauritius. This particular island is now famous for several threatened species or subspecies illustrated by van Zanen such as fruit bat (*Pteropus* sp.), pink pigeon (*Nesoenas mayeri*) and echo parakeet (*Psittacula eques echo*). Also pictured was the iconic dodo (*Raphus cucullatus*), probably extinct before 1647 (Cheke and Hume, 2008, Figure 4.2.3), but strangely considered by Figuier (1877: 368) to be no loss to humanity. Interestingly, Charles Darwin visited Mauritius in 1836 on his way home, but by that time it was already biologically impoverished. So concerned was Darwin over the decline of the giant tortoises in Mauritius and elsewhere that he eventually wrote to the governor proposing conservation (Darwin et al., 1875). In any event, Darwin's *Origin of Species* (1859) provided taxonomic and evolutionary context for the explosion of knowledge of new species in the Victorian period (Winsor, 2009) and this eventually allowed for phyletic status to be taken into consideration when assessing conservation priorities.

Of course, for conservationists it is the sheer numbers of nominal species meriting attention and the many separate and intricate details of their biology that represent the first major challenge. Linnaeus described some 12,100 species in his lifetime of which 4,400 were animals and 7,700 were plants. He confidently predicted that the complete eventual plant list would not exceed 10,000 (Reid, 2009). However, the present rate of discovery of species is extremely high. At least 2,057 new vascular plants were named in 2006, along with 8,995 insects and 486 fishes (State of Observed Species, 2008). Since 2004, to date, more than 13 new amphibian species were recognized each month, with above 6000 so far named from a total list of perhaps 9000 (Global Amphibian Assessment, 2008). There has been an extraordinary explosion of knowledge across all major taxa, with new species being discovered daily, and there is no absolutely agreed upon projected total or easy means of establishing a figure.

The Global Biodiversity Assessment (Watson et al., 1995 and subsequent updates) recognizes 1.8 million described species as probably or possibly being valid, with a final projected total of perhaps 14 million. If true, this means that less than 13% of species have scientific names and a staggering 12.2 million remain to be formally described. In conservation terms, the picture is or has been partly confused by the number of fossil species represented in the growing lists, particularly for land-dwelling vertebrates, and these are not always separately indicated (MacPhee, 1999). There are also microbiological (and ultimately conservation-related) challenges in applying traditional species concepts to bacteria, viruses, and allied microorganisms and incorporating these in biodiversity figures. Whatever the precise totals may be, we know that recent species are being described at a far greater rate than the present ability to assess their threat status using standard IUCN and other systems. Overall, on the basis of limited and approximate assessments, we know there is an ever-increasing proportion of species officially classed as threatened with extinction in the IUCN Red List and other lists.

For many Linnaean and post-Linnaean taxa of conservation concern, basic questions of identity, naming, and biology remain (Dixon and Brishammar, 2007). Without this information many prospective conservation actions are rendered difficult or impossible. Before we can select and more effectively conserve representative taxa, there needs to be a clearer perspective on species numbers, life history data, ecological, genetic, demographic, and phylogenetic composition and the threats to populations. One can only really properly conserve what is known and accepted as taxonomically valid, particularly in the context of national law and international codes. Decisions on the particular species or other units to be recognized or prioritized for conservation management purposes might literally represent life or death for the affected populations or the taxon as a whole. For example, the tadpole shrimp (*Triops cancriformis*—with a fossil pedigree dating back beyond 220 million years) is known as the rarest freshwater crustacean in Great Britain. It is confined to only a very few localities, in which it is threatened and has special protection under the law, being scheduled as Endangered and protected under Schedule 5 of the UK Wildlife and Countryside Act 1981, (UK BAP, 1999). However, continental European populations of what appear to be the same species are widespread and fairly common, thus possibly meriting far less conservation attention. But are they really the same species of tadpole shrimp and might the European populations be safely and legitimately used to supplement the British ones should they become extinct? One might well argue "no," since the UK populations are self-fertilizing hermaphrodites, whereas those in Europe commonly have separate sexes and so appear to be biologically distinct. The full genetic, taxonomic, conservation, and other biological significance of these critical differences has not so far been resolved. This example could be multiplied for the invertebrates, with a particularly good instance being ongoing attempts to resolve taxonomic difficulties in working to conserve the 58 or more species (or genetic forms) of Polynesian land snails of the genus *Partula* (Johnson et al. 1988; Lee et al., 2009). In this context, Coote (2009: 25) considers that:

> The taxonomy of many species of partulid has never been clarified, creating problems for conservation planning. The advent of widespread extinctions now makes the clarification of the taxonomy a critical exercise to aid survival rather than just an arcane scientific pursuit.

The situation is similarly challenging among threatened lower vertebrate groups generally recognized as "speciose" but where individual species identification remains problematic. In Lake Victoria, East Africa, there is, for example, a supposedly monophyletic assemblage or "flock" of perhaps 500 species of haplochromine cichlid fish. The evolutionary origins of the flock may be as recent as 12,400 (carbon-14) years BP, and research on DNA suggests extensive interspecific hybridization at some perhaps early stage (Verheyen et al., 2003). The present-day flock contains numerous "morphs" and color variants that may or may not represent separate, described, or undescribed species. Some authors consider that the Lake Victoria haplochromines demonstrate the fastest known rate of vertebrate speciation by the dramatic process of "explosive evolution" and, if so, this makes the conservation of the assemblage all the more important. In any event, comparatively recent introductions to Lake Victoria of the Nile perch (*Lates niloticus*), an alien invasive predator, have decimated the haplochromine flock and evidently led to a high rate of extinction, perhaps the highest for any comparable vertebrate taxon (Bruton, 1990). Chronic taxonomic difficulties and the large geographical scale of the lake (which at 68,000 km² is about the area of Switzerland) make it hard to know exactly what species have been lost from which part of the lake, what species remain, and what should be saved for conservation breeding programs managed ex situ. One early estimate (not repeated) puts the number of extinctions at 200 species (Witte et al., 1992).

The same sort of situation commonly occurs among higher vertebrates, even familiar and long-established taxa such as the colorful and otherwise conspicuous Amazon parrots (*Amazona* spp.) of South America. There are about 50 distinct phena within the genus; some of these may merit recognition as full species and others as subspecies, or discrete geographical races. Elucidating the *A. lilacina* species complex is proving to be particularly difficult and, from current environmental pressures, there is a greater than 50% chance of extinction of several contained phena in five generations or within twenty years (Mark Pilgrim, personal communication). To determine exactly what needs to be saved, why and how, is a race against time. It demands a sophisticated approach both in situ and ex situ, combining traditional museum-based taxonomic and phylogenetic studies with analyses of zoogeography, DNA, genetics, ecology, and behavior.

Similarly, among well-known South American wild mammals, there is a growing threat to the survival of the jaguar (*Panthera onca*), and it is placed as "Near Threatened" in the IUCN Red List 2008. The main threat is from ranchers who shoot them, ostensibly to prevent predation on their domestic cattle. Again, the taxonomic/zoogeographical situation for jaguars is not entirely clear, which affects population and habitat viability analyses and conservation planning. The nominal species has (or originally had) a wide zoogeographical range extending from Brazil to the southern United States, with distinct zoogeographical phena of questionable taxonomic status being sometimes recognized (Eizirik et al., 2001). In 1933, some sixteen subspecies were erected, reduced to eight subspecies in 1939 and to one nominal species in 1997—but all decisions were unsatisfactorily based on inadequate museum and living material. Sanderson et al. (2002) recognize fifty-one jaguar conservation units representing thirty different jaguar geographic regions. There is an insurance conservation breeding program operated by the European Association of Zoos and Aquariums; however, about 90% of jaguars in this EEP program are of unknown geographical origin. Hence, if there are geographically and genetically definable jaguar subspecies or populations in nature, then it is likely that successive generations of their zoo offspring will be hybrids to some extent or other, and this serves no "pure" or truly useful conservation purpose. For example, these offspring could never be returned to nature following the IUCN Reintroduction Specialist Group Guidelines (http://www.ssc.rsg.org).

The same arguments pertain for many other high-profile species held in zoos, including giraffes (*Giraffa camelopardalis*), where many of the 720 animals kept in 146 institutions are inter-populational or inter-subspecific hybrids, and where major challenges arise over the nature and identity of subspecies in giraffes and whether they merit full species status (Damen, 2008: 21). Again, the current level of taxonomic uncertainty invalidates the prospect of any such hybrids being used in reintroduction programs, as and when needed, and also compromises or minimizes the value of scientific studies conducted ex situ.

Other uncertainties come through assessing the impact on diversity from natural hybridization or introgression. This sometimes occurs between species or populations in nature and may result in adaptations of evolutionary significance and major and rapid evolution (Grant and Grant, 1992; Arnold, 2004). Indeed, Grant and Grant (1992) provide data to indicate that approximately one in ten species of birds is known to hybridize, including among the populations of Darwin's finches on an island in the Galápagos. From this, these authors argue that:

Hybridization presents challenges to the reconstruction of phylogenies, formulation of biological species concepts and definitions, and the practice of biological conservation. (p. 193)

If this phenomenon is taxonomically widespread in nature, as it may well be, and exists at a high frequency, then the idea of monophyly at the species level and the associated phylogenetic species concept (as advocated by Groves, 2004 and others) are both severely compromised. So, too, are the genetic criteria used for diagnosing species and managing conservation breeding programs.

Hybridization also relates to human domestication through artificial selection—a practice that Linnaeus engaged in enthusiastically for agricultural and economic purposes (Reid, 2009). Today, there can be taxonomic and biological challenges in separating and treating wild versus domestic stock, as in the case of the dingo (mentioned earlier) and the bactrian camel (*Camelus bactrianus* L., 1758) listed by IUCN as Critically Endangered. From newsletters published by the Wild Camel Protection Foundation (http://www.wildcamels.com) led by John Hare, there are only about 1,000 or so individual Bactrian camels (*Camelus bactrianus "ferus"*) left in the wild as opposed to some 1–2 million domesticated animals. This is a migratory species in the Gobi Desert of China and Mongolia that can (according to Hare) be distinguished superficially from its domestic counterpart through individuals' being shorter and leaner, with smaller humps and shorter hair. Analysis of DNA from the skins of wild and domestic forms evidently shows a genetic distance of about 3%, but, while this is perhaps diagnostically useful, cladists would not accept it as a meaningful way to determine phylogenetic relationships. Furthermore, even a cursory examination of the taxonomic literature indicates typological and geographical problems and very few skins and skulls and living animals involved to support the present diagnosis. There is little documentation on the supposed or implied genetic processes of domestication, which may or may not involve artificial selection for particular characters. Clearly, there are testable alternative hypotheses that might readily explain the putative differences, including zoogeographical, ecophenotypic, and dietary variation and, indeed, the possibility that more than one species or subspecies is involved.

If one accepts the status quo, then there are between 8 and 15 wild bactrian camels in captivity in China and Mongolia, and only domesticated stock in zoos around the world (664 individuals comprising 233 males, 412 females and 19 juveniles in 7 geographical categories—ISIS Data, 2009 http://www.isis.org). If one is persuaded by contrary evidence that the substantial zoo stock actually represents the wild species, in whole or in part, an entirely new perspective then emerges on conservation breeding ex situ to strategically preserve genetic diversity. There would, for example, be far less of a case for culling wild putative "hybrids" (thus limiting genetic diversity) and conducting well-intentioned but costly and procedurally difficult embryo transfer work to assist reproduction in wild populations (as advocated by the WCPF).

Setting aside the issue of uncertainty in the diagnoses of apparently well known species with obviously large adult individuals, there is a growing issue in the recognition of "cryptic" species across all major taxonomic groups (Condon et al., 2008). These taxa have become diagnosable only in recent years, following technological advances in chemosystematics, proteomics, DNA barcoding, and other innovative techniques. For example, African freshwater cigar fish (*Brienomyrus*) were on gross morphology formerly thought to represent only a few geographically widespread species, with adult individuals attaining only the comparatively small size of a few centimeters long. Studies by Arnegard and Hopkins (2003) on electrogenic communication signals and regional variation in electrical language or "accents" has led to large numbers of minimally diagnosable units (MDUs) being recognized, which evidently constitute a riverine species flock. There may be more than 30 putative species in Gabon alone. Part of the growing problem with such discoveries is where to draw the line in terms of formally recognizing MDUs and taking conservation action in respect of them. If we are to accept increasingly higher levels of detail in the discrimination of diversity, there will be far more taxonomic units than before that require consideration for conservation purposes. There will be attendant legal and cost implications in managing the situation. In any event, new techniques and technologies are certainly uncovering large-scale, hitherto unrealized diversity. Thus, the estimated list of 12 million species remaining to be formally recognized may, in some taxonomic form or another, expand exponentially.

Advances in techniques in artificial selection, genetic engineering, and other biotechnology mean that, for some taxa, the determination and location of individuals that are representative of the natural wild stock are becoming progressively more difficult. Hence, the idea of conserving the wild species in its pure form and the associated "natural genome" is gradually being challenged. For example, many freshwater tropical fishes held in public aquariums or in laboratories are from farmed sources where artificial selection and inter-populational or inter-specific hybridization is the norm, and such stock has often been reintroduced back to the wild, where there is a prospect of a genetic pollution of the original or a closely allied sibling species and other ecological or environmental impacts. In this area, biotechnological advances allow for the creation of trans-specific or interge-neric hybrids, where, for instance, glowfish (various teleost species) contain bacterial or jellyfish genes (Reid, 2001; Kinoshita et al., 2009). Clearly, there are large conservational, environmental, and bioethical issues concerning this taxon pollution (Millar and Tomkins, 2007). Also, one might reasonably question the basic scientific validity of conducting laboratory experiments and draw-ing "species-dependent" conclusions using artificially hybridized or genetically engineered fish or other livestock. These are commonly of a confused, uncertain, or unknown taxonomic identity and provenance. Scientific studies do take place where named species are not the primary research tar-get and hybridized or artificially modified specimens or populations become the research models. However, the correct taxonomically valid identification of the species involved is often fundamental to the integrity and reproducibility of biological research. Experimental biologists, in particular, are in increasing danger of losing reliable taxonomic "anchors" for their hypotheses.

TAXONOMIC BASIS OF CONSERVATION THREAT ASSESSMENT

Despite ongoing conceptual and interpretational difficulties, the Linnaean morphospecies remains a fundamental unit when assessing threat status, enshrining taxonomies in law, and determining conservation action. The IUCN Red List of Threatened Species (IUCN, 2008) is the global stan-dard, which currently includes as being threatened 23% of mammal species, 12% of birds, 32% of amphibians, and 43% of freshwater fishes. While the Red List provides a unique and reason-ably authoritative guide, there are not enough evaluations of threats overall and there are notable gaps and seemingly irrational biases in taxonomic coverage in some areas of conservation con-cern. Among lower vertebrates, almost every named amphibian species has been dealt with (via the Global Amphibian Assessment or GAA—http://globalamphibians.org; Mendelson et al. 2006.), but nothing like all fishes. There is evidently a strong general taxonomic bias in favor of vertebrates over most invertebrates (Clark and May, 2002), save perhaps dragonflies and butterflies (Odonata and Lepidoptera). A twenty-five-year publication analysis of *Zoo Biology* indicates a decline in descrip-tive and basic science articles and a strong bias toward the large, charismatic megafauna notably carnivores and primates, with species-rich groups of small, potentially threatened bats, rodents, and marsupials insufficiently covered (Anderson et al., 2008).

Setting these issues aside, the current Red Listing process is essentially a sophisticated taxo-nomic database, utilizing demographic, genetic, and other biological, zoogeographical, or anthro-pogenic criteria to determine threats and extinction risks (Mace and Lande, 2005). The Red List distinguishes levels of threat from Extinct in the Wild (EW) and Critically Endangered (CR) to Least Concern (LC), but these categories have undergone several phases of reevaluation and redefi-nition (Mace and Lande 2005). The accompanying IUCN standardized classification of threats incorporates familiar elements such as human-induced habitat deterioration or loss from factors such as deforestation, damming, and pollution. There are also impacts from the introduction of alien invasive species and the continuing overexploitation of many taxa, including through poaching. The causes of persecution of animals can be complex or surprising. Poaching for "bush meat" is, for instance, not simply a response by poor people to eke out a living but is actually a luxury trade with a strong demand in western European markets. Bush meat is sold at a premium above beef or pork in the habitat countries and can be found in the markets of many European cities, including Paris. Animal persecution can result when particular species become blamed, perhaps erroneously, for transmitting emergent infectious diseases to humans or other animals. For example, the epidemic spread in humans of severe acute respiratory syndrome (SARS) has been associated with, among other taxa, various subspecies of Chinese masked palm civet cat (*Paguma larvata*) (Zhong, 2004), that of auto-immune disease syndrome (HIV/AIDS) with chimpanzee (*Pan troglodytes*) (Hahn, et al. 2000), retroviruses with bushmeat (Chomel et al., 2007) and bird flu—the potentially zoonotic H5N1 strain of influenza—with various species of migratory water birds (Lui et al., 2005). More than 70 of all emergent infectious diseases are associated with wild species (Chomel et al., 2007: 5), and this may sometimes result in the irrational mass slaughter of disease-associated species, which becomes a kind of taxonomic genocide. Emergent diseases in their entirety may have something to do with the way human beings manage the planet (http://www.iwar.org./uk/news-archives/2005/05-09.htm).

There are other taxonomic databases for assessing or signalling conservation status and priorities such as the "biodiversity hotspots" established by the Alliance for Zero Extinction for species that are endemic to a single, small geographical site (http://www.zeroextinction.org/); and "EDGE" developed by the Zoological Society of London, which balances "evolutionary distinctiveness" against global endangerment (http://www.edgeofexistence.org/; Isaac et al., 2007). This form of analysis gives the giant panda (*Ailuropoda melanoleuca*) a higher conservation weighting and priority than other more generalized species of bear, which fosters the argument that the giant panda might actually have had its day and so does not merit this priority (Entwistle and Dunstone, 2001). At lower levels of generality, the incorporation of phylogenetic and evolutionary criteria fans the debate over whether we should simply be conserving evolutionarily significant units (ESUs, which might exclude traditional subspecies) versus MDUs, which would often include species or subordinate populations (Fraser and Bernatchez, 2001); and this discussion has legal and practical implications (Mace and Purvis, 2008). Academic debates continue over whether the African Northern White Rhino is a species or subspecies of Ceratotherium, yet it is now believed to be Extinct in the Wild (EW) and only eight individuals survive, all in zoos (Tonge et al., 2009). So far, millions of dollars have been spent in attempts to conserve this taxon.

Lying beyond the Red List category EW, which implies that some animals survive ex situ in a zoo or other wildlife sanctuary, lies the grim reality of the category Extinct (EX), signifying that the species is lost forever from the recent fauna. Of course, there have been frequent species extinctions throughout all of evolutionary history, taking in entire higher-level taxa such as the dinosaurs and trilobites, but some authorities consider that species extinction rates now run at 100–1000 times the natural background rate, although it is not entirely clear how this can be accurately established on temporal and spatial scales. In any event, the UN's Global Environmental Outlook Report (2000, http://www.unep.org) predicts a general pattern of species extinction over the next 30 years of 24% of the world's mammals, 12% of birds, 50% of amphibians, 20% of freshwater fishes, and 50% of the invertebrates. This ignores subspecies and other categories of biodiversity and assumes the same number of valid species as currently recognized (or realistically estimated) and threats impacting much as they do now, rather than becoming worse. For mammals and birds this means that we could lose 1,130 and 1,183 species, respectively. Nonetheless, these data provide a false sense of accuracy and reliability because anticipating or absolutely confirming extinctions remains a most difficult and scientifically fraught task. Lomborg (2001: 249, 252, 257) is particularly scathing about the insubstantial basis for the widely quoted estimate of "40,000 species lost every year" given by Norman Myers in his 1979 book *The Sinking Ark*.

Responding to generalized assertions (based on, say, ecological models that indirectly infer the loss of species per unit area of rainforest destroyed) taxonomists in the late-1990s began to ask the crucial question "what named species is extinct where, or might soon become so?" The critical scientific evaluation of recent extinctions became an exercise for the Committee for Recently Extinct

Organisms (CREO) established in May 1999 at the American Museum of Natural History in New York (Reid, 1999). This exercise resulted in the publication *Extinctions in Near Time* (MacPhee, 1999), covering 40,000 years of human history and pointing up some gross inaccuracies in contemporary Red Lists and a paucity of convincing evidence concerning many species presumed to be extinct. Also highlighted was the inadequacy of the scientific process in analyzing extinction and the need for a better system of validation, utilizing field data, museum, and living specimens and based on logical decision trees. Nonetheless, despite much positive progress, many of the identified problems and resource issues in validating extinctions remain, particularly for zoogeographically widespread taxa.

TAXONOMY AND THE STATUTORY PROTECTION OF SPECIES AND BIODIVERSITY

While this chapter focuses on the conservation of species, many governmental or inter-governmental organizations have the high aspiration to conserve biodiversity, including species. However, the concept of biodiversity widely adopted refers to more than just named species—the definition also covers genes, genomes, populations, habitats, and ecosystems. This more elaborate, difficult, and diffuse concept, taking in elements of sustainability, hardly allows for accurate generalized assessments of threats to biodiversity and meaningful measurement of an increase or decline in biodiversity. So what is it that we are actually striving to conserve, at what level of generality and in what quantities or qualities, and how do we reliably and consistently define and name the constituent parts and relate them to the whole? Again, taxonomic and associated issues in biodiversity conservation extend well beyond the 14 million or so species currently believed (on not absolutely certain grounds) to inhabit the planet, excluding much conservationally significant microbial life (Cockrell and Jones, 2009). This has international legal implications, as in the case of the Convention on the International Trade in Endangered Species. The Convention, among other things, covers animal species and products deriving from them such as meat, eggs, and skins, often in a national or geographical context, but it does not cover biodiversity in its totality as currently conceived. To regulate international trade in elements of biodiversity requires new laws, new species-related concepts, including in genomics, and new techniques in enforcement, including rapid identification of potentially illegal constituents through say DNA "fingerprinting" and forensic chemosystematics.

Restricting the discussion to species alone, it should be emphasized that there are various different vernacular, biological, taxonomic, and legal meanings to the term. While not invariably the case, individual governments often accept the IUCN Red List, or regional and national expressions of this, as the principle taxonomic listing for threatened species. They then use this as a basis for formulating and developing national schedules, conservation policies, and supportive legislation (refer to Ecolex a database that provides comprehensive international coverage of environmental law and operated jointly by IUCN, FAO, and UNEP—http://www.ecolex.org/start.php). The Endangered Species Act (ESA), 1973, in the United States (revised in 2008/2009 by the outgoing administration of President George Bush) is one high profile example of a far-reaching national law that sparked intense controversies over species and other alpha-level taxonomic concepts (Clegg, 1995). The ESA (Sect 3, 15) liberally defines "species" in a legal sense to include: "… any subspecies of fish or wildlife or plants, and any distinct population segment of any species of vertebrate, fish or wildlife which interbreeds when mature."

This definition relates to the biological species concept and clearly allows for the protection of taxa at and below the species level, but this could conceivably include hybridized stock. Elsewhere in the ESA, the need to conserve pest species, disease vectors, and parasites is discounted. This is not entirely justified, considering the potentially complex role of these organisms in supporting ecosystems, biodiversity, and individual animal health. In any event, the enactment of the ESA has engendered fierce and continuing debate over the economic costs in implementation, versus conservational or environmental, benefits in recognizing alpha-level taxa, and over the use of public funds in designating and managing protected areas. Today, companies with large developments and profits at risk are happy to challenge through the courts the systematic, ecological, geographical, or threat concepts underpinning legislation, with a view to winning individual cases or overturning entire statutes. For conservationists, to progress a case in protecting an individual species, subspecies, or distinct population can prove to be financially expensive and time-consuming. Hence, it is more than ever essential that the concept behind the name, together with associated data on life history, genetics, and geographical distribution, can survive intense scientific and legal scrutiny. The risk of a legal challenge exists at many levels, applying to individual systematic or conservation biologists through to the IUCN Species Survival Commission itself.

One classical early example is the case of survival of a species of fish (the snail darter, *Percina tanasi*) versus the construction of the Tellico Dam on the Little Tennessee River (Plater, 2009). In 1973 David Etnier, a biologist from the University of Tennessee, discovered the snail darter in the Little Tennessee River, and it was subsequently scheduled as an endangered species under the ESA. At the same time, the controversial US$150 million Tellico Dam was being constructed by the Tennessee Valley Authority with a risk of the project's being severely delayed or even aborted if the law on species protection was upheld. The Tellico Reservoir—to be created by the Tellico Dam—would, it was argued, alter the habitat of the river to the point of extirpating the snail darter by making a comparatively stagnant lake out of 33 miles (53km) of free-flowing river—the darter's only known natural habitat. Under the aegis of the ESA, a case against the construction of the dam was taken to court by opponents. Over a six-year period, a lawsuit was progressed by the environmental lobby to halt the construction of the dam. Eventually, the case was taken to courts in Washington D.C. and then the U.S. Congress, which authorized the completion of dam construction (by then 90% installed). However, before the closure of the gates of the Tellico Dam and formation of the reservoir, numerous snail darters were transplanted into the Hiwassee River (with, it should be said, uncertain consequences in terms of environmental impact). The snail darter has since been found in other locations and so the Little Tennessee River was never the sole environment in which these fish lived, as had been alleged in the original legal suit. The scientific case was found to be flawed and, following further conservation assessments, the snail darter was taken off the endangered species list in the 1980s. Nevertheless, had matters turned out otherwise, the snail darter might now be extinct.

FINANCIAL VALUE, OWNERSHIP, AND FATE OF TAXONOMIC NAMES

Conservationists have to face the hard economic fact that governments, international agencies, and environmental NGOs do not raise and provide anything like sufficient financial resources to definitively assess conservation needs and implement biodiversity action plans across the globe, nor to properly support museum and zoological institutions and their databases in the cause of conservation. But can humanity in strategic terms really afford to spend so little (by comparison with worldwide military or general economic spending or the current multibillion-dollar support of failing banks) and regard this outgoing as almost a luxury rather than a necessity? In recent years, the emerging discipline of ecosystem services has articulated the essential role of nature in terms of enabling human survival, the alleviation of poverty, and facilitation of development. The UK Government's Department of the Environment (DEFRA) is, for example, working on an Ecosystem Services Research Project to establish the economic and other bases for an ecosystems approach and how it can be used to make effective assessments of the benefits that the natural environment provides (http://www.ecosystemservices.org.uk/). This "services" model is working toward a true valuation of the real worth of species, habitats, and the wider environment in relation to human needs. At first sight, it might appear to some that a standing rainforest has no particular material value, but it actually or potentially provides very many vital and sustainable benefits including oxygen, climate, carbon sequestration, food, timber, pharmaceutical products, genes, and ecotourism. Many of these actual or potential services can now be costed, valued, ranked, and prioritized in a more-or-less standardized bio-accountancy system. Wider beneficiaries may then issue financial or carbon credits to local communities that are paid to conserve the habitat for posterity. Ultimately, most arguments in ecosystem services depend on named, definable biological entities including species. What price is there, for example, on saving a higher-level taxon or an individual name within an ecosystem?

Taxon selection and conservation prioritization (which may involve a limitation on natural genetic diversity) ultimately translates into management and maintenance costs in situ and ex situ, which can be fairly high. For example, costs to implement the IUCN-SSC Amphibian Conservation Action Plan are estimated at the equivalent of more than €257 million (http://globalamphibians. org). The price of saving individual species can be high—up to the equivalent of €32 million was expended on the program to conserve the California condor on- and off-site (Kaplan, 2002; http:// www.dfg.ca.gov/wildlife/species/t_e_ssp/condor). An individual elephant in a European zoo conservation breeding program may cost >€45, 000 per annum to maintain (Mark Pilgrim, personal communication). Nonetheless, developing and maintaining taxonomic databases for conservation and other purposes is very cheap by comparison with military, banking-support, or other government spending. The International Trust for Zoological Nomenclature (ITZN) has predicted that the cost of implementing the first phase of ZooBank will be £1 million (http://www.zoobank.org/). It may cost about $11 million/€7.4 million to fully develop the Zoological Information Management System (ZIMS) of the International Species Information System (ISIS) (http://www.isis.org/). This elaborate computerized taxonomic, animal welfare, and conservational database is used by 735 institutions in 73 countries and contains information on 2 million living zoo and aquarium animals in >15,000 higher-level taxa and >10,000 species, together with some 4 million individual historical records. It contains basic animal records, veterinary medical records, and a collection planning and studbook management system.

Linnaeus himself is associated with early experiments in animal breeding and record-keeping and traditional biotechnology (Reid, 2009). Today, zoos, museums, botanical gardens, and field conservationists can address some issues in taxonomy and conservation biology through genetic resource banks (Holt and Pickard, 1999; Lerman et al., 2009; Clark, 2009; Kouba and Vance, 2009). This includes the freezing (cryopreservation) at ultra-low temperatures of cells, gametes, and tissues, or even whole organisms and maintaining the associated taxonomic databases (Reid, 1994a; Ryder, 2005). These can be used for various supportive purposes in conservation, including genetic analysis and assisted reproduction. The international Frozen Ark Project partnership involving museums, universities, zoos, and other agencies (Clarke, 2009; http://frozenark.org) seeks to obtain and cryopreserve samples from 7,565 threatened Red Listed species across a wide range of taxa. There remain substantial issues in ethical procedures, the ownership of samples and the growing cost of this long-term exercise. The cryopreservation of individual specimens or samples, may cost above €500 per annum (Bill Holt, personal communication). Fortunately, with technological advances the costs of sequencing entire genomes have dropped from billions of dollars to tens of thousands, or less (Service, 2006).

Established practice under the International Code for Zoological Nomenclature (Jeffrey, 1973; Winston, 1999) is that the first person to apply a Linnaean latinized binomial to a species and provide a formal taxonomic description becomes the author of the species and in effect its scientific discoverer, subject to that description's being validly published. Species are often scientifically named after their morphological appearance, color, or geographical occurrence, or in honor of an eminent scientist or other favored person. Many names are conceived of and published by taxonomists working outside of the range countries of the species concerned. Indeed, the majority of taxonomists and institutions holding preserved collections are still located in the temperate zone in Europe and North America, well away from biodiversity hotspots in the tropics, from whence many species and specimens derive. So the material that should be regularly referred to is often located remote from the habitat countries and from the site of any conservation problem. Access to described material by indigenous scientists and any material benefit or kudos associated with the species for local people is thus commonly minimized, yet benefit sharing is a requirement under the Convention on Biodiversity (CBD). Issues of ownership of species names, and access and benefit-sharing in relation to traditional knowledge and the recognized economic values of particular taxa, have come to the fore in the 21st century. Local people may feel that so-called discoveries by scientists from other countries make them seem ignorant, disempowered and disinherited. A pertinent example here is the description of a new diminutive snake, *Leptotyphlops carlae* (Blair Hedges, 2008), from Barbados. The author of the species, evolutionary biologist S. Blair Hedges from Penn State University, named the snake after his herpetologist wife Carla Ann, but this act sparked heated debate in Barbados—how could someone "discover" a snake long known to locals albeit by another name? In any event, why name it scientifically after his wife rather than honor a worthy indigene of Barbados?

Clearly, some species or environments are worth more than others to human society. For example, if one is seeking funding for animal conservation or welfare projects, domesticated breeds are far more relevant to the general public than wild taxa or even designated habitats such as the Galapagos World Heritage Site that was so important to the development of Darwin's (1859) theory of evolution by natural selection, and recently the source of a putative new pink species of land iguana (Gentile et al., 2008). The welfare of the domesticated cat (*Felix catus* L., 1758, representing in effect a "pseudo-species") is, in financial terms, valued far more highly by society than the conservation of a species of gorilla or rhino (Save the Rhino International Fundraising Database). Paradoxically, through predation, domestic cats—especially when feral in Europe, North America, and elsewhere—are a major cause of general biodiversity loss among native wildlife (Clarke and Pacin, 2002). Since 1970 the population of domestic cats in the United States has more than doubled, with estimates of >100 million individuals, and with attempts to remove them from the wild often vigorously opposed by local citizenry, despite widespread and deleterious impacts on wildlife. This may be because cats and other household domestic animals often act as companions and, it seems, as child substitutes (Archer, 2003). A strong, human emotional engagement seems to be at the heart of their popularity and ability to attract large-scale support for their welfare, including financial contributions. All of this is opposed to the more abstract, remote, and general concepts of species, biodiversity, and ecosystem services that often attract far less funding. These important aspects are now being explored by conservation psychologists (http://www.conservationpsychology.org/).

The idea of raising funds for taxonomy by entrepreneurial means is not new, even if it is sometimes seen to be controversial. Since the time of Linnaeus, species, genera, and higher-level taxa have been named by taxonomists in honor of individuals. This was sometimes with the pure motive of honorific recognition of a scientific contribution and at other times with a clear view to soliciting patronage or acknowledging financial, publishing, or other support, or, through taxonomic flattery, minimizing opposition to some zoological thesis or other. There are, then, many fundraising precedents from the 18th, 19th, and 20th centuries—notably Linnaeus' own successful efforts to gain sponsorship and patronage, including from the Swedish Royal family (Reid, 2007).

A contemporary exercise by the conservation agency Amphibian Ark (http://www.amphibianark.org/) to auction naming rights for five species of frog was to raise money for the Year of the Frog public awareness campaign concerning the global amphibian extinction crisis. This engendered debate regarding selling naming rights in the face of uncertain legal or moral ownership. Auctioning names in the support of conservation can be seen to have a pure and worthy motive. Realistically, conservation has to be funded and—while acknowledging the potential risks, political, social, scientific, and otherwise of auctioning names—money can be directly applied to solving conservation problems where the species actually comes from. Also, authorship of new species names can be organized through range-country taxonomists.

In the end, conservationists need to be confident about taxonomic identifications and nomenclatural stability (Melville, 1995), and in assessments of conservation threat and status before embarking on long-term, resource-intensive species management programs. Taxonomically distinct species may often need to be managed separately and management costs will rise correspondingly, particularly ex situ where separate facilities are usually constructed, costing millions of euros in some cases. There will be major variations in total population and gene pool estimates, Red List status, and conservation prioritization, if separate entities are subsequently recognized from a species hitherto undivided. Conversely, if a species is taxonomically de-recognized (synonymized) then there are abortive or opportunity costs to consider, perhaps covering many years of conservation endeavor. Strategic wildlife meta-population management objectives in situ and ex situ may involve working to conserve 90% of genetic variability for 50 years.

It would be a mistake to imagine that the taxonomic status of all large charismatic megavertebrates recognized since the time of Linnaeus remains intact. The Asiatic elephant, *Elephas maximus* L. 1758, is a species threatened with extinction (assessed as Endangered (EN) (Choudhury et al. 2008). Formally existing in millions, there may be (on crude and optimistic assessments) a total of 41,000–52,000 individuals in the wild Asiatic elephant population, but with a continuing steep general rate of decline. One of the few remaining single continuous populations in Assam in northern India may number fewer than 2,000. However, it is still not clear whether the separate, increasingly fragmented, inbred populations of Asiatic elephant represent one or more taxonomically and genetically discrete entities, and domestication and translocation complicate matters. The genetic analysis of Fernando et al. (2003) indicates a small but distinct population of Asiatic elephants in Borneo that, it is argued, have been there since the Pleistocene with no genetic inbreeding depression. The World Wildlife Fund (WWF) Malaysia (2003) publicized this research, endorsing a view that a dwarf subspecies occurs on Borneo that requires special conservation consideration. Others have contested this (Ibbotson, 2003; Shim, 2003) on the basis of potential ancient movements of domesticated Asiatic elephant stock within the geographical region. From a detailed review of osteological and zooarchaeological material and historical records, Cranbrook et al. (2007) are broadly supportive of the researches of Fernando et al. (2003). However, they consider that, rather than complete isolation since the Pleistocene, introductions of elephants by humans have actually taken place. Consequently, the population of Asiatic elephants isolated in northeastern Borneo might well consist of remnant survivors of introduced Javan elephant (*Elephas maximus sondaicus*). This is evidently a valid subspecies, hitherto believed to be extinct since about the 12th century AD but not a dwarf population by comparison with others in Asia. If true, then this geographically displaced and potentially "pure" surviving population of *Elephas maximus sondaicus* is a viable ESU, demanding a high conservation priority and management separate from other populations.

There are radical new perspectives on the taxonomic and conservational status of primate species recognized by Linnaeus (1758). This follows analyses of actual or potential primate extinctions on different temporal and spatial scales (Harcourt, 2002) and the groundbreaking taxonomic work of Groves (2001, 2004). The orang-utan (*Pongo pygmaeus* L., 1758) has recently been confirmed as including more than one species or subspecies (Steiper, 2006). Island populations were originally designated as *P.p. abelii* from Sarawak and *P.p. pygmaeus* from Borneo. However, from a thorough examination of genetics and morphology, Groves (2001) elevated these to the status of full species: *P. abelli* and *P. pygmaeus*, respectively. Since then, three or four putative subspecies have been researched on Borneo by Jalil et al. (2008) using mitochondrial and molecular data: *P.p. pygmaeus* (Sarawak and northwest Borneo), *P.p. wurmbii* (Central Kalimantan), *P.p. morio* (East Kalimantan and Sabah—with the latter geographical population possibly representing another separate subspecies). Independent of any remaining taxonomic uncertainty, all orang-utan populations are in severe decline (Wich et al., 2008), with the total for Borneo about 54,000: 3,000–4,500 in Sabah; 34,975 in Central Kalimantan, and 4,800 in East Kalimantan. This places *P. abellii* in Sumatra as CR, alongside the putative subspecies of *P. pygmaeus* in East Kalimantan. These taxonomic and demographic data present significant challenges for the future of zoo conservation breeding populations, where orang-utan species or subspecies are not always recognized or separately managed (Becker et al., 2008). Among the 647 *"Pongo pygmaeus"* in zoos worldwide, 70 (or 11%) of current stock are hybrid; and many are of unknown subspecific status or geographical origin (ISIS Data, 2009).

There are similarly complex issues concerning the conservation of the chimpanzee (*Pan troglodytes* L. 1758). Again, subspecies have been erected, the most threatened of which is *P.t. vellerosus*, where there may be only 5,000–8,000 surviving in the Nigerian montane forest (Beck and Chapman, 2008). On the basis of genomic explorations, Goodman et al. (2002) transferred the entire species from *Pan* to the genus *Homo*, as a close (sister group) relative of humans. This raises fundamental issues concerning the "mental taxon" to which the chimpanzee might belong, as well as associated animal welfare or human rights issues (Cavalieri and Singer, 1994; Reid, 2002; Sommer, 2009). Hence, if we allow the extinction of the Linnaean chimpanzee, we may, in effect, be contemplating the death of a part of humanity.

ACKNOWLEDGMENTS

I would like to thank the editors for their constructive comments. My colleagues Roger Wilkinson, Mark Pilgrim, and Bill Holt generously provided references and personal communications for inclusion in the text. Vikash Tatayah of the Mauritius Wildlife Foundation kindly sourced a letter for me (via Daniel Glaser and Judith Magee of the Natural History Museum, London) on Charles Darwin's request to conserve giant tortoises of Aldabra Island and the Mascarenes. Claudine Gibson and Sally Reid proofread and otherwise improved the manuscript.

REFERENCES

Alves, R.J.V., Filho, V., and Machado, M.D. 2007. Is classical taxonomy obsolete? *Taxon*, 56(1): 287–288.

Anderson, M.1986. From predator to pet: Social relationships of the Saami reindeer–herding dog. *Central Issues in Anthropology*, 6.2: 3–11. American Anthropological Association.

Anderson, U.S., Kelling, A.S., and Maple, T.L. 2008. Twenty-five years of Zoo Biology: A publication analysis. *Zoo Biology*, 27: 444–457.

Anglin, J.M. 1995. Classifying the world through language: Functional relevance, cultural significance and category. *International Journal of Intercultural Relations*, 19(2): 161–181.

Angulo, E., Deves, A-L., Saint Jalmes, M., Courchamp, F., 2009. Fatal attraction: Rare species in the spotlight. *Proceedings of the Royal Society*, 276: 1331–1337.

Archer, J. 2003. Why do people love their pets? *Evolution and human behaviour*. 18(4): 237–259.

Arnegard, M.E. and Hopkins, C.D. 2003. Electric signal variation among seven blunt-snouted *Brieonmyrus* species (Teleostei: Mormyridae) from a riverine species flock in Gabon, Central Africa. *Environmental Biology of Fishes*, 67(4): 321–339.

Arnold, M. 2004. Transfer and origin of adaptations through natural hybridization: Were Anderson and Stebbins right? *The Plant Cell*, 16: 562–570.

Beck, J. and Chapman, H. 2008. A population estimate of the endangered chimpanzee *Pan troglodytes vellerosus* in a Nigerian montane forest: Implications for conservation. *Oryx*, 42(3): 448–451.

Becker, C., de Jongh, T., Vermeer, J., Bemment, N., and Pilgrim, P. 2008. Orang utans: Distribution, species status and social system—consequences for the EEP management, the future husbandry and enclosure design. Amsterdam: Executive Office of the European Association of Zoos and Aquaria. Pp. 1–10, with Annexes.

Berlin, B. 1973. Folk systematics in relation to biological classification and nomenclature. *Annual Review of Ecology and Systematics*, 4: 259–271.

Bird-David, N. 2002. Animism revisited: Personhood, environment and relational epistemology. In Harvey, G. (Ed.) *Readings in indigenous religions*. London and New York: Continuum. Pp 72–105

Broberg, G. 2006. *Carl Linnaeus*. Stockholm: Swedish Institute.

Bruton, M. N. 1990. The conservation of the fishes of Lake Victoria, Africa: An ecological perspective. *Environmental Biology of Fishes*, 27: 161–175.

Cavalieri, P. and Singer, P. 1994. *The great ape project: Equality beyond humanity*. New York: St Martins.

Cheke, A. and Hume, J. 2008. *Lost land of the Dodo. An ecological history of Mauritius, Réunion and Rodrigues*. London: T and AD Poyser.

Chomel, B.B., Belotto, A., and Meslin, F-X. 2007.Wildlife, exotic pets and emerging zoonoses. *EID Journal*, 13(1). Internet publication available from http://www.cdc.gov/ncidod/EID/13/1/6.hl.

Choudhury, A., Lahiri Choudhury, D.K., Desai, A., Duckworth, J.W., Easa, P.S., Johnsingh, A.J.T., et al. 2008. *Elephas maximus*. In: IUCN 2008. 2008 IUCN Red List of Threatened Species. <www.iucnredlist.org>. Downloaded on 12 February 2009.

Clark, J.A. and May, R.M. 2002. Taxonomic bias in conservation research. *Science*, 297: 191–192.

Clarke, A.G. 2009 (in press) The Frozen Ark Project: The role of zoos and aquariums in preserving the genetic material of threatened animals. *International Zoo Yearbook*, 43:222–230.

Clarke, A.L. and Pacin, T. 2002. Domestic cat "colonies" in natural areas: A growing exotic species threat. *Natural Areas Journal*, 22(2): 154–159.

Clegg, M.T. (Ed.) 1995. Species definitions and the Endangered Species Act. in *Science and the Endangered Species Act*. Washington D.C.: National Academy Press. Pp. 46–70.

Clutton–Brock, J. 1981. *Domesticated animals from early times*. London: British Museum (Natural History) and William Heinemann Ltd.

Cockrell, C.S. and Jones, H.L. 2009. Advancing the case for microbial conservation. *Oryx* 43 (4): 520–526.

Condon, M.A., Scheffer, S.J., Lewis, M.L., and Swensen, S.M. 2008. Hidden neotropical diversity: Greater than the sum of its parts. *Science*, 320(5878): 928–931.

Coote, T. 2009. Action Plan for the long-term conservation of French Polynesia's last surviving populations of endemic tree snails of the genera *Partula, Samoana* and Trochomorpha. 2nd revision, January 2009. IUCN Partulid Global Species Management Programme in association with La Service Direction de l'Environment de la Polynésie Française.

Cowper Powys, J. 1929. *The meaning of culture*. New York: W.W. Norton.

Cranbrook, Earl of, Payne, J., and Leh, C.M.U. 2007. Origin of the elephants *Elephas maximus* L. of Borneo. *The Sarawak Museum Journal*, 63(84): 95–125.

Cristofferson, M. 2007. Macroevolution and macroecology for the biological synthesis. *Gia Scientia,* 36 (1): 9-16.

Damen, M. 2008. Tall blondes in the picture: Giraffe EEP [Endangered Species Breeding Programme] celebrates twenty years. *EAZA News*, (62): 20–21.

Darwin, C., et al. 1875. [Letter to A.H. Gordon, Governor of Mauritius, requesting the protection of the Giant Tortoise on Aldabra]. Published in *Transactions of the Royal Society of Arts and Sciences of Mauritius* n.s., 8: 106-109.

Darwin, C. 1859. On the origin of species by means of natural selection or the preservation of favoured races in the struggle for life. London: John Murray.

Darwin, C. and Wallace, A. 1858. On the tendency of species to form varieties; and on the perpetuation of varieties and species by natural means of selection. *Proceedings of the Linnean Society of London*, 3: 45–62.

Davis, P. 1996. *Museums and the natural environment. The role of natural history museums in biological conservation.* London and New York: Leicester University Press.

Diamond, J. 1992. *The third chimpanzee: The evolution and future of the human animal*. New York: Harper Collins.

Dixon, G.R. and Brishammar, S. 2007. From Linnaeus to biodiversity. *Biologist*, 53: 149–153.

Dollinger, P., Robin, K., Weber, F. 2006. The significance of zoos for nature conservation. A contribution to the implementation of the World Zoo and Aquarium Conservation Strategy. *Rigi Symposium*, 17–19 February 2005. Landcape and Animal Park Goldau. Berne: WAZA. Pp. 1–89.

Edmonds, M. 2009. Is the brain hard-wired for religion? http://www.howstuff works.com/brain-religion htm. (Accessed 27th October 2009).

Eizirik, E., Jae–Heup, K., Menotti-Raymond, M., Crawshaw, P.G., O'Brien, S., and Johnson, W.E. 2001. Phylogeography, population history and conservation genetics of jaguars (Panthera onca, Mammalia, Felidae). *Molecular Ecology*, 10: 65–79.

Elledge, A., Leung, L.K-P., Allen, L.R., Firestone, K., and Wilton, A.N. 2006. Assessing the taxonomic status of dingoes *Canis familiaris dingo* for conservation. *Mammal Review*, 36 (2): 142–156.

Entwistle, A. and Dunstone, N. 2001 (Eds). *Priorities for the conservation of mammalian diversity: Has the Panda had its day?* Cambridge: Cambridge University Press.

Fernando, P.J., Vidya, T.N.C., Payne, J., Stuewe, M., Davison, G., Alfred, R.J. et al. 2003. DNA analysis indicates that Asian elephants are native to Borneo and are therefore a high priority for conservation. *PLoS Biology*, 1: 1–16.

Figuier, L., 1877. *Reptiles and birds. A popular account of their various orders with a description of the habits and economy of the most interesting*. New edition translated from the original French and revised by Parker Gillmore. London: Cassell, Petter and Galpin.

Fisher, J., Simon, N., and Vincent, J. 1972. *The red book. Wildlife in danger.* London: Collins.

Fortey, R. 2008. *Dry Store Room No. 1. The secret life of the Natural History Museum.* London: HarperCollins Publishers.

Fraser, D.F. and Bernatchez, L. 2001. Adaptive evolutionary conservation: towards a unified concept for defining conservation units. *Molecular Ecology*, 10: 2741-2752.

GAA 2008. *Global Amphibian Assessment.* International Union for Conservation of Nature and Natural Resources. (http://www.iucnredlist.org/amphibians).

Gentile, G., Fabiani, A., Marquez, C., Snell, H.L., Snell, H.M, Tapi, W., and Sbordoni, V. 2008. An overlooked pink species of land iguana in the Galápagos. *Proceedings of the National Academy of Science* 106 (2): 507–511.

Goodman, M., McConkey, E.H., and Page, S.L. 2002. Reconstructing human evolution in the age of genomic exploration. In: Harcourt, C. and Sherwood, B. (Eds.) *New perspectives in primate evolution and behaviour.* Otley, West Yorkshire: Westbury Academic and Scientific Publishing. Pp. 47–70.

Gosse, P.H. 1854. *The aquarium: An unveiling of the wonders of the deep sea.* London: John Van Voorst, Paternoster Row.

Graham, C.H., Ferrier, S., Huettman, F., Moritz, C., and Townsend Peterson, A. 2004. New developments in museum-based informatics and applications in biodiversity analysis. *Trends in Ecology and Evolution*, 19(9): 497–503.

Grant, P.R. and Grant, R.B. 1992. Hybridization of bird species. *Science*, 256: 193–197.

Groves, C. 2001. *Primate taxonomy.* Washington D.C.: Smithsonian Institution Press.

Groves, C. 2004. The what, why, and how of primate taxonomy. *International Journal of Primatology*, 25(5): 1105–1126.

Gunther, R.W.T. 1928 (Ed.). Further correspondence of John Ray. London: Adlard and Son. Printed for the Ray Society.

Hahn, B.H., Shaw, G.M., De Cock, K.M., and Sharp, P.M. 2000. AIDS as a zoonosis: Scientific and public health implications. *Science*, 287 (5453): 607–614.

Harcourt, A.H. 2002. A biology of primate extinction, at two temporal and spatial scales. In: Harcourt, C. and Sherwood, B. (Eds.) *New perspectives in primate evolution and behaviour.* Otley, West Yorkshire: Westbury Academic and Scientific Publishing. Pp. 309–322.

Hartley, S. and Kunin, W.E. 2003. Scale dependency of rarity, extinction risk and conservation priority. *Conservation Biology*, 17(6): 1559–1570.

Hennig, W. 1966. *Phylogenetic systematics.* Urbana: University of Illinois Press.

Holt, W.V., Pickard, A.R. 1999. Role of reproductive technologies and genetic resource banks in animal conservation. *Reviews of Reproduction*, 4: 143–150.

Ibbotson, R. 2003. Domesticated elephants in Borneo. *Sabah Society Journal*, 20: 1–6.

Isaac, N.J., Turvey, S.T., Collen, B.E. et al. 2007. Mammals on the EDGE: Conservation priorities based on threat and phylogeny. *PloS ONE*, March 2007 (3) e296.

IUCN 2008. The IUCN Red List of Threatened SpeciesTM. Online. Available at: http://www.iucnredlist.org/.

Jalil, M.F., Cable, J., Sinyor, J., Lackman-Anchrenaz, I., Bruford, M.W., and Goossens, B. 2008. Riverine effects on mitochondrial structure of Bornean orang–utans (*Pongo pygmaeus*) at two spatial scales. *Moecular Ecology*, 17: 2898–2909.

Jeffrey, C. 1973. *Biological nomenclature.* London: Arnold.

Johnson, M.S., Clarke, B., and Murray, J. 1988. Discrepancies in the estimation of gene flow in Partula. *Genetics*, 120(1): 233–238.

Kaplan, M. 2002. Condors' future on a cliff's edge. New Scientist.com news service. 8th April, 2002. Online. Available at: http://www.newscientist.com/article/dn2137–condors–future–on–a–cliffs–edge.html.

Kinoshita, M. Murata, K., Naruse, K. and Tanaka, M. 2009. *Medaka. Biology, management and experimental protocols.* Wiley-Blackwell.

Kitchener, A.C. 1997. The role of museums and zoos in conservation biology. *International Zoo Yearbook*, 35: 325–336.

Koerner, L. 1999. *Linnaeus: Nature and nation.* Harvard: Harvard University Press.

Kouba, A.J. and Vance, C.K. 2009. Applied reproductive technologies and genetic resource banking for amphibian conservation. *Reproduction, Fertility and Development* 21: 719-737.

Leakey, R. E. 1981. *The making of mankind.* London: Book Club Associates by arrangement with Michael Joseph Limited.

Lee, T., Birch, J.B., Coote, T., Pearce-Kelly, P., Hickman, C., Mayer, J-V and Ó Foighil, D. 2009. Moorean tree snail survival re-visited: a multi-island genealogical perspective. *BMC Evolutionary Biology* 9: 2004 (http://www.biomedcentral.com/1471–2148/9/204).

Lermen, D., Blömeke, B., Browne R., et al. 2009. Cryobanking of viable biomaterials: implementation of new strategies for conservation purposes. *Molecular Ecology* 18: 1030 1033.

Linnaeus, C. 1753. *Species plantarum, exhibentes plantas rite cognitas, ad genera relatas, cum differentiis specificis, nominibus trivialibus, synonymis selectis, locis natalibus, secundum sytema sexuale digestas.* Stockholm: Salvius.

Linnaeus, C. 1758. *Systema naturae per regna tria natura, secundum classes, ordines, genera, species, cum characteribus, differentiis, synonymis, locis,* editio decimal, reformata, Volume 1. Stockholm: Salvius.

Lomborg, B. 2001. *The sceptical environmentalist. Measuring the real state of the world.* Cambridge: Cambridge University Press.

Lowe, S., Browne, M., Boudjelas, S., and De Poorter, M. 2000. 100 of the world's worst invasive alien species from the Global Invasive Species Database. *Aliens* (12). Lift-out supplement, pp. 12. (December, 2000) (www.issg.org/booklet.pdf).

Lloyd, A.C. 1962. Genus, species and ordered series in Aristotle. *Phoronesis: A Journal for Ancient Philosophy,* 7(1–2): 67–90(24).

Liu, J., Xiao, H., Lei, F., Zhu, Q., Qin, K. et al., 2005. Highly pathogenic H5N1 influenza virus infection in migratory birds. *Science,* 309(5738): 1206.

MA 2003. *Ecosystems and human well-being. A framework for assessment.* Millenium Ecosystem Assessment: Island Press.

Mace, G. and Purvis, A. 2008. Evolutionary biology and practical conservation: bridging a widening gap. *Molecular Ecology* 17: 9–19.

Mace, G. and Lande, R. 2005. Assessing extinction threats: Towards a re-evaluation of IUCN Threatened Species Categories. *Conservation Biology* 5(2): 148–157.

MacPhee, R.D.E. 1999 (Ed.) *Extinctions in near time. Causes, contexts and consequences.* New York: Springer.

Mayr, E. 1969. *Principles of systematic zoology.* McGraw-Hill.

Mayr, E. 1942. *Systematics and the origin of species, from the viewpoint of a zoologist.* Cambridge: Harvard University Press.

McNeely, J.A. (ed). 1998. *Major conservation issues of the 1990's.* Results of the World Conservation Congress Workshops. Montreal: IUCN The World Conservation Union.

Melville, R.V. 1995. *Towards stability in the names of animals. A history of the International Commission of Zoological Nomenclature 1895–1995.* London: International Trust for Zoological Nomenclature.

Mendelson, J.III, Lips, K., Gagliardo, R. et al. 2006. Confronting amphibian declines and extinctions. *Science,* 313: 48.

Merker, B. 2000. Synchronous chorusing and human origins. In: Wallin, N.L. Merker, B., and Brown, S. (Eds.) *The origins of music.* Cambridge MA: MIT Press. Pp. 315–327

Millar, K. and Tomkins, S. 2007. Ethical analysis of the use of GM fish: Emerging issues for aquaculture development. *Journal of Agricultural and Environmental Ethics,* 20: 437–453.

Miller, B., Conway, W., Reading, R., Wemmer, C., Wild, D., Kleiman, D. et al. 2004. Evaluating the conservation mission of zoos, aquariums, botanical gardens and natural history museums. *Conservation Biology,* 18(1): 86–93.

Myers, N. 1979. *The sinking ark: A new look at the problem of disappearing species.* New York, Pergamon Press. xiii + 307 pp.

Pellier, F., Reale, D., Watters, J., et al. 2009. Value of captive populations for quantitative genetics research. *Trends in Ecology and Evolution* 24 (5): 263–270.

Penning, M., Reid, G.McG., Koldewey et al., 2009. *Turning the tide. A Global Aquarium Strategy for Conservation and Sustainability.* WAZA Executive Office Bern, Switzerland.

Pickett, K.M. 2005. The new and improved phylocode, now with types, ranks, and even polyphyly. A conference report from the First International Phylogenetic Nomenclature Meeting. *Cladistics,* 21: 79–82.

Plater, Z. 2009. The snail darter, the Tellico Dam and sustainable democracy lessons for the next President from a classic environmental law controversy. *Boston College Law School Lectures and Presentations.* Available at http://www.bc.edu/schools/law/ .

Popper, K. 2002. *Conjectures and refutations. The growth of scientific knowledge.* 3rd ed. London: Routledge and Kegan Paul.

Reid, G.McG. 1994a. The preparation and preservation of collections. In: Stansfield, G. Mathias, J. and Reid, G. McG. (Eds). *Manual of natural history curatorship.* London: Museum and Galleries Commission, Her Majesty's Stationery Office. Pp. 28–69.

Reid, G. McG. 1978. A systematic study of labeine cyprinid fishes with particular reference to the comparative morphology, functional morphology and morphometrics of African *Labeo* species. PhD Thesis, University of London.

Reid, G. McG. 1994b. Live animals and plants in natural history museums. In: Stansfield, G. Mathias, J. and Reid, G. McG. (Eds). *Manual of natural history curatorship*. London: Museum and Galleries Commission, Her Majesty's Stationery Office. Pp. 190–212.

Reid, G. McG. 1999 (Ed.). *Criteria for defining extinction and evaluating evidence for extinction*. Draft Report on International Workshop, Committee for Recently Extinct Organisms (CREO), 15–16 May 1999. New York: Center for Biodiversity and Conservation, American Museum of Natural History.

Reid, G. McG. 2001. Biotechnology. In: Bell, E. (Ed.) *Encyclopaedia of the world's zoos*. Vol. 1. Pp. 132–136.

Reid, G. McG. 2002. Preface. In: Harcourt, C. and Sherwood, B. (Eds.) *New perspectives in primate evolution and behaviour*. Otley, West Yorkshire: Westbury Academic and Scientific Publishing. Pp. ix–x.

Reid, G. McG. 2007. Linnaeus' fishes, past present and future. In: Gardiner, B. and Morris, M. (Eds.), *The Linnean Collections*. The Linnean Special Issue No. 7. The Linnean Society of London: Wiley–Blackwell, Oxford. Pp. 75–84.

Reid, G. McG. 2009. Carolus Linnaeus (1707–1778): His life, philosophy and science and its relationship to modern biology. *Taxon*, 58(1) (February): 1–14.

Reid, G. McG., Macdonald, A.A., Fidgett, A.L., Hiddinga, B., and Leus, K. 2008. Developing the research potential of zoos and aquaria. The EAZA Research Strategy. Amsterdam: EAZA Executive Office.

Ryder, O. 2005. Conservation genomics: Applying whole genome studies to species conservation efforts. *Cytogenetic and Genome Research*. 108: 6–15.

Sanderson, E.W., Redford, K.H., Chetkjewicz, C–L.B., Rabinowitz, A.R., Robinson, J.G., and Taber, A.B. 2002. Planning to save a species: The jaguar as a model. *Conservation Biology*, 16: 58–72.

Savolainen, V., Cowan, R.S., Vogler, A.P., Roderick, G.K., and Lane, R. 2005. Towards writing and encyclopaedia of life: An introduction to DNA barcoding. *Philosophical Transactions of the Royal Society*. Series B (Biological Sciences) 360(1462): 1805–1811.

Service, R.E. 2006. The race for the $1000 genome. *Science*, 311: 1544–1546.

Shim, P.S. 2003. Another look at the Borneo elephant. *Sabah Society Journal*, 20: 7–14.

Sommer, V. 2009. Wir sind Menschenaffen. Evolutionäre Anthropologie radikal gedacht. ["We are apes." Towards a radical evolutionary anthropology.] Keynote lecture at the 2009 Symposium: *Darwin, die Evolution und unser Bild vom Menschen*. Deutsches Hygienemuseum und Universität, Dresden.

SOS 2008. State of observed species. A report card on our knowledge of the earth's species. International Institute for species Exploration, Arizona State University, International Commission on Zoological Nomenclature, International Plant Names Index and Thomson Scientific (publisher of the Zoological Record) (www.species.asu.edu).

Steiper, M.E. 2006. Population history, biogeography and taxonomy of orang–utans (Genus: *Pongo*) based on a population genetic meta–analysis of multiple loci. *Journal of Human Evolution*, 50(5): 509–522.

Stuart, B.L., Rhodin, A.G., Grismer, L.L., and Hansel, T. 2006. Scientific description can imperil species. *Science* 312, 1137.

Sutherland, W.J. Adams, W.M, Aronson, R.B., et al. 2009. One hundred questions of importance to the conservation of global biological diversity. *Conservation Biology* 23 (3): 557–567.

Tamera, B. 2005. *The life and times of Hammurabi*. Bear: Mitchell Lane Publishers.

Thomas, K. 1983. *Man and the Natural World*. Changing attitudes in England 1500–1800. Allen Lane.

Tonge, S. Holst, B., and Dickie, L. 2009. EAZA *Statement in response to the proposed translocation of Northern White Rhino*. EAZA Executive Office Amsterdam.

Ucko, P.J. and Rosenfeld, A.1967. *Paleolithic cave art*. McGraw-Hill.

UK BAP. 1999. UK Biodiversity action plan for tadpole shrimp (*Triops cancriformis*). UK Biodiversity Group. Tranche 2. *Action Plans: Terrestrial and freshwater species and habitats*. Volume VI, page 131.

Verheyen, E., Salzburger, W., Snoeks, J., and Meyer, A. 2003. Origin of the superflock of cichlid fishes from Lake Victoria, East Africa. *Science*, 300: 325–329.

Watson, E. and Smith, R. 2009. *Why taxonomy matters*. Case studies collated by Bionet–International: The Global Network for Taxonomy (www.bionet–intl.org/why; bionet@bionet–intl.org).

Watson, R.T., Heywood, V.H., Baste, I., Dias, B., Gámez, R., Janetos, T., Reid, W., and Ruark, G. (Eds). 1995. *Global biodiversity assessment. Summary for policy makers*. United Nations Environment Program. Cambridge: Cambridge University Press.

WAZA 2005. *Building a ruture for wildlife—The World Zoo conservation and aquarium strategy*. WAZA Executive Office, Berne, Switzerland.

Wich, S.A., Meijaard, E., Marshall, A.J. et al. 2008. Distribution and conservation status of the orangutan (*Pongo* spp.) on Borneo and Sumatra: How many remain? *Oryx*, 42(3): 329–339.

Wilson, D.S. 2003. *Darwin's cathedral: Evolution, religion and the nature of society.* Chicago: Chicago University Press.

Wilson, E.O. 1992. *Biophilia: The human bond with other species.* Cambridge, MA: Harvard University Press.

Winsor, M.P. 2009. Taxonomy was the foundation for Darwin's evolution. *Taxon* 58 (1): 43–49.

Winston, J.E. 1999. *Describing species. Practical taxonomic procedure for biologists.* New York: Columbia University Press.

Witte, F., Goldschmidt, T., Wanink, J., van Oijen, M., Goudswaard, K., Witte–Maas, E., and Bouton, N. 1992. The destruction of an endemic species flock: Quantitative data on the decline of the haplochromine cichlids of Lake Victoria. *Environmental Biology of Fishes*, 34: 1–28.

WWF 2003. New subspecies of elephants discovered in Borneo: Sabah Wildlife Department and WWF's tests indicate Pygmy Elephants to be new subspecies. WWF Malaysia Press Release (http://www.wwfmalaysia.org/newsroom/pressrel/2003/pr030903.htm).

WZO 1993. *The World Zoo conservation strategy.* Basel Switzerland: World Zoo Organisation and Captive Breeding Specialist Group of IUCN/SSC. Chicago: Chicago Zoological Society.

Zhong, N. 2004. Management and prevention of SARS in China. *Philosophical Transactions of the Royal Society* (Series B) 359(144):1115–1116.

5 Engineering a Linnaean Ark of Knowledge for a Deluge of Species

Quentin D. Wheeler

CONTENTS

Our Linnaean classifications can be thought of collectively as an "Ark of Knowledge," a place where specimens, characters, natural history observations, geographic distribution records, published descriptions—all that is known of higher taxa—can be assembled as a representation of the diversity of species in nature. This ark is the most comprehensive and reliable source of information about species today. It will acquire an even more pervasive and important role in an uncertain future, as environments change and species are lost. In the here and now, it is a baseline of information against which changes in biodiversity can be detected—in the future, a legacy of information about biodiversity bequeathed to our descendants.

The original ark, the Linnaean system of classification and names, has served science and humanity well. Its fundamental structures are appropriate to a knowledge ark of any dimension, but its capacity must be immediately expanded to accommodate a much larger-scale exploration of species. Linnaean names, in the world of species diversity, were the original "unique identifiers" by which information could be retrieved or communicated about each of the distinct kinds of living things. Linnaean names continue to be potent in this regard, having been incorporated into sophisticated computer search strategies (e.g., Patterson et al. 2006, Franz et al. 2008). Because related species share evolutionary novelties, it is desirable that they be housed together in the ark, in the first instance in genera and then, in turn, in families, orders, and so forth.

Willi Hennig recognized that evolutionary history provides the one common thread around which knowledge of millions of species can be organized logically. Descent with modification results in nested sets of characters, each modified from a previously existing character. Species have similarly nested sets or relationships. Thus, the historical pattern of relative recency of common ancestry is the basis for a reference system generally applicable to the whole of biology, and that pattern is expressible through Linnaean names and ranks (Hennig 1966, Platnick 1979, Nelson and Platnick 1981). While new evidence occasionally necessitates species' changing positions, by and large our general reference system has proven reliable.

While there is nothing wrong with the fundamental structure of our ark, the tools and methods we use to populate it with specimens, characters, and observations are dated and inadequate in scale and tempo for the challenge of responding to the biodiversity crisis. We must discover and describe

53

ten or more million new species while continuing to test a rapidly expanding inventory of "known" species. Adopting methods of work and tools appropriate to such a massively scaled task is essential if we are to learn our world's species and their history before much evidence is lost.

The absence of support for the taxonomic agenda from the ecological community is difficult to understand and much more difficult to justify. Ecologists have complained about the "taxonomic impediment"—that is, the inability to identify many if not most species at many study sites—yet many have shown little enthusiasm for supporting taxonomists in the exploration of species and little respect for taxonomy even when done to exceptional levels of excellence. One sometimes gets the impression that discovering and describing species simply does not matter, that ecological theorizing can overcome the simple inconvenience of lacking knowledge of most of the moving parts in the large and complex machinery of ecosystems. I cannot name another area of science where ignorance is preferred over knowledge and common sense would dictate that the more we know of the unique components of ecosystems the more we can learn about those ecosystems themselves. My guess is that, as species can be identified and knowledge of their unique combination of morphological, behavioral, and physiological characters become known, clever ecologists will make good use of them.

Linnaean classifications and nomenclature are superbly suited to communicating about species and clades. Continuing the traditions of Linnaeus and Hennig, cyber-taxonomy now offers an opportunity to modernize taxonomy's infrastructure and practices, and to vastly accelerate its work (Wheeler 2008a). Were it not for the biodiversity crisis there would be no great urgency attached to the goals of taxonomy. We now recognize that if we are to assemble collections representative of earth's species that it must be done now. So that our knowledge approaches comprehensiveness, we must accelerate taxonomy ultimately by orders of magnitude. So that our knowledge is reliable, we must do so without abandonment or dilution of the core of proven taxonomic theory and practice.

It is critical to decision makers that taxonomists establish baseline knowledge of our planet's species and to science that taxonomists pursue answers to their fundamental questions (Cracraft 2002, Page et al. 2005, Wheeler 2008b). Without such knowledge, how are ecologists, conservationists, and resource managers to detect changes in the occurrence and distribution of species? How are we ever to have a deep and reliable knowledge of evolution if we allow most of the evidence of its history (in the form of species and their characters) to disappear without a trace? And if we are to continue refining our knowledge of phylogeny—a process that can continue unabated even after many species are extinct—then we must make our inventory of species and their characters as complete as possible.

It is important to emphasize that while taxonomic information is essential for credible environmental science, that taxonomy is first an independent, curiosity-driven science in its own right. Those questions (Cracraft 2002, Page et al. 2005) are deserving of answers in and of themselves. Were we to discover some level of biological diversity on Mars or some other "earth-like" planet, these would be the first questions asked and answered by the National Aeronautics and Space Administration (NASA). Species are not more interesting simply because they have evolved on some other world; in all likelihood the most we will ever learn or understand of evolution and evolutionary history will result from a careful, detailed study of earth's species. The sooner we acknowledge this simple reality and get on with taxonomy's agenda, the better science, society, and the environment will be.

Our society places too much emphasis on the latest methods and technologies at the price of keeping our focus on compelling scientific questions such as those at the core of taxonomic research. As a result, we invest disproportionate funds and energy in studies whose results and impact are short-lived. A moleculo-phenogram based on a few sequenced genes is soon replaced by a subsequent analysis of different or additional gene sequences. But a revision or monograph done well and based on carefully structured hypotheses about characters (Platnick 1979) will be a useful reference a century or more from now, even factoring in accumulated specimens, species, characters, and nomenclatural changes, and collections of specimens are of even more enduring value, particularly those representing times and places where species can never be re-collected from nature.

Many biologists seem to regard the mission of taxonomy as impossibly great, but there is abundant evidence to the contrary. For example, with limited funding and essentially no coordination of efforts, entomologists have managed to describe about 1,000,000 species (Foottit and Adler 2009) representing an estimated quarter of the most species-rich taxon (Grimaldi and Engel 2005). It is also encouraging that although the needs of descriptive taxonomists have been hugely neglected in both the United States and Europe that entomologists are currently describing new species of insects at a rate twice the average for the period 1758–2008. These facts should give us confidence in the achievability of our prize. Given improved access to research resources, improved tools and research environments, increased funding, and deliberate coordination of efforts (Wheeler 2008b), there are no excuses for not accelerating the work of taxonomists by orders of magnitude.

For the sake of argument, let us assume that about ten million species await discovery and description. By information-age standards this is not an especially large number. Many cities update telephone directories annually for millions of residents, and astronomers are busy mapping an estimated two hundred *billion* stars in the Milky Way. It is time for taxonomists to think big.

Taxonomy stands at one of the most consequential crossroads in its centuries-old history. It can reverse the neglect that has impoverished its progress in recent decades by reasserting its independence and importance, and demanding the infrastructure and support that it needs and deserves. Or it can acquiesce to those who propose to replace its rigorous theories with quick-and-dirty, but vastly inferior practices such as arbitrary genetic distances in the place of explicitly testable species hypotheses. The mistaken acceptance of similarly untestable phenetic distances in the 1970s was tragic in the sense that it wasted a decade of time and funding. Today, with recognition of the biodiversity crisis, the costs of such a mistake are understood to be much greater. For hundreds of thousands of species there will be no second chance if we fail to do the job right. It is urgent and essential that taxonomists step up to the challenge, reclaim the scientific questions that are rightfully theirs, and insist that species exploration progress to the high standard of excellence already achieved by taxonomists. Opinions of some advocates of DNA barcoding notwithstanding, the purpose of taxonomy is not simply to identify species for other biologists but rather to explore the transformational histories of species and characters (Wheeler 2007).

WELCOMING A DELUGE OF SPECIES

Currently, about 20,000 species are described each year. Were we to increase that rate by just one order of magnitude to 200,000 species per year, we could double our existing knowledge of species in just a decade to about 4 million species. In four decades we could potentially raise that number to about 10 million total species. This number is a little optimistic but approximately right. As the number of "known" species increases, however, so too do the number of species hypotheses that must be tested with newly collected specimens and newly discovered characters. In other words, recognizing new species is a function of testing limits of existing species. As total knowledge of species increases, so too does the challenge of testing and corroborating existing species hypotheses as a necessary component of the search for new species.

Is it realistic to increase the rate of species description by an order of magnitude? How quickly might such an acceleration of descriptive taxonomy be achieved? I believe that one order of magnitude is easily achieved almost immediately by a combination of the following factors. Given a more aggressive approach, I believe that significantly more speed is possible.

First is a change of attitude. Were the several thousand existing taxonomists allowed to devote full attention to revisions, monographs, floras and faunas, we could have an instant rise in numbers of new species reported. Many taxonomists face significant constraints from employers, funding agencies, and peer pressure. Within some natural history museums, newly hired scientists are expected to do something "more" (e.g., ecology, population, or molecular genetic micro-evolution, etc.) than just descriptive work. This detracts from their productivity and means that the vast research resource embedded in our herbaria and museums is underutilized. As taxonomists retire from universities, they are seldom replaced in kind but instead with some alternatively popular biological sub-discipline or at best with a hybrid researcher such as described for museums. Similarly, it is very difficult to get a grant to fund a taxonomic revision or monograph. Instead, it must be put into some broader context, often diluted, with much funding and energy going instead to a related set of questions that slow taxonomic progress. There are more than 5,000 taxonomists in the world who would constitute a formidable workforce if allowed to do so, especially coupled with student training.

It is understandable that granting agencies want to objectively prioritize awards. Nearly every taxon is in need of revision, so why should a study of any particular taxon be funded at the exclusion of others? It is therefore reasonable to prioritize such funding in relation to what improved taxonomic status could make possible. Thus, taxa of interest to ecology, micro- or macro-evolution, agriculture, disease, or other topics could be funded first. It is important, however, to acknowledge that every species, every clade, has a unique story to tell us and that supporting curiosity-driven taxonomy is in itself important.

Second is a change in working practice. U.S. National Science Foundation-funded Planetary Biodiversity Inventory awards have demonstrated that rapid progress can be made by putting together teams of relevant experts from around the world to focus on a single taxon (e.g., Page 2008). Such teamwork has the potential to significantly increase efficiency and to speed progress (Knapp 2008). (See Figure 5.1.)

Finally are technological advances. The fusion of advances in cyberinfrastructure with the needs of taxonomists and museums has the potential to realize enormous gains in the productivity of taxonomy (Wheeler 2008b). Access to microscopes and other instruments remotely, comprehensive 3-D image archives of type specimens, detailed and annotated image libraries of characters, access to specimen-associated data through the Global Biodiversity Information Facility and other portals, imaging and visualization tools, software tailored to accelerate preparation of descriptions, online publishing, international video conferencing, and related tools can have a profoundly positive impact.

I believe that any combination of these three factors can deliver, very quickly, the needed one-order-of-magnitude increase in the rate of species discovery to realize 200,000 new species per year. Reasonable arguments could be made that any one of these has the potential of reaching such a goal, suggesting that a combination is absolutely capable of it.

While this plan is reasonable, inexpensive, and almost certain to succeed, it is in itself insufficiently bold. What we really need is a societal commitment to intensively invest in species exploration for the next fifty years. Why not aim for 200,000 species per year within two years, ramping up to 500,000 species "treatments" per year within a decade? Over this same period, the work would gradually shift from primarily describing new species to critically testing existing species and expanding our detailed knowledge of each. If we end up with ten million species and the capacity to describe or test half a million each year, then no species hypothesis need ever be more than five years out of date. This is quite a contrast to the status quo, where species in large groups, such as insects, are tested only once or twice each century.

This taxonomic "moon shot" would be repaid many times over in three fundamental ways. First, as a relatively complete and reliable baseline of data, information, and knowledge against which changes in biodiversity can be detected and measured. Second, as a comprehensive research resource for macro-evolution, allowing us to refine our understanding of character transformations, speciation, and phylogenetic diversification far into the future. And third, by providing scientists access to biological diversity, we can continue to discover new approaches to solving scientific, technical, and societal problems. Organic substances, genes, and natural models continue to guide discovery of such solutions.

Success must include investment in the education of the next generation of descriptive taxonomists, engineering a cyberinfrastructure tuned to meet the specific needs of taxonomists, expanding

FIGURE 5.1   Cybertaxonomy will be based on international teams of experts who work collectively to both discover and describe new species and continuously test "known" species. This is an artist's vision of one such knowledge community. The orbiting "electrons" represent individual taxonomists; the "nucleus" represents cyberinfrastructure that gives them access to instruments, specimens, and all that is known of the taxon. Existing data, information, and knowledge are drawn upon as needed, and improved and expanded data, information, and knowledge deposited in a shared virtual- or electronic-monograph or taxon knowledge bank. Image: Graphic artist Frances Fawcett. After Wheeler (2008b).

natural history collections and their infrastructure, and strengthening links between taxonomy and its many user communities. The current trend toward expecting every individual taxonomist to be engaged in a wide range of approaches must be reversed as an absolute requirement if we are to avoid mediocrity. A doctoral student cannot be expected to gain sufficient experience or expertise in the many aspects of modern taxonomy. How can a student in five years truly master comparative morphology, the latest molecular techniques, phylogenetic theory, phylogenetic analysis software, effective field work, biogeographic theory, paleontology, and ontogeny, among others? Even those taxonomists who do become such polymaths in their careers are best served by starting with a solid and deep foundation in one or a few areas and then building upon that strength. With billions of facts and millions of species in urgent need of discovery, does it not make more sense to educate and employ a diverse workforce of specialists, each bringing unique knowledge, skills, and passion to their piece of the puzzle? There are more than enough characters to warrant lifetimes spent on simply doing comparative morphology, paleontology, ontogeny, or molecular studies to their highest level of excellence. More than enough species await discovery to warrant training and supporting an army of collectors, preparators, and technicians. More than enough theoretical and technical challenges warrant supporting an additional army of theoretical taxonomists, cyberinfrastructure engineers, and programmers to subtend taxonomy. We are thinking too small. Let's aim for a much larger, richer, and diverse workforce.

Regardless of the combination of approaches adopted by any single student in her or his degree, it is important that the next generation of taxon specialists be given trans-disciplinary "literacy," that is, that they are trained to work effectively as part of a local or distributed team of taxon experts who openly collaborate and share their particular expertise. It is critical that all relevant subdisciplines of taxonomy be done. It is not necessary that they be done by every individual taxonomist.

ENGINEERING THE ARK

We are extremely fortunate that taxonomic theory is in relatively good shape thanks to Hennig (1966) and subsequent advances (e.g., Nelson and Platnick 1981). As Nelson (2004) has observed, the Hennigian theoretical revolution was cut short by a de-emphasis of the study of characters and the flood of molecular data that is treated as though it revealed phylogeny. While essentials are in place, it is important that Hennig's revolution be re-engaged so that we learn as much as possible from an intensive exploration of species and their characters.

Although mobilizing data remains an important, necessary, and enormous goal (e.g., Polaszek et al. 2008), we must move beyond thinking of cybertaxonomy as nothing more than databases. Databases are necessary for the success of cybertaxonomy of course, but are in themselves not sufficient. We must think of new kinds and uses of instrumentation, such as telemicroscopy access to type and rare specimens. We must think of new ways to analyze and visualise hypotheses about complex characters and their diversification. In total, cybertaxonomy should deliver to taxonomists every instrument, research resource, software, hardware, and colleague needed to make rapid progress in describing new species and improving and corroborating descriptions of known species (Figure 5.2).

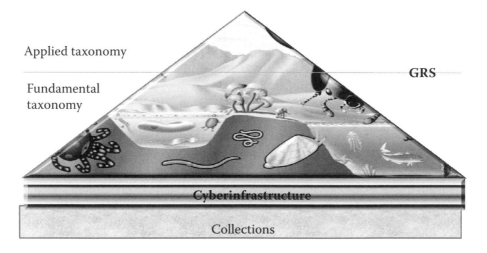

FIGURE 5.2 Cybertaxonomy envisions a foundation of collections of specimens upon which is superimposed a taxonomy-specific cyberinfrastructure that gives access to specimens, instrumentation, data, and other research resources. Fundamental curiosity-driven taxonomy results in the general reference system (GRS) that, through Linnaean names and classifications, opens access for the application of taxonomy by many user communities.

Monophyletic taxa are not constrained by geopolitical boundaries or ecosystems, and thus we need to see a new era of open international collaboration and communication. Knowledge of taxa should be seen as the property of humanity, since no nation can learn or understand its flora or fauna in isolation. In addition to the gained efficiency of such international collaborations, perhaps we will achieve enough political clout to reverse the short-sighted laws and regulations that prohibit the collecting needed to discover our world's species. This is possible given two principles. First, open access to collecting must be met with reciprocal open access to what is learned. And second, any patented discoveries resulting from such species exploration must result in an equitable and pre-agreed distribution of profits back to the country of origin. The odds of such commercially valuable discoveries are, alas, slim. Yet the fear of losing rights to a valuable substance or discovery resulting from one rare species is now condemning nations to ignorance about the vast majority of their species.

The more we know of the world's species the more intelligent and effective our planning and prioritization for species exploration. This depends in part on much improved coordination of efforts among museums and botanical gardens, both to share existing information associated with collections and to avoid unnecessary redundancy of efforts. Species exploration should no longer be seen as "ecology light" but rather as hard taxonomy. It is inherently inefficient to approach species exploration on a local and provincial basis; much can be gained by supporting teams of taxon specialists to send collecting teams where they are needed (Wheeler 1995).

The taxonomic and collections communities need to come together, explicitly define the details needed to make the new Ark effective and efficient, link such specifications for infrastructure with explicit goals (such as the short-term increase to 200,000 new species per year), and speak with one voice to demand the funding needed to educate a new generation of experts and support them, their infrastructure, and their collections to meaningfully and rapidly address the core questions of taxonomy.

CONCLUSIONS

We have enormously important contributions to make and little time in which to do so. It is time to demand the new Ark, founded on expanded herbaria and museums and a newly constructed worldwide cyberinfrastructure for taxonomy. We must reassert the excellence of the theory and practice of descriptive taxonomy and no longer allow the prejudices and ignorance of some experimental biologists who don't know good descriptive science when they see it. We must as a community set an ambitious ten-year agenda for taxonomy and species exploration and detail what is needed to achieve it. We can make contributions to evolutionary and environmental biology that no future generation will be able to do. This is of such consequence to science, society, and the environment that we simply cannot fail in our mission.

ACKNOWLEDGMENTS

It will be through the vision of, and communication facilitated by, international gatherings of taxonomists that a new infrastructure, newly articulated goals, and new era of efficient collaborative research will become a reality for taxonomy. I thank Andrew Polaszek for the vision to organize *Systema Naturae* 250 and the local organizers in Paris at the Museum National d'Histoire Naturelle for such a productive and encouraging meeting (Figure 5.3).

FIGURE 5.3   Some of the attendees of the *Systema Naturae* 250 symposium standing under the watchful gaze of Jean-Baptiste Lamarck in the Jardin de Plantes, Paris. From left to right: Quentin Wheeler, Alessandro Minelli, Sandra Knapp, Hans Dieter Sues, Andrew Polaszek, Hans Odöö (who delivered an introductory lecture in the persona of Carl von Linné), and E. O. Wilson.

REFERENCES

Cracraft, J. 2002. The seven great questions of systematic biology: An essential foundation for conservation and the sustainable use of biodiversity. *Annals of the Missouri Botanical Garden* 89: 127–144.
Foottit, R. and P. Adler (Eds.) 2009. *Insect Biodiversity*. New York: Blackwell.

Franz, N., R. K. Peet, and A. S. Weakley. 2008. On the use of taxonomic concepts in support of biodiversity research and taxonomy. In Q. D. Wheeler (Ed.), *The New Taxonomy*. Systematics Association Special Publication Vol. 76. Boca Raton, FL: CRC Press. 41–86.

Grimaldi, D. A. and M. S. Engel. 2005. *Evolution of the Insects*. New York: Cambridge University Press.

Hennig, W. 1966. *Phylogenetic Systematics*. Urbana: University of Illinois Press.

Knapp, S. 2008. Taxonomy as a team sport. In Q. D. Wheeler (Ed.), *The New Taxonomy*. Systematics Association Special Publication Vol. 76. Boca Raton, FL: CRC Press. 33–53.

Nelson, G. 2004. Cladistics: Its arrested development. In D. M. Williams and P. L. Forey (Eds.), *Milestones in Systematics*. Systematics Association Special Publication Vol. 68. Boca Raton, FL: CRC Press. 127–147.

Nelson, G. and N. Platnick. 1981. *Systematics and Biodiversity: Cladistics and Vicariance*. New York: Columbia University Press.

Page, L. M. 2008. Planetary biodiversity inventories as models for the new taxonomy, In Q. D. Wheeler (Ed.), *The New Taxonomy*. Systematics Association Special Publication Vol. 76. Boca Raton, FL: CRC Press. 55–62.

Page, L. M., H. L. Bart, Jr., R. Beaman, L. Bohs, L. T. Deck, V. A. Funk et al. 2005. *LINNE: Legacy Infrastructure Network for Natural Environments*. Illinois Natural History Survey, Champaign, IL.

Patterson, D. J., Remsen, D., Marino, W. A., and C. Norton. 2006. Taxonomic indexing: Extending the role of taxonomy. *Systematic Biology* 55: 367–373.

Platnick, N. I. 1979. Philosophy and the transformation of cladistics, *Systematic Zoology* 28, 537–46.

Polaszek, A., Pyle, R., and D. Yanega. 2008. Animal names for all: ICZN, ZooBank and the new taxonomy, 129–141, In Q. D. Wheeler (Ed.), *The New Taxonomy*. Systematics Association Special Publication Vol. 76. Boca Raton, FL: CRC Press.

Wheeler, Q. D. 1995. Systematics, the scientific basis for inventories. *Biodiversity and Conservation* 4: 476–489.

Wheeler, Q. D. 2007. Invertebrate systematics or spineless taxonomy? *Zootaxa* 1668: 11–18.

Wheeler, Q. D. (Ed.) 2008a. *The New Taxonomy*. Systematics Association Special Publication Vol. 76. Boca Raton, FL: CRC Press.

Wheeler, Q. D. 2008b. Taxonomic shock and awe. In Q. D. Wheeler (Ed.), *The New Taxonomy*. Systematics Association Special Publication Vol. 76. Boca Raton, FL: CRC Press. 211–226.

6 Historical Name-Bearing Types in Marine Molluscs

An Impediment to Biodiversity Studies?

Philippe Bouchet and Ellen E. Strong

CONTENTS

INTRODUCTION

Molluscs are the second largest animal phylum in terms of the global number of described species, and they have the largest known diversity of any marine group. We currently know about 82,000 valid described mollusc species (53,000 marine), with a yearly increment of about 580 new species descriptions (350 marine) (Bouchet 2006). Marine molluscs account for a quarter of all described marine biota. Gastropods make up roughly 85% of these and, specifically, the caenogastropods (comprising the vast majority of what people call "seashells") account for around two-thirds of total marine molluscan diversity. Seashells have attracted the interest of naturalists, amateurs, and collectors since well before Linnaeus, and the result has been both a blessing and a curse for biodiversity exploration. This chapter is mainly concerned with the impediments of historical and current taxonomic practices in documenting marine shelled gastropod diversity, and exploring ways to move forward in view of the daunting diversity that remains to be described. Other groups of molluscs (e.g., marine slugs) and certainly several other phyla of marine invertebrates suffer from similar types of problems, and would probably also benefit from the solutions explored here.

Linnaeus knew of roughly 870 species (H. Dekker, pers. com.) of "Vermes Testacea," a higher category that included the shells of molluscs and brachiopods, the tubes of serpulid polychaetes, and the plates of goose barnacles, while the shell-less molluscs (e.g., nudibranchs) were classified under "Vermes Mollusca," together with holothurians and annelid polychaetes. Leaving aside the phylogenetic rearrangements that have taken place since *Systema Naturae*, the significance of Linnaeus's classification is that species of "Testacea" were described and named exclusively based on their shells. This dichotomy between shell and animal was pursued to nomenclatural absurdity by several late 18th-century/early 19th-century zoologists. For example, Giuseppe Poli (1746–1825) used a parallel nomenclature for the shells and for the animals that occupied them. An example would be the animal *Callista* Poli, 1791 living in the shell *Callistoderma* Poli, 1795 (Bivalvia, Veneridae). Analogous systems have been developed in taxa with complex life cycles (e.g., hydroid and medusa stages of Hydrozoa), but Poli's approach differed in that different names were applied to different (and coexisting) parts of the same animal.

Although, very early on, authors such as Georges Cuvier (1769–1832), John Edward Gray (1800–1875), Franz Hermann Troschel (1810–1862), Otto Mörch (1828–1878), Paul Fischer (1835–1893), and Eugène Louis Bouvier (1856–1944) used anatomical characters as a basis for classification, and the radula and digestive tract anatomy, respiratory and excretory organs, nervous and circulatory systems are considered to be essential for the recognition of families, superfamilies, orders, and subclasses, by contrast the shell alone was considered adequate for species identification. The naming of shelled molluscs based exclusively on their shells has remained standard practice to this day, and over 80% of new descriptions of shelled marine gastropod species published in 2006 contained a description of the shell only. However, this time-honored system is compromised by the poor quality of much type material, and the resulting complexity poses a significant impediment to modern biodiversity studies.

MOLLUSCAN SHELLS IN BIODIVERSITY STUDIES

The shells of molluscs are a significant asset for biodiversity exploration because they make robust post mortem remains that potentially persist for a long time after the death of the animals that secreted them. This, of course, facilitates fossilization, and the fossil record of molluscs is exceptionally good. It also facilitates sampling and conservation of specimens; shells of molluscs do not require sophisticated preservation protocols, and sailors, missionaries, explorers, and travelers historically returned to their native countries with natural history collections, of which shells and corals, because of their hard parts, made up a significant proportion. In museum collections, the conservation of shells barely requires any maintenance at all and, except for Byne's disease (de Prins 2005) and—in the case of microshells—"glass disease" (Geiger et al. 2007), shells remain for centuries in museum drawers unaltered, whereas herbaria and insect collections require constant attention to prevent fungal and insect damage. Another advantage of the shells of molluscs is that they provide an important window on actual molluscan diversity, even if living animals are so rare, seasonal, or elusive that they are rarely or never sampled. After 25 years of intensive exploration in New Caledonia, as many as 73% of all 1,409 turrid gastropod species documented are represented only by empty shells, and 34% by a single empty shell (Bouchet et al. 2009).

But the shells of molluscs also have significant limitations. These external exoskeletons are susceptible to erosion after death, and even during the life of the animal. Shells are gradually eroded biologically (by the attack of boring and encrusting organisms), physically (by the mechanical movements of the waves), and chemically (by dissolution or diagenetic alteration of the minerals). But how these processes affect individual specimens is not uniform in space and time and there is no direct equivalence between the age of a shell and the quality of its preservation. Even ancient fossils hundreds of millions of years old may be exquisitely preserved, complete with the microsculpture of their protoconch (Bandel et al. 2002, Yoo 1994, Fryda 1999). By contrast, many shells collected on seashores may be (severely) worn—collectors use the expression "beach worn" to qualify seashells polished by sand and wave action so that details of sculpture and color are lost. However, such shells have been and occasionally still are described as new species (Turton [1932], for instance, did so on a massive scale).

The role of name-bearing types is to provide an anchor for the virtual world of names in the real world of animals. In common with many—perhaps most—other animal taxa, the historical descriptions of new recent shelled molluscs lack many characters that are today considered essential for proper (super) family assignment and species discrimination. But as progress is made in the understanding of characters and character-states, entomologists can still return to name-bearing types to, for example, mount sclerotised pieces from the internal genital anatomy, acarologists can examine the position of setae with the scanning electron microscope (SEM), ichthyologists can do meristic counts from x-rays of preserved specimens, etc. All this is feasible because name-bearing types of insects, mites, or fish were (with a few rare, exceptions) alive when they were collected. By contrast, the name-bearing types of shelled molluscs were mostly empty shells at the time they were collected. In such cases it is not only impossible to examine standard anatomical characters, including the radula, but very often the shell itself lacks important diagnostic features such as the protoconch, color, and microsculpture.

THE PROTOCONCH REVOLUTION

With their accretionary growth, gastropod shells typically exhibit at their apex a protoconch secreted by the embryo or veliger larva before hatching or metamorphosis, whereas the rest of the shell (the teleoconch) is secreted by the post-metamorphic snail. The morphology of the protoconch reflects the larval ecology of the animal that secreted it (Jablonski and Lutz 1980, 1983). A protoconch with a large initial nucleus, consisting of 1–2 whorls (i.e., "paucispiral"), and a single protoconch/teleoconch discontinuity, is characteristic of species with non-planktotrophic larval development (also called lecithotrophic or (improperly) direct development). A protoconch with a small initial nucleus and two distinct accretionary stages, including protoconch I (typically comprising less than one-half whorl secreted by the embryo in the egg-case) and protoconch II (secreted by the swimming veliger typically consisting of 3–5 whorls, i.e., "multispiral"), is characteristic of species with planktotrophic larval development. Protoconch II is also frequently adorned with complex sculpture patterns that provide additional morphological characters. The size of protoconchs is usually in the range of 0.5–2 mm, and they are thinner—thus less solid—than subadult and adult teleoconch whorls. These characteristics mean that the protoconch is the most vulnerable part of the shell and is often the first to become damaged or corroded, even during the lifetime of the animal.

The value of the protoconch as a taxonomic character was first highlighted by Dall (1924), and Powell (1942) even attributed genus-level importance to the paucispiral vs. multispiral protoconchs of Turridae (but see Bouchet 1990). With the advent of SEM, malacologists were prompt to understand its potential for examining and illustrating protoconchs (e.g., Robertson 1971, Thiriot-Quiévreux 1972). Indeed, protoconchs are so essential in the systematics of certain shelled gastropod families that Marshall (1983) wrote:

> I cannot emphasize too strongly that under absolutely no circumstances should further new species [of Triphoridae] be proposed unless a complete, unworn protoconch can be illustrated. Protoconchs should always be illustrated by scanning electron micrographs.

It has been demonstrated that the mode of larval development (i.e., planktotrophic vs. non-planktotrophic), is a species-specific character of shelled molluscs (Hoagland and Robertson 1988, Bouchet 1989). The two modes of larval development often coexist in the same genus, and there are numerous examples of species pairs/complexes, with very similar or even indistinguishable teleoconchs, but distinct—multispiral vs. paucispiral—protoconchs (e.g. Boisselier-Dubayle and Gofas 1999, Véliz et al. 2004) (Figure 6.1). Historically, such species pairs/complexes were considered to be one species. When two species are recognized, the question that arises is to which of the two species the name in current use should be applied. In principle, the answer should be straightforward: examine the name-bearing type, and confirm its protoconch type. Many times, however, the name-bearing types of shelled molluscs, especially those described before the 1970s, lack protoconchs, are fragmentary, or lack both the tip of the shell and its base, obliterating characters now considered important or essential.

FIGURE 6.1 Two species of *Cerithium* from the Mediterranean formerly included in the highly variable species *Cerithium vulgatum* Bruguière, 1792. Based on protoconchs, spawn and allozyme data, two species are recognized, one with multispiral protoconch (left), one with paucispiral protoconch (right) (see Boisselier-Dubayle and Gofas 1999). *Cerithium vulgatum* is the oldest name and there are 38 available names currently included in its synonymy. When type material is extant, it lacks the discriminating protoconch that is always eroded in the subadults and adults, and it is conjectural which name applies to each species. The minor differences apparent between the two specimens are less than the range of variation within each species. Scale bars: 10 mm (adults) and 100 μm (protoconchs). (Photos courtesy of S. Gofas.)

   Examination in the Muséum National d'Histoire Naturelle (MNHN) and the National Museum of Natural History (USNM) (Figure 6.2) of the name-bearing types of 64 and 61 nominal species of Triphoridae, respectively, shows that only roughly one-third have a protoconch in good condition, and another third have only remains that allow some "guesswork," while the remaining third are in an atrocious condition, i.e., beach worn (see, for example, Figure 6.2A); rarely do such historical types rise to the level of modern, live-collected shells in excellent condition. Buried in the taxonomic literature on molluscs, not infrequently one finds observations such as: "The holotype of *Murex mundus* Reeve, 1849 [...] is beachworn, probably subadult, and lacks the protoconch and the first teleoconch whorl. [Its] identity remains thus uncertain" (Houart 2003). Common sense would perhaps recommend that such name-bearing types be discarded and replaced by a neotype. However, a replacement must be approved by the International Commission on Zoological Nomenclature (ICZN) following an application, on a case-by-case, name-by-name, basis. Furthermore, the judgment of one author who regards a specimen as unrecognizable can be challenged by another author, even purely on principle. Also, the eyes of two taxonomists with different training and experience may not "see" exactly the same thing on the same shell. As a result, such an application would be considered ineffectual by many zoologists, except in the case of emblematic or biologically or ecologically important species. But what about an insignificant species such as *Murex mundus*? It is symptomatic that so few molluscan petitions in the *Bulletin of Zoological Nomenclature* are for replacement of existing type material by a neotype, whereas such petitions are not rare for arthropods or vertebrates. The implicit significance is that malacologists are perhaps either more pragmatic or more cynical about the taxonomical value of types.

   Taxonomists are thus often left to guess from subtle differences in teleoconch sculpture to which of two or more biological species a name-bearing type belongs. After reasonable guesswork and

FIGURE 6.2 Types of triphorid species described by Paul Bartsch and deposited in the USNM; A to F, worst to best. A. *Triphoris montereyensis* Bartsch, 1907, figured holotype, USNM 32216. B. *Triphoris stearnsi* Bartsch, 1907, figured holotype, USNM 32259. C. *Triphoris pedroanus* Bartsch, 1907, figured syntype, USNM 152206. D. *Triphoris peninsularis* Bartsch, 1907, figured holotype, USNM 106424. E. *Triphoris africana* Bartsch, 1915, figured holotype, USNM 186804A. F. *Triphoris dalli* Bartsch, 1907, figured holotype, USNM 195375. Scale 1 mm.

inference from type localities (when the two species in a pair have partly non-overlapping distributions), the type can often be pinned down to a certain species. But for how long? Only until one discovers there is not one, but a species complex, that matches the fragmentary type? The following quote was written by Marshall (1983) with Triphoridae in mind, but could apply just as well to Cerithiopsidae and many other caenogastropods and shelled heterobranchs:

> Many species based on imperfect type specimens will be impossible to identify, at least until the faunas of particular type localities are sufficiently well known for topotypes to be identified with confidence. In cases where the type locality is unknown, or when a type specimen comes from a locality where two or more species with indistinguishable teleoconchs occur, species based on specimens lacking the protoconch or lacking even the first whorl may be permanently unrecognizable.

By leaving so much open to personal interpretation and appreciation, name-bearing types clearly do not fulfill their function to anchor nomenclature in verifiable statements.

NO RADULA OR GENETIC MATERIAL FROM EMPTY SHELLS

The situation becomes even worse when anatomical characters are essential to discriminate between species or to allocate species to the correct genus and family. The genital anatomy of seaslugs is routinely used for species-level discrimination, but the anatomy of shelled molluscs is considered to be informative primarily at higher taxonomical levels—genus, family, or superfamily (e.g. Strong 2003; but see Reid 1996). Most anatomical characters are time-consuming to observe, interpret, and illustrate, especially on large suites of specimens, but the radula is an exception. Radulae are relatively easy to extract (also from dried soft parts) and, starting with Solem (1972), their examination and illustration has become greatly facilitated by SEM. The radula is a ribbon situated in the buccal bulb of a gastropod; it consists of a chitinous membrane carrying rows of minute mineralized or sclerotised teeth, each measuring a few microns to several tens or even hundred microns, coming in all sorts of shapes and numbers depending on the taxon considered. Radulae provide superb taxonomical characters at family, genus, or even species level.

Not infrequently, species or genera with featureless shells that are indecisively attributable to one of several possible families can be unambiguously classified in the correct family based on their radula (e.g. *Onoba bassiana* Hedley, 1921, the type species of *Botelloides* Strand, 1928, originally described as a rissoid, and transferred to Trochidae by Ponder [1985]; *Daphnellopsis lamellosus* Schepman, 1913, originally described as a turrid, and transfered to Muricidae by Houart [1986]). Species of Triphoridae with indistinguishable shells can have strikingly different radulae, indicative not only of species-level, but also of genus-level differences. For instance, *Marshallora adversa* (Montagu, 1803) and *Similiphora similior* (Bouchet and Guillemot, 1978) have virtually indistinguishable protoconchs and teleoconchs, but very different radulae (Bouchet 1985) (Figure 6.3). When name-bearing types are empty shells, even with well-preserved protoconchs, none of these characters are available for scrutiny. This is another situation where name-bearing types fail to fulfill their taxonomical and nomenclatural function.

Finally, there is the issue of molecular characters. Historical sea-fan, insect, or crustacean types may have faded colors, broken legs or branches, yet they were alive (or at least freshly dead) at the time they were collected or preserved, and in principle their preserved or dried remains contain genetic material. Similarly, types of vertebrates often contain useable DNA in the bones and teeth, although not necessarily alive when originally collected. The DNA may be degraded and technically difficult to extract and sequence with current routine techniques (e.g., Hajibabaei et al., 2006), but the fact is that name-bearing types of octocorals, insects, or crabs do contain tissues with potentially sequenceable DNA (e.g., Berntson and France, 2001; Hylis et al., 2005; Fisher and Smith, 2008; Zhang et al., 2008). By contrast, classical molluscan types do not contain any DNA, and no technological advance will improve the situation. As molecular sequencing, especially COI, is just becoming a routine taxonomic character in marine molluscs (e.g., Kelly et al., 2007; Mikkelsen et

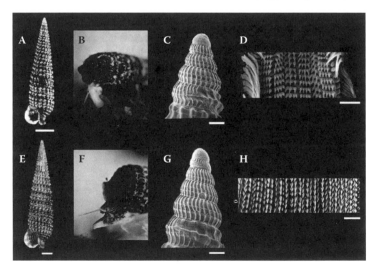

FIGURE 6.3 *Marshallora adversa* (Montagu, 1803) (A–D) and *Similiphora similior* (Bouchet and Guillemot, 1978) (E–H). Note the indistinguishable teleoconchs (A, E, scale 1 mm) and protoconchs (C, G, scale 100 μm) (the very minor differences between the two specimens are of the same magnitude and nature as differences between two conspecific specimens), but very different head–foot pigmentation and radulae (D, H, scale 10 μm). The name *Murex adversus* Montagu, 1803, was resurrected from the synonymy of *Trochus perversus* Linnaeus, 1758, and, in the absence of extant type material, a live-taken neotype was designated. *Triphora similior* was described as a new species, and a live-taken holotype was designated. After Bouchet (1985).

al., 2007; Johnson et al., 2008), we are already facing cases of barely distinguishable shells that can be recognized as distinct species only by molecular and radular characters (Kantor et al., 2008) (Figure 6.3).

Freshly collected specimens can, with some difficulty, be attributed to one of the species, based on subtle characters of the teleoconch, but older, beach-worn specimens—including name-bearing types—remain in limbo. In the example discussed by Kantor et al. (2008) (Figure 6.4), the authors stated:

> The syntypes of *Pleurotoma cingulifera* Lamarck, 1822 are badly-worn specimens that render their identification difficult. [...] Although some ambiguity persists, we apply the name *cingulifera* to the form with "semi-enrolled" marginal radular teeth. We are of the opinion that this is nomenclaturally more stable than leaving *cingulifera* as a *nomen dubium* and describing the form with "semi-enrolled" marginal radular teeth as a new species. Ideally, the current name-bearing types of *Pleurotoma cingulifera* Lamarck, 1822 and *Xenuroturris legitima* Iredale, 1929 should be replaced by live-taken neotypes with known radular and molecular characteristics. This can be done only by a decision of the International Commission on Zoological Nomenclature, but we are of the opinion that this is unnecessary as long as the systematic and nomenclatural conclusions of this paper are accepted by zoologists.

As molecular barcoding becomes routine in marine molluscan taxonomy, such cases are bound to become more and more common, and this will be yet another situation where name-bearing types fail to fulfill their nomenclature stabilizing function.

In this chapter we have focused on those situations where species are distinguishable by trivial differences in teleoconchs, but significant differences in protoconchs, radulae, living animal features, or molecular characters, a situation particularly profound among the multitudes of similar species of gastropods in the five most diverse molluscan families—Cerithiopsidae, Triphoridae, Pyramidellidae, Turridae, and Eulimidae. Some may argue that all marine molluscs do not fall into this situation, and that there are many instances of clear-cut species with highly distinct teleoconch characters. In these cases, the name-bearing types are unambiguous and hence the taxonomy is also unambiguous and uncontroversial. In such cases, the existence of types may

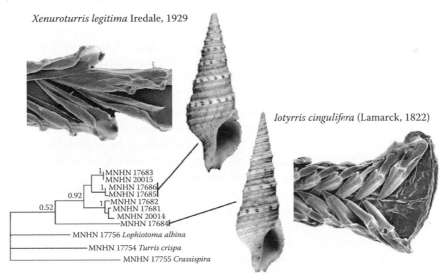

FIGURE 6.4 *Xenuroturris legitima* Iredale, 1929 (above) and *Pleurotoma cingulifera* Lamarck, 1822 [now *Iotyrris cingulifera*] (below). Closely resembling turrid species distinguished by molecular and radular characters, but with almost identical teleoconchs (the minor differences apparent between the two specimens are mainly due to size). The badly worn syntypes of *P. cingulifera* are not attributable with absolute certainty to one of the two biological species, and remain in limbo. After Kantor et al. (2008).

not be absolutely essential for the stability of nomenclature. For example, we may not need to have a name-bearing type for the dogwhelk *Nucella lapillus* (Linnaeus, 1758) as long as nobody suspects it hides a species complex. But even the iconic common cowrie *Cypraea tigris* Linnaeus, 1758 has been shown to harbor cryptic diversity based on molecular characters (Meyer, 2003), and taxonomic ambiguity of nomenclatural types may be more widespread than we currently are prepared to accept.

DOCUMENTING AND DESCRIBING THE MOLLUSCAN FAUNA OF THE WORLD

The description of marine molluscs currently continues unabatedly at the pace of roughly 350 new species per year. Not only is there no sign of leveling off, but there was, in fact, a steady increase of 68% in the naming of new marine molluscs between the 1960s and the 1990s (Bouchet 1997), and this trend is sustained. Even in European seas, where the shallow-water mollusc fauna was reputed to be completely inventoried since the beginning of the 20th century, in the late 1960s new species started again to be named and described, as a result of attention given to micromolluscs and nudibranchs. However, the main reservoirs of undescribed marine molluscs are in the tropics and the deep sea. Inspired by the work of entomologists in the canopies of tropical rainforests, recent large-scale marine expeditions are returning from the field with fantastic collections of molluscan specimens—with postlarvae, juvenile, subadult, and adult—live-taken, with digital images of living animals and clips of tissue for sequencing. Whereas "seashell" families (cowries, cones, volutes, miters, etc.) are fairly well known, with few new shallow-water species remaining to be discovered, the percentage of new species in hyper-diverse families of micromolluscs can be as high as 80–90% (Anders Warén, pers. com. cited by Bouchet et al. 2002). In deeper water, the Turridae (*s.l.*) is represented in New Caledonia alone by 1,409 morphospecies, with estimates that over 80% of these are undescribed (Alexander Sysoev, pers. com.). Preliminary examination of collections from other South Pacific island groups suggests that an estimate of 5,000, mostly undescribed, deep-water species of turrids is not beyond the realm of possibility (Bouchet et al. 2009).

A bottleneck shared by entomology and malacology is the assignment of names to the morphospecies or operational taxonomic units (OTUs) generated by such massive sampling exercises. There is justifiable concern for the broadening gap between discovering and documenting the diversity of the world and backing this exercise with sound nomenclature (Thompson 1997). Admittedly, molluscan systematics is not the only branch of zoology that suffers from the instability of names in taxonomic limbo, but with taxonomic inflation fueled by shell collecting, the magnitude of the problem is certainly unique to malacology. No one has ever evaluated the magnitude of the bank of nomenclaturally available names for Recent molluscs, but we believe that it is on the order of 300,000–500,000 species-group names; of these, 82,000 designate species currently considered valid, with an estimated 150,000–250,000 that have been evaluated as synonyms, and possibly another 100,000 in taxonomic purgatory (not evaluated, not in use, *nomina dubia*, etc.). How many are based on an existing name-bearing type in a publicly accessible institution? How many of these are based on specimens that defy modern standards as acceptable biological vouchers? Nobody knows. Without question, there is a critical need to evaluate the quality and quantity of names in the existing "name bank." Gittenberger (1993) used the expression "digging in the graveyard of synonymy" to describe the unrewarding task of confronting modern taxonomic hypotheses to nominal species currently considered invalid. We can extend the notion of a "graveyard" to encompass not only the names that are currently considered synonyms but also the taxonomic purgatory of names that have yet to be assessed. There are probably 100,000–150,000 molluscan species still to be named, and although we respect the history of our discipline and the work of our predecessors, too much is at stake in the face of a sixth extinction to waste our efforts digging in the graveyard of synonymy and matching modern live-taken specimens with ineffective name-bearing types.

DO WE NEED A MALACOLOGICAL "REVOLUTION"?

Biodiversity studies gain nothing by trying to resurrect or preserve ambiguous nominal species. Given this burden, the artisanal pace of the documentation and description of molluscan diversity can never hope to keep up with the industrial scale at which new species are discovered based on gene sequences alone. Not surprisingly, some researchers are advocating the development of name-free taxonomy for molecular OTUs (Klee et al., 2007), and molluscan systematics is effectively rooted in quicksand if (tens of) thousands of names in current use are backed by name-bearing types that do not fulfil their nomenclature stabilizing function.

In 250 years of morphology-based nomenclature, taxonomists have had time to accept that taxonomy is fallible and open to disputes and mistakes. But through reference to name-bearing types deposited in publicly accessible collections, the rules of nomenclature that have been in place since the latter half of the 19th century have permitted the correction of such mistakes and have brought remarkable stability. By contrast, much of molecular sequencing is effected by individuals who have no background in species identification, and who tend to view the historical heritage of taxonomy, including nomenclature, as a burden and a thing of the past. However, there is a growing body of literature (e.g., Vilgalys 2003) that documents the unreliability of many identifications associated with sequences deposited in public DNA databases. Before we declare traditional specimen-based nomenclature obsolete, modern sequence-based taxonomy must accept binding nomenclatural rules that guarantee stability and permit the correction of errors. The molecular systematists of today must develop a culture of vouchering specimens in long-term publicly accessible collections. Only then will there be a verifiable link between a specimen, a sequence, and a name. We praise the Data Submission Protocol for the Barcode of Life Database (BOLD), which requires vouchers and voucher information for all submissions of new molecular sequences (http://www.barcodinglife. org/docs/BOLD%20Data%20Submission.pdf), and we advocate that this should be mandatory practice in other molecular databases and for publication of research papers in molecular systematics journals.

A malacological revolution could take valuable lessons from other fields of biodiversity, namely bacteriology and botany, that have faced these challenges and already implemented solutions. In the 19th century and first half of the 20th century, bacteriologists tried to follow the provisions of the Botanical Code of Nomenclature, because bacteria had traditionally been considered fungi, the Schizomycetes. Methods of study were, however, very different. Also, much emphasis was placed on cultural characteristics, so that type cultures were of critical importance. In the 1960s and 1970s, bacteriological systematics was making fantastic progress thanks to new molecular approaches that facilitated discrimination between species and less ambiguous species definitions, while at the same time bacteriological nomenclature was being hampered by a very large percentage of published names that could not be used, due to lack of good descriptions and type cultures. A decision was then made to make a completely new start for nomenclature of bacteria on January 1, 1980. Lists were made of names that could be satisfactorily associated with known bacteria, and these formed the foundation document, the *Approved Lists of Bacterial Names* (1980). Names not on these lists lost standing in nomenclature (although provisions were made to revive old names subject to certain safeguards).

In botany, the concept of epitype (initially termed "protypus"), was proposed by Barrie et al. (1991) with words that could be applied almost without change to the situation we have described above:

Inadequate type material is frequently encountered when working with older names. All too often the surviving original material consists of poorly preserved or poorly collected specimens, vague drawings or sketches lacking the characters crucial to precise identification. The [Botanical] Code requires that a name be typified by original material if it exists, but if the obligate element is ambiguous the resulting typification does little to promote nomenclatural stability.

This new category of type, adopted in the Tokyo Code (1994) of Botanical Nomenclature, is intended to provide the necessary diagnostic characters to unambiguously fix the identity of the primary type. The epitype does not displace the primary type it is to interpret, but rather supplements it.

Just as the *Approved Lists of Bacterial Names* laid new foundations for the modern era of prokaryote taxonomy, an ideal hypothetical "*Approved List of Molluscan Names*" would contain all 82,000 molluscan names in current use, backed by name-bearing types that can be confronted to modern morphological and molecular screening; this would necessitate the establishment of thousands of neotypes/epitypes. Such an ideal approved list would have no sympathy for wobbly nominal species backed by beach-worn empty shells.

If molluscs were birds, mammals, or frogs, such a revolution might perhaps not be completely unrealistic, but the level of scientific effort devoted to the systematics of molluscs and other invertebrates is two orders of magnitude less than that devoted to tetrapod vertebrates and one order of magnitude less than that devoted to plants (Gaston and May 1992). The vast purgatory of nominal species is undoubtedly the main reason preventing a world register of mollusc species. Most of the hyper-diverse and most taxonomically challenging molluscan families are essentially without a single expert, and there simply is not the workforce to build an approved list of molluscan names that would stand the test of time, let alone to evaluate the contents of the "name bank." Rather than being elevated by the fantastic opportunities made possible by SEM, digital photography, and the molecular revolution, it is unacceptable that species-level systematics in the most diversified marine phylum is progressing at a snail's pace because of the burden of this historical legacy.

ACKNOWLEDGMENTS

Henk Dekker, Serge Gofas, Roland Houart, and Yuri Kantor kindly provided examples and illustrations. Bruce Marshall, Winston Ponder, and Anders Warén acted as a sounding board on an earlier version of this chapter. Sandra Knapp and Valery Malécot introduced us to the botanical "epitype" concept and its application. Philippe Maestrati and Gilberto Marani assisted with preparing Figures 6.3 and 6.4 for publication.

REFERENCES

Bandel, K., A. Nützel, and T.E. Yancey. 2002. Larval shells and shell microstructures of exceptionally well-preserved Late Carboniferous gastropods from the Buckhorn Asphalt deposit (Oklahoma). *Senckenbergiana Lethaea*, 82(2): 639–689.

Barrie, F.R., J.L. Reveal, and C.E. Jarvis. 1991. Proposals to permit the designation of a discriminating element to allow the precise interpretation of an ambiguous type. *Taxon*, 40(4): 667–668.

Berntson, E.A. and S.C. France. 2001. Generating DNA sequence information from museum collections of octocoral specimens. *Bulletin of the Biological Society of Washington,* 10: 119–129.

Boisselier-Dubayle, M.C. and S. Gofas. 1999. Genetic relationships between marine and marginal-marine populations of *Cerithium* species from the Mediterranean Sea. *Marine Biology*, 135(4): 671–682.

Bouchet, P. 1985. Les Triphoridae de Méditerranée et du proche Atlantique. *Lavori della Societa Italiana di Malacologia*, 21: 5–58.

Bouchet, P. 1989. A review of poecilogony in gastropods. *Journal of Molluscan Studies*, 55: 67–78.

Bouchet, P. 1990. Turrid genera and mode of development: The use and abuse of protoconch morphology. *Malacologia*, 32: 69–77.

Bouchet, P. 1997. Inventorying the molluscan diversity of the world: What is our rate of progress? *The Veliger*, 41(1): 1–11.

Bouchet, P. 2006. The magnitude of marine biodiversity. In: C.M. Duarte (Ed.). *The exploration of marine biodiversity. Scientific and technological challenges.* Bilbao: Fundación BBVA, 31–62.

Bouchet, P., P. Lozouet, and A. Sysoev. 2009. http://www.sciencedirect.com/science?_ob=ArticleURL&_udi=B6VGC-4WF4J21-2&_user=10&_rdoc=1&_fmt=&_orig=search&_sort=d&_docanchor=&view=c&_searchStrId=1068383053&_rerunOrigin=google&_acct=C000050221&_version=1&_urlVersion=0&_userid=10&md5=4a f72f07a091b62c51a3499b13da5d42. *Deep Sea Research Part II: Topical Studies in Oceanography,* 56: 1724–1731.

Dall, W.H. 1924. The value of the nuclear characters in the classification of marine gastropods. *Journal of the Washington Academy of Sciences*, 14: 177–180.

De Prins, R. 2005. Deterioration of shell collections: causes, consequence and treatment. *Gloria Maris*, 43: 1–75.

Fisher, B.L. and M.A. Smith. 2008. A revision of Malagasy species of *Anochetus* Mayr and *Odontomachus* Latreille (Hymenoptera: Formicidae). *PLoS ONE*, 3: e1787.

Frýda, J. 1999a. Higher classification of the Paleozoic gastropods inferred from their early shell ontogeny. *Journal of the Czech Geological Society*, 44: 137–153.

Gaston, K.J. and R.M. May. 1992. Taxonomy of taxonomists. *Nature*, 356: 281–282.

Geiger, D.L., B.A. Marshall, W.F. Ponder, T. Sasaki, and A. Warén. 2007. Techniques for collecting, handling, preparing, storing and examining small molluscan specimens. *Molluscan Research*, 27: 1–50.

Gittenberger, E. 1993. Digging in the graveyard of synonymy, in search of Portuguese species of *Candidula* Kobelt, 1871 (Mollusca: Gastropoda Pulmonata: Hygromiidae). *Zoologische Mededelingen*, 67: 283–293.

Hajibabaei, M., M.A. Smith, D.H. Janzen, J.J. Rodriguez, J.B. Whitfield, and P.D.N. Hebert. 2006. A minimalist barcode can identify a specimen whose DNA is degraded *Molecular Ecology Notes*, 6: 959–964.

Hoagland, K.E. and R. Robertson. 1988. An assessment of poecilogony in marine invertebrates: phenomenon or fantasy? *Biological Bulletin*, 174: 109–125.

Houart, R. 1986. Mollusca Gastropoda: Noteworthy Muricidae from the Pacific Ocean, with description of seven new species. *Mémoires du Muséum National d'Histoire Naturelle, Paris*, ser. A, 133: 427–455.

Houart, R. 2003. Description of a new muricopsine species (Gastropoda: Muricidae) from the Southwestern Indian Ocean. *Iberus*, 21(2): 91–98.

Hylis, M., J. Weiser, M. Obornik, and J. Vavra. 2005. DNA isolation from museum and type collection slides of microsporidia. *Journal of Invertebrate Pathology*, 88: 257–260.

Jablonski, D. and R. Lutz. 1980. Molluscan larval shell morphology. Ecological and paleontological applications. pp., I: Skeletal growth of aquatic organisms (D. Rhoads, R. Lutz, Eds). New York: Plenum. 323–377.

Jablonski, D. and R. Lutz. 1983. Larval ecology of marine benthic invertebrates: paleobiological implications. *Biological Reviews*, 58: 21–89.

Johnson, S.B., A. Warén, and R.C. Vrijenhoek. 2008. DNA barcoding of *Lepetodrilus* limpets reveals cryptic species. *Journal of Shellfish Research*, 27: 43–51.

Kantor, Yu., N. Puillandre, and P. Bouchet. 2008. Morphological proxies for taxonomic decision in turrids (Mollusca, Neogastropoda): A test of the value of shell and radula characters using molecular data. *Zoological Science*, 25: 1156–1170.

Kelly, R.P., I.N. Sarkar, D.J. Eernisse, and R. Desalle. 2007. DNA barcoding using chitons (genus *Mopalia*). *Molecular Ecology Notes*, 7: 177–183.

Klee, B., I. Hyman, A. Wiktor, and G. Haszprunar. 2007. Cutting the Gordian knot of a taxonomic impediment. A plea for MOTU–numbers (Molecular Operational Taxonomic Units). Abstracts, World Congress Malacologia July 20–25, Antwerp, Belgium. 115.

Marshall, B.A. 1983. A revision of the Recent Triphoridae of southern Australia. *Records of the Australian Museum*, suppl. 2: 1–119.

Meyer, C.P. 2003. Molecular systematics of cowries (Gastropoda: Cypraeidae) and diversification patterns in the tropics. *Biological Journal of the Linnean Society*, 79: 401–459.

Mikkelsen, N.T., C. Schander, and E. Willassen. 2007. Local scale DNA barcoding of bivalves (Mollusca): A case study. *Zoologica Scripta*, 36: 455–463.

Ponder, W.F. 1985. A revision of the genus *Botelloides* (Mollusca: Gastropoda: Trochacea). *South Australia Department of Mines and Energy, Special Publication*, 5: 301–327.

Powell A.W.B. 1942. The New Zealand Recent and fossil Mollusca of the family Turridae. *Bulletin of the Auckland Institute and Museum*, 2: 1–188.

Reid, D. 1996. *Systematics and evolution of* Littorina. London, The Ray Society. 463 pp.

Robertson, R. 1971. Scanning electron microscopy of larval marine gastropod shells. *The Veliger* 14:1–12.

Solem, A. 1972. Malacological applications of scanning electron microscopy II. Radular structure and functioning. *The Veliger* 14, 327–336.

Strong, E.E. 2003. Refining molluscan characters: Morphology, character coding and a phylogeny of the Caenogastropoda. *Zoological Journal of the Linnean Society*, 137(4): 447–554.

Thiriot-Quiévreux, C. 1972. Microstructures de coquilles larvaires de Prosobranches au microscope électronique à balayage. *Archives de Zoologie Expérimentale et Générale*, 113: 553–564.

Thompson, F.C. 1997. Names: The keys to biodiversity. In: Reaka-Kudla, M.L., D.E. Wilson and E.O. Wilson (Eds.), *Biodiversity II*. Washington D.C.: Joseph Henry Press, 199–211.

Turton, W. H. 1932. *The marine shells of Port Alfred, S. Africa.* London: Oxford University Press. 331 pp.

Veliz, D., F.M. Winkler, and C. Guisado. 2004. Developmental and genetic evidence for the existence of three morphologically cryptic species of *Crepidula* in northern Chile. *Marine Biology,* 143(1): 131–142.

Vilgalys, R. 2003. Taxonomic misidentification in public DNA databases. *New Phytologist,* 160: 4–5.

Yoo, E.K. 1994. Early Carboniferous Gastropoda from the Tamworth Belt, New South Wales, Australia. *Records of the Australian Museum,* 46(1): 63–120, pls. 1–23.

Zhang, D.-Z., B.-P. Tang, and S.-L. Yang. 2008. Improved methods of genome DNA extraction and PCR for old crab samples including lectotype of museum. *Sichuan Journal of Zoology,* 27: 31–34.

7 Flying after Linnaeus
Diptera Names since Systema Naturae (1758)

Neal L. Evenhuis, Thomas Pape,
Adrian C. Pont, and F. Christian Thompson

CONTENTS

> In the manifestation on earth of life which connects the past with the future, and comes between the seen and the unseen, the two-winged fly has a most important part, whether it be in service to flowers, in consumption of matter in a transition state, in sustaining the life of birds and of fishes, or in maintaining the balance of life by appropriating its appearance in other forms, and thus adjusting the number of moving creatures to their required use or work.
>
> **—Francis Walker, 1874**

Flies are beautiful, diverse, and important. While many are small, drab, and inconspicuous, some flies are large, colorful, and eye-catching, usually seen on flowers or foliage. Flies come in a multitude of shapes, habits, and sizes, and can be considered the ecologically most varied of the insect orders. Because of this diversity, flies affect man in more ways than any other group of animals. For agriculture, some flies are major pests of crops, but others are beneficial by killing such pests either as parasites or predators. Flies are also essential for the successful pollination of many crops, so, for example, without flies there would be no chocolate (Young 1994, National Research Council Committee on the Status of Pollinators 2007). For the environment, flies clean up after us by decomposing our wastes and recycling critical nutrients. Dipteran mouthparts continuously filter out detritus and microorganisms from large quantities of waste water in our sewage treatment systems. Flies decompose dead bodies, excrement, and rotting vegetation. To the detriment of our health, flies carry diseases that cripple us and bring casualties and suffering beyond imagination. The major fly-borne diseases today remain malaria (close to half a billion cases and 1.2 million deaths per year) and dengue (tens of millions of cases per year) (WHO 2003), but there are many other devastating diseases and new ones continue to emerge that are transmitted by biting flies. But flies also improve our health. Blow fly maggots provide expert treatment of bedsores and diabetes-related gangrene, and *Drosophila*, the vinegar fly and the most widely used organism to study genes and genomes, has provided valuable insights into our hereditary disorders (Ashburner 2006). Even for fields like law enforcement and forensic science, flies are critical. Flies are usually the first to find a dead body and there is a range of species that develop at various rates and under different circumstances on those bodies, so knowing the fly species can help determine the time and place of death and even sometimes what the person did before dying (Goff 2001).

Flies are ancient. The sister to modern flies arose some 250 million years ago in the Upper Permian, and flies began diversifying in the Upper Triassic (225 million years ago) (Grimaldi and Engel 2005). The result of this diversification, especially a rapid radiation in the late Cretaceous or early Tertiary, is one of the largest groups (clades) of organisms on Earth. Flies today constitute about 10% of the known (published) biodiversity (some 154,000 species) and are distributed among 161 families and almost 20,000 genera. But estimates suggest that we know only 10% of what is actually living on our planet; there are some one and a half million species of flies occurring world-wide, 90% of them not yet named.

The past is a prologue as we build on the knowledge of our predecessors and learn from their mistakes. The official start of the modern understanding of flies and their classification has been deemed to be the 10th edition of Linnaeus' *Systema Naturae* in 1758. Linnaeus (Figure 7.1) produced the first comprehensive database of the natural world. He was first to apply a uniform set of names for all known organisms and to place those names into a "natural" hierarchical classification. He divided nature into three kingdoms, many classes, orders, genera, and species. Flies were placed in the order Diptera (there were a couple of exceptions) of the class Insecta. Linnaeus derived both his classification of flies and their name from Aristotle. For the order, the principle character Linnaeus used was the presence of only two wings, not four or none as in most insects, but he also noted the

FIGURE 7.1   Linnaeus at the age of twenty. (From Fée 1832.)

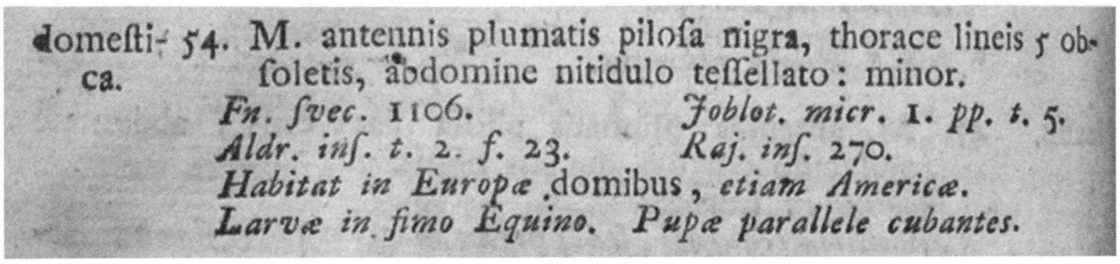

domesti- 54. M. antennis plumatis pilofa nigra, thorace lineis 5 ob-
ca. foletis, abdomine nitidulo teffellato: minor.
Fn. fvec. 1106. *Joblot. micr.* 1. *pp. t.* 5.
Aldr. inf. t. 2. *f.* 23. *Raj. inf.* 270.
Habitat in Europa domibus, *etiam America.*
Larva in fimo Equino. Pupa parallele cubantes.

FIGURE 7.2 The entry for *Musca domestica* in *Systema Naturae*, 10th edition (Linnaeus 1758).

halteres, the modified second pair of wings, which is the most conspicuous unique autapomorphy of the clade. He divided the order into 10 genera and 191 species. For each species, he provided a single word, an epithet, then a diagnosis, distribution statement, summary of the biology, and references to where further information could be found. For flies, Linnaeus cited the works of 24 different authors. His entry for *Musca domestica* (Figure 7.2) provides an example of his method.

Because Linnaeus was deemed the first, his work was perfect, without any missing names or works. This is also what is desired today: a single comprehensive authoritative reference work. The only things missing from *Systema Naturae* were images. As knowledge of nature increased, the *Systema Naturae* quickly became difficult to maintain. Linnaeus's last edition of his *Systema* was the 12th (Linnaeus 1767), and he declined to produce another. A German, Johann Friedrich Gmelin, took on this task and produced a 13th edition (the portion on flies is in Gmelin 1790). But even before this, the students of Linnaeus had already decided to divide the task among themselves. Of those, Johann Christian Fabricius took responsibility for the insects and produced his *Systema Entomologiae* (Fabricius 1775). His system was identical in form to *Systema Naturae*, but was restricted to just the insects and included many more of them. However, as the number of species and the knowledge about them increased over time, the original "system" of all insects could no longer be maintained. Fabricius realized this and began a series of works devoted to one order each. Toward the end of his life, he provided the *Systema Antliatorum* (Fabricius 1805), devoted just to flies, and in that work Fabricius changed the name of the order from Diptera to Antliata. Again, the format was the same as in Linnaeus, but with much more knowledge summarized. In 1805 Fabricius knew 1,151 species of flies, distributed among 78 genera and based on his own work and that of 46 other authors (Table 7.1). Unfortunately, this did not include all that was known at the time, as there were some 1,767 missing names represented in the work of 49 missing authors. After Fabricius, more and more workers began to study flies (Figure 7.3), and the number of species and names increased exponentially (Figure 7.4).

Today we no longer have comprehensive works like the *Systemae* of Linnaeus or Fabricius. We do understand what we have lost, and using modern technologies (computers and the Internet), we have begun to build their modern equivalents (e.g., *Encyclopedia of Life*, *Global Names Architecture*). The impediments that made the continued updating of the *Systema* impossible were the inflexibility

TABLE 7.1
Number of Valid Genera and Valid Species as well as Total Number of Names for the Three Major Diptera Nomenclators through Time (Fossils Excluded)

	Linnaeus 1758	Fabricius 1805	BDWD 2008
Genera	10	78	11,672
Species	191	1,151	156,668
Names	201	1,242	273,974
Authors	24	46	4,680

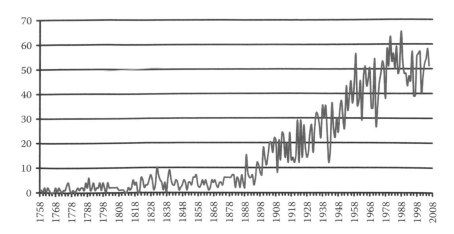

FIGURE 7.3 Growth in the number of authors over time. (Authors plotted as per year of their first description of a dipteran taxon, which means that the number of actively publishing authors in any particular year may be considerably higher.)

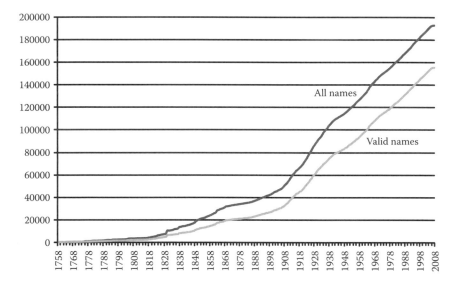

FIGURE 7.4 Accumulation curves for the number of species-group names of Diptera through time (fossils excluded).

of printing and the increased cost of disseminating knowledge by printing with fixed type reproducing text in ink on paper. Today, computers are taking over the physical aspects of printing and provide an easy means for integrating the past with current knowledge, and they also allow for alternative dissemination media beyond paper. The Internet with the Web is a relatively new medium, allowing anyone anywhere with a computer and online access to receive information in real time from anywhere in the world. The challenge is now neither the production nor the media, but what society wants of our science: systematics.*

What do people want to know and need to know about flies? People may have a strange fly, so they want to know: "What is it?" "What does it do?" and "Where did it come from?" Or they may have a problem, for example with rotting oranges. That leads them to ask about the cause; and when

---

* The accepted term for taxonomy and nomenclature and other aspects of our science today is systematics, a term that is directly derived from *Systema Naturae*. That is, the science of inferring the system inherent in the natural world.

they hear that a fly maggot is the problem, they also want to know what fly it is. The resolution of each starts with the identification of the fly, which leads to a name. With the name, people can then access knowledge (Thompson 1993). Knowledge, which was first locked up in *Systema Naturae*, is now dispersed across hundreds of thousands of works, but maybe some day soon it will again be centralized or rather interconnected into an *Encyclopedia of Life* (http://www.eol.org).

To build such an encyclopedia, we must first assemble the critical pieces of the bioinformatic infrastructure. Just as what we need for life in a modern society is transported on a system of airways, highways, seaways, and railways, biodiversity information must also be disseminated via a critical infrastructure. That infrastructure starts with names in nomenclators and species in catalogues. Information on identification and classification is disseminated through revisions and monographs. Most of the infrastructure of systematics remains in the traditional printed medium, but some has been or is being migrated to the online, digital medium of the Internet. For flies, the first critical component of this infrastructure is the *BioSystematic Database of World Diptera* (BDWD; http://www.diptera.org), a combination of a nomenclator and a species catalogue.

Years ago a Chinese philosopher noted that knowledge begins with applying the correct names. So, to deliver and decipher biodiversity information about flies, we are first developing the BDWD as an authority for information about the names of all flies. The names are then organized into a classification (or a taxonomy) just as Linnaeus and Fabricius did in their *Systemae*. For flies, the BDWD provides a single authoritative classification developed by consensus among contributors and derived from a more comprehensive taxonomy, which includes information on the characters used to generate the scientific hypotheses underlying the classification. The classification and names are shared with global solutions such as the *Integrated Taxonomic Information System* (http://www.itis.gov/) and *Species 2000 Annual Checklist of Life* (Bisby et al. 2009; http://www.sp2000.org/). Virtually all the taxonomic information remains in the traditional print medium, but an increasing amount is now appearing online, such as in the serials *Zootaxa* (http://www.mapress.com/zootaxa/) and *ZooKeys* (http://pensoftonline.net/zookeys/index.php/journal) as well as the massive amount of legacy data being digitized through the *Biodiversity Heritage Library* (http://www.biodiversitylibrary.org/), *Google® Books* (http://books.google.com/), AnimalBase (http://www.animalbase.org), and *Gallica* at the Bibliothèque National de France (http://gallica.bnf.fr).

The basic unit of the Linnaean and modern classifications is the species. Classifications are merely hierarchical groupings of these basic units. Panzer (1792–1813), at least for entomology, was the first to recognize that if information dissemination was focused on this unit so that each unit was separate and independent from others, then new information could be easily integrated and new classifications easily generated. He did this in the format of small booklets, species-by-species, with text and image on facing pages (Figure 7.5). Today this is recognized as the concept of the online species page, some of the first examples of which were placed online at the Diptera Website (http://www.diptera.org) in April 1996 and are now incorporated into the *Encyclopedia of Life*.

FIGURE 7.5 An example of the "species pages" found in Panzer (1792–1813); here his entry for *Musca arcuata* Fabricius, 1781.

The last but most important set of components of the biodiversity infrastructure for users is identification tools. These range from the early diagnoses provided by Linnaeus and Fabricius to modern interactive, image-rich expert systems that run on hand-held iPods—but these last systems are only for the birds. For the flies, there are still only a few examples, such as the *Fruit Fly Expert Identification System* (Thompson 1999) for the identification of pestiferous fruit fly species and *On-The-Fly* (Hamilton et al. 2006) for the identification of the families of Australian flies.

BIOSYSTEMATIC DATABASE OF WORLD DIPTERA: TODAY AND TOMORROW

The BDWD is designed as a comprehensive online information source for all the critical information about scientific names of flies and the basic information about species of flies. This system grew out of a vision of a group of dipterists who wanted to capitalize on the knowledge that had been generated in preparing a series of regional catalogues of Diptera that began in the 1960s with a catalogue of Nearctic Diptera that involved Canadian and U.S. fly specialists (Stone et al. 1965). In 1984, at the International Congress of Entomology in Hamburg, dipterists gathered to celebrate the start of the effort to catalog the Palaearctic Region (Soós and Papp 1984), the largest and historically most complex region. Subsequent years focused on completing the cycle with an Australasian/Oceanian catalogue and starting a series of world catalogues. Funding was successfully obtained from the U.S. National Science Foundation to do that last regional catalog (Evenhuis 1989), and private funding contributed to the production of the world fossil fly catalog (Evenhuis 1994), but the U.S. Department of Agriculture (USDA) declined to fund further world catalogs. Instead, USDA provided pilot-test project funds to develop new technologies for an expert identification and information system for fruit flies (Thompson et al. 1993). This project provided the basis for the current BDWD. Unfortunately, no additional funds were provided for the BDWD directly, and what support was made available came from outside sources such as the Schlinger Foundation and the Global Biodiversity Information Facility. The original database software used was based on a Wang proprietary COBOL data-management system, and later migrated to FileMaker® Pro, the current software system.

The BDWD is a totally online system containing all the critical information about the system and contents. What follows is merely a snapshot of what was available online in August 2008 at the time of the *Systema Naturae* 250 symposium. Irregularly, the BDWD is archived (originally to CD-ROM and then to DVD) via the *Diptera Data Dissemination Disk* series (Thompson and Evenhuis 1999, Norrbom and Thompson 2004). As segments of information are completed and peer-reviewed, they have been published in traditional print format in the series *Myia* (see, for example, Woodley 2001). Today, the components online are the nomenclator and reference files and all the appropriate supporting documentation for the system. The only major components not yet online are the species interface and online editing facilities for specialists.

The nomenclator and reference files contain all the essential nomenclatural details as well as minimal species information. For each name, information is provided about the original source and format of the name, type information if the name is available, the nomenclatural and taxonomic status of the name, the distribution and proper spelling if the name is valid, and a link to the original reference. The BDWD was built incrementally so as to provide useful information more quickly than having to wait until it was complete at optimal standards (the sources from which the BDWD was built are documented online). Each record includes a quality assurance standard indicator (these are also documented online) telling users how complete the record is especially in respect to our ultimate status of taxonomic and nomenclatural peer review by assigned specialists. Records meeting the ultimate level are identified by the name(s) of the specialist(s) and date of review. Currently only about 6% meet this highest level, but in reality most records are as good as those already published (that is, the source from which they derived) or better (Table 7.2).

The species interface will differ from the nomenclator only in the way the user can query the information. At present, a user enters a name and the nomenclator returns nomenclatural and

---

TABLE 7.2
BDWD Statistics as of August 2008 Indicating Number of Records and the Proportion Reaching the Quality Assurance Level at which They Are Ready for Pre-Publication Peer Review

Number of Records:
193,974 species (156,668 valid)
22,879 genera (11,672 valid)
25,770 references

Records Compared with Original Literature:
29,493 species (~15%)
6,138 genera (~27%)

Records Nomenclaturally and Taxonomically Reviewed:
11,509 species (~6 %)
2,462 genera (~11 %)

---

taxonomic information about that name. The species interface will allow queries about the species and some of its other attributes, such as distribution and biology. So, one can ask, for example, for a list of all the fruit flies known from Costa Rica or for a list of all the species that are known to attack a certain fruit. The challenge of the species interface is to determine which attributes users want to query (Conservation status? Distribution? Economic importance? Hosts? Morphology?) and then encoding that information. Today, the nomenclator includes only minimal distributional data for species.

The most important aspect of the whole BDWD enterprise is the team—the people who have contributed their expertise and labor to build the BDWD. Today, some 60 people have contributed to the BDWD but the team is always growing, so check online (http://www.diptera.org) for any changes to our team.

The final aspect of the BDWD is its legal status, which is documented online under copyrights and citation. The critical fact is that BDWD is a community enterprise built by dipterists for themselves and for all people. So the information is without copyright and is freely available to all. While at various times the master database may have resided physically in some institution, that master was always a product of the BDWD team and belonged to those people. When the BDWD first went online, it was hosted by the Smithsonian Institution; later it was transferred to the USDA, and most recently it is served by the Zoological Museum, Natural History Museum of Denmark. In the future it will keep migrating to the best place that is willing to properly maintain and improve it.

ACKNOWLEDGMENTS

We express our thanks to the entire BDWD team of contributors and to all others who have supplied information over the years, and to the Schlinger Foundation, the Global Biodiversity Information Facility, the United States Department of Agriculture, and our home institutions, all of which provided financial support at various times.

REFERENCES

Ashburner, M. 2006. *Won for all: How the* Drosophila *genome was sequenced*. New York: Cold Spring Harbor Laboratory Press, 107 pp.

Bisby, F.A., Roskov, Y.R., Orrell, T.M., Nicolson, D., Paglinawan, L.E., Bailly, N., et al. (Eds.) 2009. *Species 2000 and ITIS catalogue of life: 2008 annual checklist.* [CD-ROM] Species 2000, Reading, U.K.

Evenhuis, N.L. (Ed.) 1989. *Catalog of the Diptera of the Australasian and Oceanian regions.* Honolulu: Bishop Museum Press and E. J. Brill 1,155 pp.

Evenhuis, N.L. 1994. *Catalogue of the fossil flies of the world (Insecta: Diptera).* Leiden: Backhuys Publishers, 600 pp.

Fabricius, J.C. 1775. *Systema entomologiae, sistens insectorum classes, ordines, genera, species, adiectis synonymis, locis, descriptionibus, observationibus.* Kortii, Flensbvrgi et Lipsiae [= Flensburg and Leipzig]; [32] + 832 pp.

Fabricius, J.C. 1805. *Systema Antliatorum secundum ordines, genera, species adiectis synonymis, locis, observationibus, descriptionibus.* xiv + 15–372 + [1] + 30 pp. C. Reichard, Brunsvigae [=Brunswick]

Fée, A.L.A. 1832. Vie de Linné, rédigeé sur les documens autographes laissés par ce grand homme, et suivie de l'analyse de sa correspondance avec les principaux naturalistes de son époque. *Mémoires de la Société des Sciences, de l'Agriculture et des Arts de Lille* 1832(1), xi + [1] + 379 + [1]

Gmelin, J.F. 1790. *Caroli a Linne, systema naturae per regna tria naturae, secundum classes, ordines, genera, species; cum caracteribus, differentiis, synonymis, locis. Editio decima tertia, aucta, reformata* [= Ed. 13.] Vol. 1: *Regnum Animale,* Pt 5. G.E. Beer, Lipsiae [= Leipzig]. 2225–3020.

Goff, M. L. 2001. *A fly for the prosecution: How insect evidence helps solve crimes.* Cambridge, MA: Harvard University Press. 225 pp.

Grimaldi, D. and Engel, M.S. 2005. *Evolution of the insects.* New York: Cambridge University Press. 755 pp.

Hamilton, J.R., Yeates, D.K., Hastings, A., Colless, D.H., McAlpine, D.K., Bickel, D., et al. 2006. On The Fly—The Interactive Atlas and Key to Australian Fly Families. [CD-ROM] Australian Biological Resources Study (ABRS) and Centre for Biological Information Technology (CBIT).

Kohler, R.E. 1994. *Lords of the fly.* Drosophila *genetics and experimental life.* Chicago: Chicago University Press. 344 pp.

Linnaeus, C. 1758. *Systema naturae per regna tria naturae, secundum classes, ordines, genera, species, cum characteribus, differentiis, synonymis, locis.* 10th ed., Vol. 1. Laurentii Salvii, Holmiae [=Stockholm], 824 pp.

Linnaeus, C. 1767. *Systema naturae per regna tria naturae, secundum classes, ordines, genera, species, cum characteribus differentiis, synonymis, locis.* 12th ed. (revised.). Laurentii Salvii, Holmiae [= Stockholm]; Vol. 1 (2): 533–1327 + [37] pp.

National Research Council Committee on the Status of Pollinators. 2007. *Status of pollinators in North America.* Washington, D.C.: National Academies Press. 322 pp.

Norrbom, A.L. and Thompson, F.C. (Eds.) 2004. *Diptera data dissemination disk,* Volume 2. 2 April 2004. [CD-ROM].

Panzer, G.W.F. 1792–1813. *Favnae insectorum germanicae initia oder Devtschlands Insecten.* Felseckerschen, Nürnberg.

Soós, Á. and Papp, L. (Eds.) 1984–1993. *Catalogue of Palaearctic Diptera.* Vols 1–14. Akadémiai Kiadó and Hungarian Natural history Museum, Budapest; Amsterdam: Elsevier Science.

Stone, A., Sabrosky, C.W., Wirth, W.W., Foote, R.H., Coulson, J.R. (Eds.) 1965. A catalog of the Diptera of America north of Mexico. *United States Department of Agriculture, Agricultural handbook* 276, 1969 pp.

Thompson, F.C. 1996. Names: The keys to Biodiversity. Chapter 13, pp. 199–216. In Reaka-Kudla, M.L., Wilson, D., Wilson, E.O. (Eds.), *Biodiversity II. Understanding and protecting our biological resources.* Washington, D.C.: Joseph Henry Press. v + 551 pp.

Thompson, F.C. 1999. Introduction [to Fruit Fly Expert Identification System and Systematic Information Database]. *Myia* 9: 1–6.

Thompson, F.C. and Evenhuis, N.L. (Eds.) BioSystematic Database of World Diptera. Diptera Names Working Data Files. *Diptera data dissemination disk,* Volume 1. 22 February 1999. [CD-ROM].

Thompson, F.C., Norrbom, A.L., Carroll, L.E., White, I.M. 1993. The fruit fly biosystematic information database. In Aluja, M. and Liedo, P. (Eds.), Proceedings of the International Symposium on Fruit Flies of Economic Importance 1990, Antigua, Guatemala, October 14–20, 1990. New York: Springer-Verlag. 3–7.

WHO. 2003. *The world health report: 2003 : Shaping the future.* World Heath Organization, Geneva; 204 pp.

Woodley, N.E. 2001. A world catalog of the Stratiomyidae (Insecta: Diptera). *Myia* 11, 473 pp.

Young, A.M. 1994. *The chocolate tree, A natural history of cacao.* Washington, London: Smithsonian Institution Press. 200 pp.

8 e-Publish or Perish?

Sandra Knapp and Debbie Wright

CONTENTS

INTRODUCTION

Dissemination of results is one of the critical steps in the practice of science, and today's scientists do this in many forms: through presentations at conferences, invited lectures, and most importantly through publication. Publication in its simplest form means merely the dissemination of results—communication. Scientific publishing is fundamentally different from publication in that sense in that it involves an element of review and permanency. This difference between scientific publishing and general publishing was established by Henry Oldenburg, who, as the Secretary of the Royal Society, established the oldest scientific journal in the English language, the *Philosophical Transactions of the Royal Society* (only the French *Journal des sçavans* is older in any language, first published on January 5 1665). The first issue of the *Philosophical Transactions* was edited by Oldenburg and published on March 6 1665. In it he laid out four functions for publication: (1) dissemination, (2) registration, (3) certification, and (4) archiving (what is called the "Minutes of Science"). How these intersect with the needs and desires of the authors of such publications is laid out in Table 8.1. Oldenburg sent papers submitted to the *Philosophical Transactions* to other scientists for review, the first instance of peer-review, today the cornerstone of scientific publishing. Publishing in the 17th century was set in the context of concern over what today we might call intellectual property; authors were concerned with receiving credit for work they had done (Guédon, 2001), and for its dissemination to others beyond the reach of local scientific gatherings. These issues are still relevant today, but in the present research environment authors also expect rapid publication and that necessarily means rapid peer-review.

Scientific (or scholarly) publishing has been characterized as having long periods of stasis and apparent stability punctuated by bursts of intense change and upheaval (Guédon, 2001). Today's world of increasing electronic access to scholarly work, the rise of demands for open access to

---

TABLE 8.1
The Functions of Scientific Publishing as Laid Out by
Henry Oldenburg for the Royal Society in 1665

Oldenburg's Functions	What an Author Wants
Dissemination	Recognition
Registration	Ownership
Certification	Reputation
Archive (the "minutes of science")	Renown

---

information, and increased interconnectivity is a time of massive change in all communication, not only for publishing. Drivers for change in scientific publishing include a massive increase over the decades since World War II in global research and development funding and a number of key technological advances such as the Internet. The result has been an increasing output in both volume and complexity of scientific publications, more types of output, and an increase in the numbers of outlets for scientific publications. These factors have led to a period of great change in the reality and perception of how work is published in the scientific realm, and can be considered as significant a shift as that which occurred when paper replaced parchment, or when mass printing first became available. Change like this usually results in uncertainty and a certain degree of confusion and apprehension that sometimes, but not always, delays progress.

Publishing has a special place in nomenclature (the scientific naming of organisms), and "published work" in nomenclatural terms has a slightly different definition from that for science at large. All of the Codes of nomenclature (Cultivated Plants—Brickell et al., 2006; Zoological—ICZN, 1999; Bacteriological—LaPage et al., 1992; Botanical [including fungi]—McNeill et al., 2006) have valid/effective publication as one of the central pillars of the naming of organisms (terminology in the Codes is subtly different, for translations and synonymy see draft proposals for a BioCode; Greuter and Nicholson, 1996; Greuter et al., 1996; Hawksworth and McNeill, 1998). Nomenclature as governed by the Codes is possibly the best and longest-lasting example of voluntary adherence to standards in science; the Codes of nomenclature have been in use since the 19th century and the need for them was considered by scientists as early as Lamarck (see Knapp et al., 2004 for a history of the Codes). The Codes, though legalistic (and therefore quite difficult to understand for anyone not already immersed in them), are not legally binding; there is nothing other than the disapprobation of the community to prevent taxonomists from setting off on their own and doing whatever they want.

None of the Codes currently recognize the publication of nomenclatural acts (see next section for definition) in electronic form, something that has led some in the scientific community to regard the taxonomic community as backward and Luddite in the extreme. At the International Botanical Congress in St. Louis in 1999 (Greuter et al., 2000), proposals for registration (seen as a preliminary step by some to allowing e-publication) were roundly defeated by both mail and show-of-hand votes, causing some acrimony within the community and misconception outside the community as to reasoning. At the 2005 Congress in Vienna, proposals to allow electronic publication of nomenclatural acts were again defeated, but suggestions about e-publishing were voted into the Vienna Code (McNeill et al., 2006; Knapp et al., 2006). Proposals to alter the International Code of Zoological Nomenclature (ICZN, 1999) and allow electronic publishing in zoology were presented in Paris in August 2008 and have recently been published (ICZN, 2008a, b, c) in order to allow broad consultation across the zoological community as is necessary for any change in the Code. The pace of change in the Codes of nomenclature is much slower than the change taking place in the publishing world, thus potentially creating a mismatch in objectives and delivery.

Real issues do exist with electronic publishing; concern over archiving, for example, is not confined to the taxonomic community. In this chapter we will address some of the perceived problems with electronic, as opposed to paper, publication of nomenclatural acts. They fall into four broad categories: (1) archiving and the retention of permanent records, (2) accessibility of the works, (3) dates of publication of works, and (4) the types of media used in e-publishing today. We do not intend to address issues associated with open access to scientific publication; this complex and fast-moving field is beyond the scope of this chapter, nor will we specifically address the many and varied mechanisms by which e-publishing might be introduced into the zoological or any other code of nomenclature. Here we outline some of the principal objections (impediments or perceived impediments) raised over the last few years to the publication of nomenclature by electronic means and raise some real challenges both taxonomists and publishers will need to address in order to allow the business of naming new taxa to enter the electronic age.

NOMENCLATURAL ACTS IN ZOOLOGY

Published work in zoological nomenclature is defined very specifically in the Code (ICZN, 1999). It is: (1) issued for the purpose of providing a public and permanent scientific record (Article 8.1.1), (2) obtainable, when first issued, free of charge or by purchase (Article 8.1.2), and (3) produced in an edition containing copies simultaneously obtainable by a method that assures numerous identical copies (Article 8.1.3). In addition, works published on CD-ROM (introduced into the 4th edition, ICZN, 1999) must state the archives in which the work will be housed (Article 8.6). Publishing in the sense of the Zoological Code then is confined to the publishing of *nomenclatural acts*: defined as "a published act which affects the nomenclatural status (*q.v.*) of a scientific name or the typification of a nominal taxon" (ICZN, 1999: 99). The definition of nomenclatural status is found in the glossary and is as follows:

> Of a name, nomenclatural act or work; its standing in nomenclature (i.e., its availability or otherwise, and in the case of a name its spelling, the typification of the nominal taxon it denotes, and its precedence relative to other names (ICZN, 1999: 111).

Thus, our discussion here of publication for the purposes of zoological nomenclature (and more or less in the other Codes as well) is confined to the coining of new names and the designation of types.

Type specimens are not typical, but serve as pointers to fix the application of names. A type specimen is an object to which a name is tied, and decisions of synonymy or identity hinge on comparison of type specimens in relation to overall variation in a putative taxon (see any modern monograph of any taxonomic group, e.g., Peralta et al., 2008). The designation of types is today an integral part of describing a new species in all the organism groups (Brickell et al., 2007; ICZN, 1999; LaPage et al., 1992; McNeill et al., 2006). This, however, was not always the case, and many names coined by taxonomists in the last centuries need to have type specimens designated so to fix application of the names. The type method came into common use only in the mid-20th century (see Knapp et al., 2004), so the many designations of type specimens for names coined without specification of a type are treated as nomenclatural acts under the Code (see Jarvis, 2007 for Linnaean examples in botany).

That the Codes of nomenclature govern the publication of new names is linked to the Principle of Priority, where the first published name included in the circumscription of a taxon has priority over later published names and thus becomes the accepted name of that taxon. Publication of new names underpins the Principle of Priority that allows decisions on what name to use (but not on what constitutes a "true" species circumscription, see below). Date of publication thus is critical for the operation of the Principle of Priority, and keeping track of dates can be challenging (see next section).

The ICZN does *not* specify publication methods for taxonomic decisions such as synonymy, contrary to some assumptions (e.g., implicit in Godfray, 2002). In fact, all of the codes of nomenclature specifically exclude scientific decisions (except the PhyloCode, see Knapp et al., 2004). The Zoological Code (1999: xix) states:

The Code refrains from infringing on taxonomic judgment, which must not be made subject to regulation or restraint"…"Nomenclature does not determine the inclusiveness or exclusiveness of any taxon, nor the rank…, but, rather, provides the name that is to be used for a taxon whatever taxonomic limits and rank are given to it (ICZN, 1999: xix).

A hypothesis (see Knapp, 2008b) as to species identity or inclusiveness is not governed by the Codes, thus the publication of taxonomy (as opposed to nomenclature) can take place in any way a scientist wishes—on paper or electronically. There are good examples of taxonomy on the Internet, where new synonymies are presented as testable hypotheses on project Websites (e.g., Solanaceae Source—http://www.solanaceaesource.org, Plant Bug PBI—http://research.amnh.org/pbi/, the All Catfish Inventory PBI—http://silurus.acnatsci.org/, CATE, http://www.cate-project.org/; see also Mayo et al., 2008); the future of these experimental efforts to broaden the base of taxonomy to increase accessibility and potentially speed up taxonomic documentation of life on earth (see Godfray, 2002; Knapp, 2008a) is another issue and warrants serious and careful consideration by the community.

IMPEDIMENTS TO ELECTRONIC PUBLISHING OF NOMENCLATURAL ACTS

ARCHIVING

There is an abiding assumption that print archives are safe and with the stewardship of copyright libraries, such as the British Library, this is to a large extent true. However, deposit of print publications into curated archives, whether by law or custom, is generally nationalistic or regional at best. There is no one library or print archive that collects and curates all print publications on a global scale. Nor can we say that there is an identifiable group or federation of libraries that has achieved this. We do not have an international system in place. The world lost many unique works with the burning (perhaps as many as three times) of the Alexandria library and still, in recent years, libraries and their holdings have been destroyed as a result of wars and rebellions. The sheer volume of the printed literature of science is such that its physical storage and archiving in the future is increasingly seen as problematic (Müller, 2008, pers. comm. reported to SK by M. Scoble).

The building of institutional and subject repositories for electronically published information is on the increase; at the time of writing it is estimated that about 1,300 repositories now exist. Some subject repositories, such as PubMed Central (see http://www.pubmedcentral.nih.gov/), have mirror sites. Although institutions and funding bodies are generally shy of publishing their compliance figures, it has been widely reported that repositories that do not mandate deposit have very low compliance. Compliance for those that do mandate deposit is better but still low. It is not clear how many of these repositories have commitments and strategies in place for long-term funding and long-term preservation of their content.

With the advent of online publication, publishers and libraries have recognized the importance of archiving electronic publications. Both groups have been working together to build safe long-term archives designed not just to store e-publications but also to ensure that documents can be retrieved and read. Some publishers have e-archiving agreements with the Royal Dutch Library to "ensure permanent access to digital objects" (see http://www.kb.nl/dnp/e-depot/e-depot-en.html). In the United Kingdom and the Republic of Ireland, legal deposit of published material into six national archives has been law for 400 years; the British Library is one of these. Content is available either through Web access, inter-library loan, or through contacting the library directly. Legal deposit is being introduced for electronic publications. Some publishers have been working in a pilot scheme with the British Library to ensure a robust system of deposit, archiving, and retrieval, necessary for long-term preservation (see http://www.bl.uk/aboutus/stratpolprog/legaldep/index.html).

The best known and most overarching archiving initiative is CLOCKSS (Closed Lots of Copies Keep Stuff Safe), whose philosophy is summed up in the quotation used on its Website:

… let us save what remains; not by vaults and locks which fence them from the public eye and use consigning them to the waste of time, but by such multiplication of copies, as shall place them beyond the reach of accident." (Thomas Jefferson, 18 February 1781).

This collaboration of research libraries and scholarly publishers is focused on building and maintaining a sustainable and distributed archive of digital scholarly publications (see http://www.clockss.org/clockss/Home). Publishers deposit their electronic content into CLOCKSS; this is then held in dark archives using LOCKSS technology at ten libraries around the world. Should a "trigger event" occur, content is released on two host sites and is available worldwide at no cost. This has already been shown to be successful; when online access to a journal published by SAGE Publications was discontinued, this acted as a "trigger event" and the content was released through CLOCKSS (for a history of the event see the CLOCKSS Website, http://www.clockss.org/clockss/Auto/Biography#CLOCKSS.27_Latest_Trigger_Event.2C_Auto.2FBiography).

Archiving and curation of electronic publication has the potential to be, and could be argued is already, more secure than for print publications. Concerns that the ephemeral nature of the electronic environment gives rise to a lack of permanency are addressed by a plethora of initiatives involving both libraries (public sector) and publishers (mostly the private sector). Archiving in reality no longer represents a major impediment to electronic publishing of nomenclatural acts, but quality control and compliance must be monitored carefully (see below).

ACCESSIBILITY

E-publication is now more accessible than print. As library budgets are squeezed more, more libraries are replacing their journals print subscriptions with online-only subscriptions. These can be as traditional institutional subscriptions or through licenses in the so called "Big Deals." This is not just to limit expenditure on holdings but also on library overheads such as staffing and space; an online subscription does not need physical shelf space. This is a continuing trend, and with the recent financial crises this may well escalate. At some point in the future some journals will meet a tipping point where print publication is no longer viable and they may then move to online-only publication.

Accessibility goes hand in hand with find-ability. Behavior in finding literature and information has changed dramatically over the last few decades. In a report published by the Research Information Network (RIN, 2006), 36% of respondents from life sciences disciplines claimed to never browse library shelves or use a librarian to find information. The increase in use of online search engines, databases, or portals was quite dramatic. Accessing publications by individuals through interlibrary loans or through direct contact with the author has been made even easier with e-publications; anyone can write or e-mail a request and receive either a print copy or an electronic PDF in return.

What about institutions in the developing world? The access to e-publication for the scientists working in developing world institutions with limited infrastructure and resources is often cited as an impediment to the e-publication of nomenclatural acts. Most publishers deliver free or low-cost access to qualifying countries through philanthropic programs run as collaborations between major publishers and international bodies such as the United Nations and the World Health Organization (WHO). These bodies determine the level of support and whether a country qualifies as a "developing world country." A few examples of these programmes are: (1) HINARI, set up by the WHO together with major publishers, enables qualifying countries access to over 6,200 journals covering biomedical and health literature (http://www.who.int/hinari), (2) AGORA, set up by the Food and Agriculture Organization of the UN (FAO) together with major publishers, enables qualifying countries access to over 1,278 journals covering agriculture and life sciences (http://www.aginternetwork.org), (3) Online Access to Research in the Environment (OARE), set up by the United Nations Environment Program (UNEP) together with major publishers, enables qualifying countries access to over 1,300 journals covering environmental sciences (http://www.oaresciences.org/

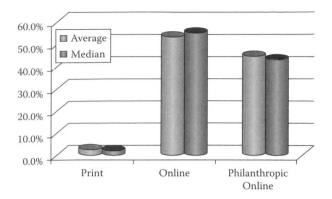

FIGURE 8.1   How institutions access scientific journals, from a sample of over 30 life science journals (range for total 6,000 to 9,000). Shown are percentages of institutional access for print-only, online and philanthropic online from the total data set. These data show that even in the developing world online access far outstrips access for print versions of scientific journals where both types of access are available. Data from Wiley-Blackwell.

en), and (4) The International Network for the Availability of Scientific Publications (INASP), set up by the International Council for Science (ICSU) together with major publishers with the mission to "enable a sustainable network of stakeholders that owns and drives access, use, dissemination and communication of research information." Part of the work of INASP is on the infrastructure and training to support dissemination and accessibility (http://www.inasp.info). Other programs for provision of access exist via individual libraries, institutions, and publishers. In general (see Figure 8.1) institutional access to scientific journals is skewed toward the electronic versions, showing the importance of e-publication as the access route to literature in both the developed and developing worlds.

Scientists in the developing world are increasingly dependent on e-publications for access to literature. The various philanthropic programs providing access, the work of INASP in the infrastructure enabling access and communication, the ability to easily access literature through interlibrary loan or direct request to the author, all these contribute to how literature can be accessed and are highly dependent on an e-publishing model. There is a real risk that reliance on print-only publication, rather than empowering, will disenfranchise scientists working in the developing world.

DATE OF PUBLICATION

The date of publication is a critical factor in recognition of nomenclatural acts due to the Principle of Priority, a pillar of all the Codes (see Knapp et al., 2004; ICZN, 1999, Chapter 5: 22–24). Decisions as to the name with priority can sometimes come down to days or months rather than years (for a botanical example involving description of the collections of the great naturalist Alexander von Humboldt, see Hiepko, 2006).

A potential problem with the current online early model for publication of e-journals lies in the stipulation of the Code that "multiple, identical" copies of publications containing nomenclatural acts be available (ICZN, 1999, Article 8.1.3, see Table 3). Once an article is paginated, it is no longer identical to the online early version, usually released with only within article (i.e., each article begins on page 1) pagination. That pagination of online versions is not identical makes it difficult to publish nomenclatural acts in a wide range of journals, in essence limiting taxonomists to a few journals. Nomenclatural acts are often part of wider taxonomic papers (especially typification decisions) and the requirement for print publication thus restricts taxonomic work, or forces taxonomists to separate their results in an unnatural manner in order to both gain broad dissemination and high impact and comply with the Codes.

E-publications can be easily date stamped. Some journals, although publishing in both the print and online, do this within the paper, as can be seen in a number of journals (e.g., *Evolutionary Applications, Naturwissenschaften, New Phytologist, Zootaxa*). In general, publications that appear online before print do not bear identical pagination (although there are exceptions). If a registration system were to be established (see ICZN, 2008a, b, c), whether voluntary or mandatory, the thorny issues of date of publication would need to be established relative to such a system. Many papers that are published online have a Digital Object Identifier (DOI). The DOI does not of itself provide a date of publication but, as shown in Harris 2004, will provide a link to the content. In Harris, the link, http://dx.doi.org/10.1007/s00114-002-0353-8, takes the reader to the published paper, and the date of online publication can be clearly seen within the paper itself. Harris (2004) proposes to the Commission that the Code be modified and that

> … documents published electronically with DOI numbers and that are followed by hard-copy printing and distribution be exempt from Article 9.8 and be recognized as valid, citable sources of zoological taxonomic information and that their electronic publication dates be considered definitive.

Not all scholarly publications are published online using the DOI system. Will these be around in the next century? We argue that, provided date of online publication for an article can be clearly shown within the paper, then this can be considered the definitive date of publication, irrespective of whether the article is published in print at some later date or whether it has a DOI or other similar registration number.

Medium Type

The medium on which scientific (and other) work is published has changed radically since publication began. Early works were handwritten on vellum; paper itself was considered ephemeral when first introduced. Paper quality has varied considerably over the centuries; look at many scientific journals published in Germany in the period of financial instability and hardship between the two world wars to see the effects such constraints had on paper quality and thus "permanency" of printed works. Here we have selected four main media for scientific publication: CD-ROM, Websites, print peer-reviewed editions, and electronic or online peer-reviewed editions. In turn, we examine how these current publishing media satisfy the criteria of Oldenburg's original four functions (see Table 8.2) and the requirements of the Zoological Code (see Table 8.3).

TABLE 8.2
How Different Media of Scientific Publication Satisfy Henry Oldenburg's 1665 Criteria

Medium of Publication	Dissemination	Registration	Certification	Archive
CD-ROM	√	√	√	X
Website	√ [but can be hard to find]	√	√ [but only if peer-reviewed]	X [no commitment]
Print peer-reviewed edition	√ [less so now]	√	√	√ [but not long-term]
Electronic/online peer-reviewed edition	√	√	√	√

TABLE 8.3
How Different Media of Publication Satisfy the Requirements of the Fourth Edition of the International Code of Zoological Nomenclature

Medium of Publication	Public (8.1.1)	Permanent (8.1.1)	Obtainable (8.1.2)	Multiple Copies (8.1.3)	Statement (8.6)
CD-ROM	√	X	√	√	√
Website	√ [but can be hard to find]	X [no commitment]	√	X	√
Print peer-reviewed edition	X [less so now]	√ [but not long-term]	√	√	√
Electronic peer-reviewed edition	√	√	√	√	√

CD-ROM

These fall down on permanence and archiving: they have a limited life span and are dependent on hardware (which rapidly goes out of date) to be read. Images of the historic Sloane herbarium archived as high-resolution (low-resolution images still available on the Internet, see http://www.nhm.ac.uk/research-curation/research/projects/sloane-herbarium/) onto CD-ROM at the Natural History Museum in the mid-1990s are today largely corrupted and thus unrecoverable. A classic case often cited is that of the *Domesday Book* edition for the 20th century. Originally written by monks on vellum in the 11th century, the 1086 *Domesday Book* is held in the United Kingdom's Public Record Office at Kew and can still be consulted, but the 20th-century edition compiled in 1986 to mark the 900th anniversary of the original was stored on video discs for specially designed computers that 15 years later were obsolete; the information thus was not accessible (see http://www.guardian.co.uk/uk/2002/mar/03/research.elearning, although the information has now been retrieved at considerable cost, see http://www.nationalarchives.gov.uk/domesday/). In 1999 the Zoological Code allowed publication on CD-ROM (ICZN, 1999, Article 8.6) with some provisos, but the botanical community has steadfastly avoided the mention of particular media (or even reproduction methods such as PDF) in the International Code for Botanical Nomenclature (McNeill et al., 2006).

Websites

Internet Websites are instantly and widely available, but can be hard to find even using sophisticated search engines. Anyone can set up a Website, and there is no permanent archive of all Websites (although the Web itself is indexed regularly, at least for now, by most search engines). Literature published only on Websites can be lost as people move between institutions, institutions delete inactive sites or sites that no longer have grant-funding, and URLs change or no longer resolve. An additional potential problem with using the Web as the primary method of e-publishing nomenclatural taxonomic work is that the open Web does not have a stamp of validation or quality control. To be part of the scholarly, scientific literature, content ultimately needs to be fully peer reviewed and validated; to be citeable, any given piece of information must be retrievable over the long term (unchangeable information is by implication able to be trusted). Models such as that being developed by the CATE initiative (see Godfray et al., 2007; http://www.cate-project.org/) may go some way to solving this peer review and validation problem. The value of peer review is questioned by some as not relevant to nomenclature; this we feel is short-sighted in the extreme, as naming organisms should be as rigorously and carefully done as any other science.

Institutional and subject repositories as Websites are not necessarily safe for any period of time; a funding body has as much right to take away as to give resourcing. A very real and recent example of this is the Arts and Humanities Data Service funded by the Arts and Humanities Research Council. This service closed in 2008, although some resources are still available (see http://en.wikipedia.org/wiki/Arts_and_Humanities_Data_Service). Reorganization of funding agencies can often impact on the sustainability of Web-only available information (e.g., the potential loss of the long-term datasets held at the Biological Records Centre in the now closed Monks Wood NERC Centre in the United Kingdom), although plans are usually put in place for the preservation of such data (see http://www.nerc.ac.uk/press/releases/2006/cehdecision.asp); finding such information later though can be a problem.

Print Peer-Reviewed Editions

In today's fund-restricted world, especially in public institutions, fewer libraries take print publications, and therefore dissemination of work published only in print on paper is becoming increasingly problematic. It is not easily accessible now except in the better-funded institutions.

Current efforts to digitize the literature of biodiversity (i.e., taxonomy) have thrown into relief the usefulness and reliability of digital versions of publications. Efforts such as the Missouri Botanical Garden's Botanicus (http://www.botanicus.org), the University of Göttingen's AnimalBase (Welter-Schultes, 2008, http://animalbase.uni-goettingen.de) and the multi-institutional Biodiversity Heritage Library (http://www.biodiversitylibrary.org/) mean that much of the heritage literature of taxonomy is now available in electronic form. Copyright restrictions, however, mean that much recent (and relatively recent, from 1922 to present) literature is not included (see http://plazi.org for another model).

There exists an assumed permanency of paper as a medium. Given that we are concerned with the preservation of the "minutes of science," we necessarily need to take a very long-term view. Paper does not last forever and if not kept in a controlled environment, can be completely destroyed within a short time frame. Complete holdings have been destroyed through accident or war. Digitization of heritage literature is often suggested as a solution to the vulnerability of print publications, but the apparent contradiction in reliance on digitized copies and not allowing e-publication is usually not addressed. In addition, printing and digitization both incur substantial costs—why pay twice, when e-publication could offer potentially the same level of security?

Electronic Peer-Reviewed Editions

This medium of publication complies with all four of Oldenburg's functions, but not yet with the strictures of the Codes of nomenclature, although the proposals published for revision of the Zoological Code will potentially change this (ICZN, 2008a, b, c) and the botanical community is recommending such a peer-reviewed path for the future (McNeill et al., 2006). Librarians and publishers have recognized the importance of online publication for many years and have been working closely together on many fronts to ensure that this meets the requirements of scholarly publication.

SOME CHALLENGES FOR THE FUTURE

As the taxonomic community moves into the 21st-century realm of access and publication, focus needs to be concentrated on solving some real problems associated with moving the nomenclatural publication model to keep pace with developments elsewhere. These can be divided, we feel, into those challenges to the taxonomic community, and challenges the publishing and taxonomic community should work together to solve. These are by no means the only challenges that will face taxonomy as we move into new models for both doing and presenting taxonomic works, but we believe that if communities focus on *real* challenges rather than rehashing old problems that are being addressed elsewhere there is a better chance for the vigorous rebirth of taxonomy as a 21st-century discipline. Rather than ignoring e-publication, we must set realistic requirements for nomenclatural e-publication so chaos does not ensue.

Challenges for the taxonomic community:

- Accept and value peer review. A recent study (Ware, 2008) of over 3,000 academics found that 90% of respondents thought that peer review improved the quality of their output, 60% that peer review determines the importance of the findings and the originality of the paper.
- Enable and encourage rapid publication; rapid review and recognition of online publication as valid publication, to encompass electronic publication ahead of issue or print publication.
- Don't confuse nomenclatural e-publishing with "everyday" scientific e-publishing, or scientific publishing with the generic activity of making information available.

Challenges for publishers and taxonomists to approach together:

- Establish a method of reliably and consistently date-stamping articles.
- Develop methods for pagination at every stage of online publications so that confusion in the future over versions does not ensue.
- Statements of deposition of hard copies for additional archiving need standard format and position in publications—for posterity, these need to be easy to locate.
- Archive and curate source data—taxonomic toolkits work at the molecular and gene level, why not for the taxonomic community? Too often the Online Supplementary Material is the first thing lost or inaccessible; these raw data are the stuff of hypothesis testing and need proper archiving and curation as well as do synthetic publications.

Publishers and librarians take the obligation to archive and curate the online "minutes of science" very seriously; it is critical that taxonomists require this careful curation and archiving as part of the nomenclatural publication process. We are sure there are additional issues that will surface or come to the fore in the coming years of rapid change—we only hope that, by working together, the taxonomic communities and their publishers can solve these issues as they arise, rather than pushing them to one side and ignoring them. Focusing on solutions rather than problems will ensure that taxonomic and nomenclatural publishing remains vibrant and useful "forever," however long that may be.

ACKNOWLEDGMENTS

We thank Andrew Polaszek for inviting us (a botanist and a publisher) to present our (still-developing) ideas at the "*Systema Naturae* 250—The Linnaean Ark" symposium at International Congress of Zoology in Paris in August of 2008; Bob Campbell, senior publisher with Wiley-Blackwell for sharing his extensive knowledge of scholarly publication through various seminars; the various journal editors DW has had the pleasure of working with; colleagues, especially Malcolm Scoble, Charles Godfray, and, not least, the officers and fellows of the Linnean Society of London for vibrant and challenging discussions on the regimes of taxonomic publication. SK is supported by the National Science Foundation's Planetary Biodiversity programme (DEB-0316614).

REFERENCES

Brickell, C.D., B.R. Baum, W.L.A. Hetterscheid, A.C. Leslie, J. McNeill, P. Trehane, F. Vrugtman, and J.H. Wiersema (Eds.) 2005. *International Code of Nomenclature for Cultivated Plants. 7th edition*. Regnum Vegetabile 144.
Godfray, H.C.J. 2002. Challenges for taxonomy. *Nature* 417: 17–19.
Godfray, H.C.J, B.R. Clark, I.J. Kitching, S.J. Mayo, and M.J. Scoble. 2007. The Web and the structure of taxonomy. *Systematic Biology* 56: 943–955.
Greuter, W. and D.H. Nicholson. 1996. Introductory comments on the Draft BioCode, from a botanical point of view. *Taxon* 45: 343–372.

Greuter, W., J. McNeill, D.L. Hawksworth, and F.R. Barrie. 2000. Report on botanical nomenclature, St. Louis 1999. *Englera* 20: 5–253.

Guédon, J.-C. 2001. *In Oldenburg's Long Shadow: Librarians, Research Scientists, Publishers, and the Control of Scientific Publishing*. Annapolis Junction, MD: Association of Research Libraries.

Harris, J. 2004. "Published works" in an electronic age: Recommended amendments to Articles 8 and 9 of the Code. *Bulletin of Zoological Nomenclature* 61(3): 138–148.

Hawksworth, D.L. and J. McNeill. 1998. The International Committee on Bionomenclature (ICB), the Draft BioCode (1997), and the IUBS resolution on bionomenclature. *Taxon* 47: 123–149.

Hiepko, P. 2006. Humboldt, his botanical mentor Willdenow, and the fate of the collections of Humboldt and Bonpland. *Botanische Jahrbücher für Systematik, Pflanzengeschichte und Pflanzengeographie* 126: 509–516.

ICZN. 1999. *International Code of Zoological Nomenclature*. 4th edition. International Trust for Zoological Nomenclature, London.

ICZN. 2008a. Proposed amendment of Articles 8, 9, 10, 21, and 78 of the International Code of Zoological Nomenclature to expand and refine methods of publication. *Zoological Journal of the Linnean Society* 154: 848–855.

ICZN. 2008b. Proposed amendment of Articles 8, 9, 10, 21, and 78 of the International Code of Zoological Nomenclature to expand and refine methods of publication. *Biological Journal of the Linnean Society* 95: 874–881.

ICZN. 2008c. Proposed amendment of the International Code of Zoological Nomenclature to expand and refine methods of publication. *Zootaxa* 1908:57–67.

Jarvis, C. 2007. *Order out of Chaos: Linnaean Plant Names and Their Types*. London: Linnean Society of London and the Natural History Museum.

Knapp, S. 2008a. Taxonomy as a team sport. In Wheeler, Q. (Ed.) *The New Taxonomy*. Systematics Association Special Volume 76. London: CRC Press. 33–53

Knapp, S. 2008b. Species concepts and floras—what are species for? *Biological Journal of the Linnean Society* 95: 17–25.

Knapp, S., G. Lamas, E. Nic Lughadha, and G. Novarino. 2004. Stability or stasis in the names of organisms: The evolving Codes of nomenclature. *Philosophical Transactions of the Royal Society, series B, Biological Sciences* 359: 611–622.

Knapp, S., A. Polaszek, and M. Watson. 2007. Spreading the word. *Nature* 446: 261–262.

Knapp, S., K. Wilson, and M. Watson. 2006. Electronic publication. *Taxon* 55: 2–3.

LaPage, S.P., P.H. Sneath, E.F. Lessel, V.B.D. Skerman, H.P.R. Seelinger, and W.A. Clark (Eds.) 1992. *International Code of Nomenclature of Bacteria (1990 revision)*. Washington DC: American Society for Microbiology.

Mayo, S., R. Allkin, W. Baker, V. Blagoderov, I. Brake, B. Clark, R. Govaerts, C.H.J. Godfray, et al. 2008. Alpha e–taxonomy: Responses from the systematics community to the biodiversity crisis. *Kew Bulletin* 63: 1–16.

McNeill, J., F.R. Barrie, H.M. Burdet, V. Demoulin, D.L. Hawksworth, K. Marhold, et al. 2006. *International Code of Botanical Nomenclature (Vienna Code)*. Regnum Vegetabile 146. Liechtenstein: A.R.G. Gantner Verlang KG.

Müller, E. 2008. Long term preservation of digital information, the challenges and solutions. Abstract from Stockholm Biodiversity Informatics Symposium 2008 at http://artedi.nrm.se/sbix2008/TheBookofAbstracts2.pdf.

Peralta, I.E., D.M. Spooner, and S. Knapp. 2008. Taxonomy of wild tomatoes and their relatives (*Solanum* sections *Lycopersicoides, Juglandifolia, Lycopersicon*; Solanaceae). *Systematic Botany Monographs* 84: 1–186.

RIN. 2006. Researchers and discovery services: Behaviour, perceptions and needs. Research Information Network, London. Available online at http://www.rin.ac.uk/researchers–discovery–services.

Velterop, J. 1995. Keeping the Minutes of Science. *Proceedings of the Electronic Libraries and Visual Informatics (ELVIRA) Conference* (2–14 May 1995). ASLIB, London.

Ware, M. 2008. Peer review: Benefits, perceptions and alternatives. Publishing Research Consortium, London.

Welter-Schultes, F. 2008. AnimalBase: Digitizing early zoological literature for before 1830. Abstract from Stockholm Biodiversity Informatics Symposium 2008 at http://artedi.nrm.se/sbix2008/TheBookofAbstracts2.pdf.

9 Reviving Descriptive Taxonomy after 250 Years

Promising Signs from a Mega-Journal in Taxonomy

Zhi-Qiang Zhang

CONTENTS

INTRODUCTION

The tenth edition of Linnaeus's (1758) monumental *Systema Naturae* marked the beginning of zoological nomenclature and taxonomy as a modern branch of the natural sciences. In the last 250 years, the binominal nomenclatural system that Linnaeus developed has remained a fundamental principle for biological taxonomy and has guided generation after generation of taxonomists in the description of about 1.5–2.0 million species on earth (Cracraft 2002, May 2002, Mace et al. 2005). As we celebrate 250 years of taxonomy and its achievement, the need for descriptive taxonomy, and its service to the rest of biological science, is greater than ever in this age of elevated rates of extinction and biodiversity crisis. Herein I discuss the revival of descriptive taxonomy with some encouraging data from the phenomenal growth of a mega-journal in taxonomy—*Zootaxa*—and its impact. In particular, I will discuss factors that have contributed to the success of *Zootaxa* and its implications for the revival of descriptive taxonomy.

REVIVAL OF A FUNDAMENTAL SCIENCE IN CRISIS

Taxonomy is a fundamental science (Wilson 2004). Discovering and describing how many species inhabit the Earth remains a fundamental quest of biology, especially as more than 90% of the world's species are still waiting to be discovered, named, and described. Most biologists would agree that taxonomy is important and fundamental to credible biology and documentation of biodiersity, and descriptive taxonomy is the most important task of taxonomy (Wheeler 2007).

Descriptive taxonomy had its great advancement and achievements in the first 200 years or so after Linnaeus, but experienced an unexpected turn of fortune after the advent of the "New Systematics," which emphasized experimental studies at the species level and below on the one hand, and phylogenetic studies on the other (Wheeler 2008). Descriptive taxonomy has been marginalized since the mid-1950s and has sustained serious losses in funding and academic positions in universities and museums around the world, especially since phylogenetic and molecular studies became popular in the last twenty years. The Conference of the Parties (COP) to the Convention on Biological Diversity (CBD) has recognized the so-called "taxonomic impediment," which refers to the lack of taxonomists to handle the enormous task of identifying and naming the majority of the biodiversity unknown to science (Hoagland 1996).

THE OTHER "TAXONOMIC IMPEDIMENT"

Overcoming the "taxonomic impediment" is more than addressing the need for more taxonomists. The PEET program (Partnerships for Enhancing Expertise in Taxonomy) of the National Science Foundation (NSF) of the United States has helped to train a new generation of taxonomists, leading to the declaration "Taxonomic Impediment Overcome" (Rodman and Cody, 2003). However, a recent survey of PEET graduates showed that 47% surveyed were no longer working on taxonomy one to six years after graduation and an additional 9% found employment where taxonomy plays a minor role (Agnarsson and Kuntner 2007). Evenhuis (2007) also found that some of the taxonomists on his Fiji Terrestrial Arthropod Survey project were far less active in doing taxonomy. Thus, there is the other "taxonomic impediment" to overcome—to enable some of the existing taxonomists to do and do best what they are trained to do (Evenhuis 2007). This impediment involves all steps of taxonomic research from collecting specimens and literature at the beginning to final publication of papers and monographs. The final step—the publication of the results of taxonomic research—is extremely important. "A discovery is not a discovery until it is made public" (Evenhuis 2007); likewise, a new species is not known to science until its name and description are published.

THE RISE OF A MEGA-JOURNAL IN TAXONOMY

Zootaxa was established as a rapid journal at the start of this century to overcome this last aspect of the taxonomic impediment. Although thousands of journals may publish taxonomic papers, it is increasingly difficult to publish papers on descriptive taxonomy in a timely and cost-effective manner. It is common for a taxonomist to wait for eight to ten months and sometimes years to get a paper published. Unless there is access to an institutional monograph series, it is even more difficult to publish a large taxonomic revision or monograph, not only because of costs, but because most journals are of a fixed size and have limits on the length of papers. Also during the last two decades, there has been an important historical trend in taxonomic publishing—many journal publishers and editors have been making increasing demands on authors to provide phylogenetic analysis, molecular systematics, and other modern types of information in taxonomic papers.

This impediment in publishing has a huge negative impact on taxonomy—the delay and difficulty in getting taxonomic works published can discourage taxonomists who have worked for years, and unpublished works (often publicly funded) are a huge waste of talent and resources. *Zootaxa* was founded to provide a much-needed outlet for descriptive taxonomic papers and monographs that are difficult to publish elsewhere, and as a result has received tremendous support from taxonomists worldwide, despite the fact that it is a grass-roots project without support from governments, professional societies, or institutions (Zhang 2006a). *Zootaxa* satisfied the publishing need of many zoological taxonomists, and sustained a period of rapid growth during 2001 to 2006 (Zhang 2006b). During the last two years, it has continued to grow in size, and especially in its impact, and has become a significant force in reviving descriptive taxonomy on a global scale. The following are some encouraging data on the growth of *Zootaxa* and its impact.

ENCOURAGING SIGNS FROM *ZOOTAXA*

GROWTH IN SIZE AND FREQUENCY

The growth of *Zootaxa* was extremely rapid during the first six years: from 302 pages of 20 papers in 2001 to 22,052 pages of 1,020 papers in 2006 (Figure 9.1). Thereafter, the number of papers and pages continued to increase, but at a much slower rate: e.g., the rate of increase in papers was on average 126% per year during 2002 to 2006 but only 6.5% during 2007 and 2008.

It should be noted that during 2002–2006, the greatest rate of increase in papers occurred during the second and third years (435% and 150%, respectively), due to the low base numbers of papers in the first and second years when the journal was just launched. The rate of increase in number of papers dropped to 49% and 44%, respectively, in the fourth and fifth year. The inclusion of *Zootaxa* in ISI's database *Science Citation Index Expanded* in 2004 seemed to have attracted a sudden increase in the submissions of papers in late 2004 and early 2005, which resulted in the rate of increase of papers in 2006 bouncing back to 78% (although still much lower than those of the second and third year).

The frequency of publication increased from 15 per year in 2001 to 103 per year in 2004 (Figure 9.2). It remained at about two per week (based on 50 weeks per year) during 2004 to 2006, and then increased gradually to about three per week in 2008. It should be noted that when *Zootaxa* was started, each paper was separately issued and printed until the end of the fifth year. In May 2005, short papers of fewer than 60 pages were no longer published separately; instead, they were grouped to form issues of 60, 64, or 68 pages to save paper, printing, and mailing costs. Since then, *Zootaxa* has been publishing 5–6 issues of a total of 467 pages at the frequency of two or three times each week.

INCREASING CONTRIBUTION TO DESCRIPTIONS OF NEW TAXA

The foremost task of descriptive taxonomy is the discovery and description of the other 90% of taxa not yet known to science (Wilson 2003). The number of new taxa published in *Zootaxa* per year increased from 51 in 2001 (i.e., about 1 per week) to 2,547 in 2007 (i.e., about 51 per week) (see Figure 9.3A). The total number of new animal taxa per year from all journals and books indexed in *Zoological Record* fluctuated between 17,548 and 19,495, with an average of 18,643 per year or about 373 per week* (Figure 9.3a); overall, there was no obvious increase or decrease in the total number of new taxa per year, although the variation was greater during 2004–2007 than during 2001–2003. Because the number of new taxa in *Zootaxa* was increasing each year, while the world total was more or less the same each year, *Zootaxa*'s contribution to the total new taxa described increased each year and by 2007 *Zootaxa* alone contributed 14% of all new taxa indexed in *Zoological Record* (Figure 9.3b).

---

* It should be noted that a relatively small number of journals/books in 2007 might have been delayed in reaching Zoological Record's office and new taxa therein have yet to be indexed; also it should be noted that Zoological Record aims to be exhaustive but does not have a complete coverage of all publications.

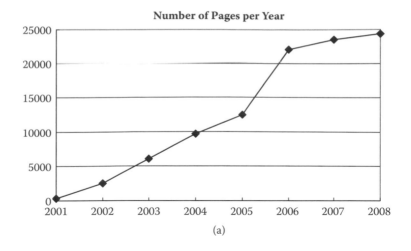

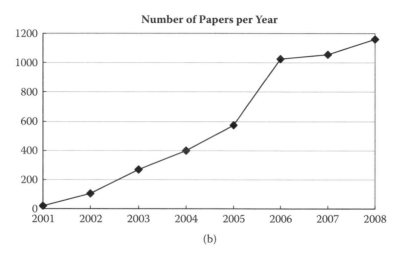

FIGURE 9.1   Growth of *Zootaxa* in size from 2001 to 2008: (a) number of pages; (b) number of papers (short notes fewer than four pages are not counted as papers).

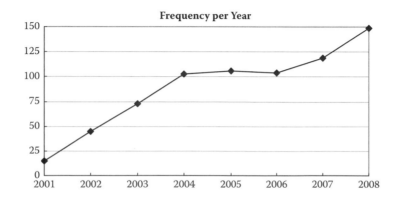

FIGURE 9.2   Frequency of publication (per year) of *Zootaxa* from 2001 to 2008.

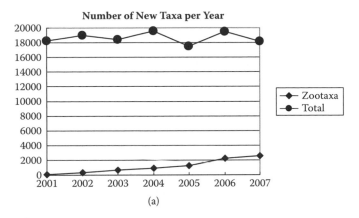

(a)

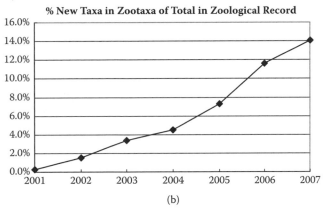

(b)

FIGURE 9.3 Growth of *Zootaxa* in the number of new taxa published from 2001 to 2007: (a) total number of new taxa published per year in *Zootaxa* versus total of new taxa of animals per year in all publications indexed in *Zoological Record*. (b) *Zootaxa*'s new taxa per year as a proportion of the total new taxa per year in *Zoological Record*.

It seems obvious that *Zootaxa* has filled an important gap in the decline of descriptive taxonomic papers published in many small journals during the last few years. One author and editor—Neal Evenhuis of the Bishop Museum in Hawaii—wrote: "As for the usefulness of *Zootaxa*—your journal is filling in one of the major gaps in the "taxonomic impediment": publishing the new taxa!"

DEVELOPING IMPACT AS A MEGA-JOURNAL IN TAXONOMY

Zootaxa became a mega-journal in 2006 when the number of papers it published per year reached 1,000 and the number of pages per year exceeded 20,000 (Zhang 2006b). In addition to its size, Zhang (2006b) emphasized two points: (1) it represented and involved the majority of scientists working in the discipline (over 2,700 in 2006), and (2) it published a significant number of the most important works in the subject area (58 monographs in 2006—more than one monograph each week).

At the end of 2001, *Zootaxa* had only 40 authors. The number of authors reached 500 by the end of 2003 and 1,000 on January 27, 2005. It exceeded 2,000 on May 15, 2006 and 3,000 on April 9, 2007. The number of *Zootaxa* authors exceeded 4,000 on January 14, 2008 and 5,000 on November 12, 2008. It should be noted that Wilson (2004) estimated that there are about 6,000 taxonomists at work worldwide on all groups of organisms combined. This may be an underestimate. Also, not all authors of *Zootaxa* are taxonomists, as some co-authors of papers are other kinds of biologists collaborating with taxonomists. Nevertheless, *Zootaxa* has represented and involved the majority of taxonomists currently working in the discipline.

The number of monographs (papers with 60 or more pages) increased rapidly from 2001 to 2006 and then leveled off during 2006 to 2008, with an average of over one monograph per week (Figure 9.4). It should be noted that from 2007, the printed area of *Zootaxa*'s pages increased from 14 × 20 cm to 17 × 24 cm, which effectively makes a 100-page book in pre-2007 format to be about 80 pages in the 2007 format. Thus, large monographic works in fact continued to increase in 2007 and 2008, although not as fast as during 2001–2006.

A commonly used measure of a paper's impact is the number of citations by other papers. Inclusion of *Zootaxa* in the *Science Citation Index Expanded* began in 2004 and the number of citations of *Zootaxa* papers rapidly increased from 27 in 2004 to 2,072 in 2008 (Figure 9.5). It should be noted that the number of papers and monographs published during 2006–2008 (Figure 9.1b; Figure 9.4) was increasing relatively slowly compared with the rapid increase in the number of citations during the same period (Figure 9.5). That translates into a greater rate of increase in the number of citations per paper, which is directly correlated with journal impact factor.

In 2008, *Zootaxa* was named a Rising Star three times consecutively (May, July, and September) by ScienceWatch.com based on its highest rate of increase in the number of citations among 392 journals in the ISI Animal and Plant Science Field (this large field contains many high-impact journals). In August 2008, *Zootaxa* was a Featured Journal from Essential Science Indicators (http://sciencewatch.com/inter/jou/2008/08aug-jou-Zoo/) for its citation achievements. In January 2009, it

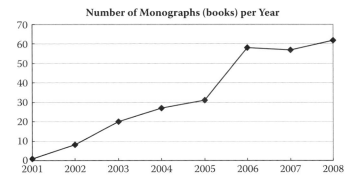

FIGURE 9.4   Number of monographs/books (papers of 60 or more pages, or special volume with collected papers) published in *Zootaxa* from 2001 to 2008.

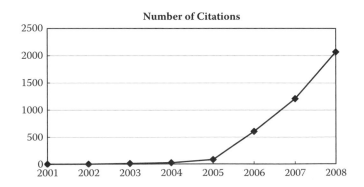

FIGURE 9.5   Number of citations of *Zootaxa* papers in ISI's *Science Citation Index Expanded (SCIE)* during 2001 to 2008. *Zootaxa* has been indexed by SCIE from 2004. The numbers of citations during 2001–2005 were obtained by the Cited Reference Search option in Web of Science.

was again named a Rising Star by ScienceWatch.com for its highest rate of increase in the number of citations among journals in the ISI Animal and Plant Science Field.

Another measure of impact is the number of times the papers are accessed online by readers. Many papers in *Zootaxa* are very popular and far more so than many non-taxonomic papers in other more trendy comprehensive journals in zoology. For example, the cybertaxonomy exemplar paper by Pyle et al. (2008) describing five new species of fish—published on January 1, 2008 on the occasion of the 250th anniversary of the officially recognized date of publication of *Systema Naturae* (Linnaeus, 1758)—was accessed 12,559 times during the first month of its publication. In comparison, the top paper in *Frontiers in Zoology* was accessed only a few hundred times each month (see http://www.frontiersinzoology.com/mostviewed).

Zootaxa has become the first choice for many authors to publish their important new species discoveries. Among the Top 10 new species of all taxonomic groups published in 2007 and also in 2008 (http://www.species.asu.edu/Top10), four were published in *Zootaxa*.

FACTORS CONTRIBUTING TO THE SUCCESS OF *ZOOTAXA*

Many factors have contributed to the success of *Zootaxa*. The most important one is that there is a great demand for an efficient outlet such as *Zootaxa* for publishing descriptive and revisionary papers in taxonomy. One author—Greg Evans of the U.S. Department of Agricultute (USDA)—wrote: "I am very impressed with the quality, ease, speed and accessibility of your journal." This summarizes most of the important features of *Zootaxa* that are attractive to authors and contribute to its success.

SPEED

Zootaxa aims to publish an average paper within one month after acceptance. It was, in general, faster than this during the first couple of years and then became slightly slower in the last few years due to the large number of accepted papers. This contrasts well with most other journals (Figure 9.6 for comparison with two other journals in taxonomy for data in 2008). One author—Fabio Moretzsohn—wrote: "*Zootaxa* is certainly a step in the right direction to help the taxonomic process move along faster."

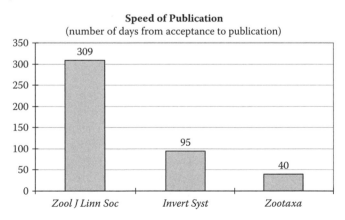

FIGURE 9.6   Speed of publication (measured as number of days from date of acceptance to date of publication) in three major journals of zoological systematics. Samples and sample sizes for three journals: *Zoological Journal of the Linnean Society*—one issue from each quarter in 2008 was sampled and the number of papers counted was 24; *Invertebrate Systematics*—all six issues of vol. 22 (2008) were sampled and the number of papers counted was 38; *Zootaxa*—one issue of each month of 2008 was sampled and the number of papers counted was 55.

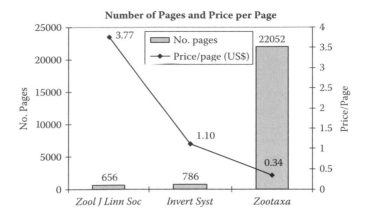

FIGURE 9.7 Cost of taxonomic journals (institution print+online price in 2006). Data of *Zoological Journal of the Linnean Society* and *Invertebrate Systematics* from Eigenfactor.org.

QUALITY

Speed of publication, however, is not achieved at the expense of quality. All manuscripts submitted to *Zootaxa* are subjected to strict peer review organized by subject editors. The participation of a large number of editors who are volunteers is important to the success of the journal; they are practicing taxonomists themselves as well as being editors, ensuring the quality of papers in *Zootaxa*. This creates a community of people with shared values that is very important to the success of our science.

EASE AND ACCESSIBILITY

No page charge is required for publication of papers or monographs, making it possible for all authors of different economic backgrounds to publish in *Zootaxa*. Color photos can be included in the online edition without cost to authors. A free e-reprint is also provided for authors' personal use (including exchanging with other scientists). Authors with funding for publication can opt to pay a low open-access fee of US$20 per printed page to make their paper free online for anyone to read. Subscription rates for *Zootaxa* are only a fraction of those of other major taxonomic journals (Figure 9.7).

FLEXIBILITY

Unlike many other journals, *Zootaxa* does not restrict the length of manuscripts. A paper of a few pages describing a new species is not too short and a monograph of a few hundred pages is not too long. The only requirement is that peers and editors consider it a quality paper that is well-presented and makes a good contribution to taxonomy. This is extremely important. Knowing that *Zootaxa* is there to publish their quality work regardless of the length of papers or number of papers encourages taxonomists to be more productive than they were before *Zootaxa*'s availability—a taxonomist, even with time and expertise, would be reluctant to prepare a large revision or book when he or she knows there are few or no outlets for publication.

CONTRIBUTION TO THE REVIVAL OF DESCRIPTIVE TAXONOMY

Zootaxa's contribution to descriptive taxonomy is self-evident in its size, its increasing share of the total literature in this branch of the discipline, and its growing impact. In addition to these, several benefits could be more important than the numbers themselves.

A major benefit of *Zootaxa* is the concentration of a vast body of papers and new taxa in a single easy-access journal that otherwise would be scattered in hundreds or even thousands of small journals or books, many of which may be expensive, difficult to access, or available only in large research libraries in developed countries. The effect is not just that of "n + n = 2n." but that of "n + n > 2n." in terms of both access and impact.

Another significant outcome of *Zootaxa* is the coming together as a community of a diverse group of specialists who were often perceived as too individualistic and fragmented into diverse subdisciplines. Within a short span of eight years, several thousand taxonomists from over a hundred countries have joined *Zootaxa* as authors, reviewers, and editors (Figure 9.8). This grass-roots development has received no institutional or government funding, nor has any money been spent in advertising or promoting this journal in other journals. This coming together is natural and completely voluntary. It creates a community of people with shared values, which is very important to the success of this journal and indeed to our science. It has even been proposed that they form an international society to facilitate collaboration among themselves and promote taxonomy on a global scale.

A really encouraging development seen in *Zootaxa* is the emergence of mega-biodiversity in developing countries such as Brazil, China, Argentina, and Mexico also as strongholds of human capacity in taxonomy (Figure 9.8). These and other developing countries accounted for about 44%

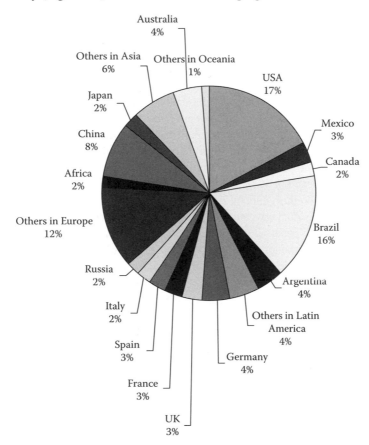

FIGURE 9.8 Geographic composition of 5,148 authors who published in *Zootaxa* during 2001–2008. Countries with fewer than 100 authors are not listed separately; countries are grouped for North America (22%), South/Latin America (27%), Europe (29%), Africa (2%), Asia (16%), and Australasia/Oceania (5%). There is a rounding error of 1%.

of 5,148 authors of *Zootaxa* at the end of 2008. Since authors of *Zootaxa* comprise the majority of active taxonomists, this estimate of proportion is thus based on a very large and reliable sample and well reflects the changing pattern in the geographic distribution of human capacity in taxonomy in the world (Figure 9.8). The developed countries (e.g., United States and Canada, and those in Europe, Australasia, and Japan) still have more taxonomists than developing countries as seen in *Zootaxa*, especially when the amount of biodiversity per country is taken into consideration. Africa and South/Southeast Asia are most deficient in taxonomic capacity. It should be noted that the above estimate is slightly biased against some non-English-speaking countries such as China, where some taxonomists are unable to write and publish in English.

Last, but not least, there is this often neglected effect of the growth of *Zootaxa* on lifting the impact factors of taxonomic journals and the consequent benefits it brings to authors and their research.

The ISI journal impact factor for a particular year is calculated by dividing the number of current year citations in ISI Science Citation Index Expanded (SCIE) journals by the number of source items published in that journal during the previous two years. In general, taxonomic journals have relatively low impact factors because, at least in part, the way the impact factor is calculated is biased against taxonomic journals (Krell 2000; Agnarsson and Kuntner 2007). Many taxonomic papers (especially descriptive taxonomic papers) are published in journals or monographs that are not indexed in *SCIE* and thus their citations are never counted because ISI includes only citations in ISI-indexed journals in the calculation of impact factors; this is in addition to the fact that taxonomic authorities for species are rarely cited in non-taxonomic papers and sometimes not even in taxonomic papers. Taxonomic papers are historical documents that are cited for a long period of time, but often not as quickly as papers in other branches of biology, especially when taxonomic research is a long process (especially revisionary work) and also publishing taxonomic papers is traditionally slow. This results in a relatively small number of papers being cited in the second or third year of publication.

The entry of *Zootaxa* into ISI SCIE in 2004 has had corrective effects on both of these two aspects discussed above. Unlike most other journals that are published in a fixed frequency/number of pages per year, *Zootaxa* has neither limits on the numbers of issues nor the number of pages published per year. In 2004, *Zootaxa* was already a large journal publishing 398 papers, which accounts for 5.5% of all SCIE-indexed papers from over 120 journals in the Zoology category. During 2004–2007, the total number of papers indexed in SCIE Zoology category increased along with the increase in number of papers published in *Zootaxa* and, by 2006–2007, *Zootaxa* alone accounted for one-eighth of all papers indexed in the SCIE Zoology category (Table 9.1). The flexibility of *Zootaxa* thus allows the rapid increase of taxonomic papers indexed in the SCIE Zoology category, which in turn increases the number of citations for taxonomic papers. Without *Zootaxa*, many of these papers would have been published in many small journals outside the ISI SCIE and the number of papers cited would have had no contribution to the calculation of impact factors of cited taxonomic journals.

TABLE 9.1
Growth of *Zootaxa*'s Share of All Papers Indexed by ISI's *Science Citation Index Expanded* in the Zoology Category (120+ Journals).

Year	Zoology (120+ journals) No. Papers	*Zootaxa* No. Papers	*Zootaxa*'s % 120+ Journals
2004	7153	398	5.5%
2005	7200	575	8.0%
2006	8216	1016	12.8%
2007	9043	1128	12.5%

Note: Indexing of *Zootaxa* by ISI's *Science Citation Index Expanded* began in 2004.
Source: Data from *ISI Journal of Citation* Reports.

Likewise, *Zootaxa* published an increasing number of large taxonomic revisions and mono-graphs (see Figure 9.4) that are not suitable for small journals with a fixed number of pages per issue. Without *Zootaxa*, some of these would have never been written or published, and many of them might have been published as individual books not indexed by SCIE. Thus, large papers with more systematic treatments and broader coverage of subjects that can attract more citations than small taxonomic papers covering single or a few species are excluded from the system, contributing to the under-estimation of impact factors of taxonomic journals. The importance of monographs can be seen in the following example: the impact factor for the taxonomic monographic series—*Bulletin of the American Museum of Natural History*—jumped to 16.385 in 2007, which is higher than those of top ecology journals such as *Trends in Ecology and Evolution* (14.797) and a premier science journal such as *Proceedings of National Academy of Sciences of the United States* (9.598).

Also, *Zootaxa*'s rapid turnover (see Figure 9.6) made it possible for many papers that cite other papers published in the previous two years to be published within the second or the third year of these cited papers, and thereby makes possible their contribution to the calculation of impact factors.

Although journal impact factors have been misused for evaluation of scientists and funding allocation, they do affect the employment of many scientists, their ranking among peers, their promotion, and funding allocation and prospects, and this includes taxonomists (Agnarsson and Kuntner 2007). It will be interesting to follow up the effect of the growth of *Zootaxa* on lifting the impact factors of taxonomic journals and the consequent benefits it brings to authors and their research.

CONCLUSIONS

Linnaeus captured the imagination of his generation with clever and provocative classifications and a sense of discovery. He opened the world of biological diversity to an ever-widening audience and sparked an age of species exploration. As society's need for reliable information about species expands, and threats of extinction accumulate, the time has arrived to revive taxonomy.

—Quentin Wheeler (2007: 17)

Reviving taxonomy, especially descriptive taxonomy, is essential for the completion of the Encyclopedia of Life (Wilson 2003). We need more taxonomists and we need them to be more effective if we want to achieve this task before it is too late (Gonález-Oreja 2008).

With the current global financial and economic crisis, it will be unrealistic to expect many new positions for taxonomists in museums, universities, and other institutions—it will be, in fact, lucky if we do not lose many current positions. As stated above, a recent survey of PEET graduates, for example, showed that half were no longer working in taxonomy one to six years after graduation. It is important for the taxonomic community to find clever ways to engage, encourage, enable, and collaborate with more courtesy taxonomists* to overcome the taxonomic impediment of the lack of taxonomists to handle the enormous task of identifying and naming the majority of the biodiversity unknown to science.

We are fortunate to live in an exciting era when cyber-infrastructure and technology offer tax-onomy an opportunity to meet its challenges in ways not foreseen in human history (Wheeler 2008). Taxonomy is not only a branch of biological sciences, but at the same time also essentially an infor-mation science. Taxonomists observe and analyze information (molecular, morphological, biologi-cal, etc.) stored in living or fossil species during the course of evolution in order to classify, name, and group them in an efficient information retrieval system that preferably reflects the phylogenetic history of the group. Cyber-infrastructure and technology can facilitate virtually every aspect of taxonomic research, such as distant viewing of specimens at museums in other countries, capture

---

* A term I use for taxonomists not in employment as taxonomists; I prefer this term over "amateurs" as the latter term has a negative connotation and is not a proper description of these taxonomists.

of images and processing of information, analysis of data and preparation of manuscripts, rapid and low-cost publication of, and online access to, authoritative taxonomic information. We are fortunate to have major initiatives such as the Global Biodiversity Informational Facility (GBIF) and the Biodiversity Heritage Library.

In addition to integrating taxonomy with cyber-technology, it is important for taxonomists to enhance coordinated international collaboration among themselves to achieve the synergistic effect of "n + n > 2n." We have a glimpse of the success of team efforts in the making of a mega-journal in taxonomy, even as a grassroots movement without the support of major institutions. For taxonomy to realize its dream to be a big science, coordinated international collaboration is essential (Knapp 2007).

In multiple issues of *Zootaxa* published each week, we are beginning to see the transformation of the 250-year-old zoological taxonomy into a dynamic branch of zoology in this new era. In *Zootaxa*, zoological taxonomy is becoming bigger, faster, and hotter than the other branches of zoology. This is a very promising sign for the revival of taxonomy as a whole.

ACKNOWLEDGMENTS

I thank Andrew Polaszek of the Department of Entomology, Natural History Museum (London, UK) for an invitation to contribute to this volume celebrating 250 years since the tenth edition of *Systema Naturae*. Some of the data on *Zootaxa* presented in this chapter were from an editorial previously published (Zhang 2008; *Zootaxa* 1968: 65–68), which benefited from reviews and comments by Dan Bickel of the Australian Museum and Andrew Polaszek, and also data provided by Nigel Robinson of *Zoological Record* (Thomson-Reuters, UK). I thank Rosa Henderson (Landcare Research) for reviewing the draft of this manuscript. I also greatly appreciate the support of over 150 *Zootaxa* editors and 5,000 authors who have made *Zootaxa* a success. My attendance at the *Systema Naturae 250* Symposium in Paris was partly funded by the Foundation for Research, Science, and Technology, New Zealand and partly by a Global Biodiversity Information Facility (GBIF) Seed Fund project.

REFERENCES

Agnarsson, I. and Kuntner, M. (2007) Taxonomy in a changing world: Seeking solutions for a science in crisis. *Systematic Biology*, 56, 531–539.

Cracraft, J. (2002) The seven great questions of systematic biology: An essential foundation for conservation and the sustainable use of biodiversity. *Annals of the Missouri Botanical Garden*, 89, 127–144.

Evenhuis, N.L. (2007) Helping solve the "other" taxonomic impediment: Completing the eight steps to total enlightenment and taxonomic Nirvana. *Zootaxa*, 1407, 3–12.

Gonález-Oreja, A.J. (2008) The Encyclopedia of Life vs. the Brochure of Life: Exploring the relationships between the extinction of species and the inventory of life on Earth. *Zootaxa*, 1965, 61–68.

Hoagland, K.E. (1996) The taxonomic impediment and the Convention of Biodiversity. *Association of Systematics Collections Newsletter*, 24(5), 61–62, 66–67.

Knapp, S. (2008) Taxonomy as a team sport. In Wheeler, Q.D. (Eds.) *The New Taxonomy*. Systematics Association Species Volume. Boca Raton, FL: CRC Press, 33–53.

Krell, F.-T. (2000) Impact factors aren't relevant to taxonomy. *Nature*, 405, 507–508.

Linnaeus, C. (1758) *Systema Naturae per regna tria naturae, secundum classes, ordines, genera, species, cum characteribus, differentiis, synonymis, locis. Editio decima, reformata. Tomus* I. Laurentii Salvii, Holmiae, 828 pp.

Mace, G., Masundire, H., and Bailie, J. (Coords.) (2005) Biodiversity. In *Millennium ecosystem assessment: Current state and trends: Findings of the Condition and Trends Working Group. Ecosystems and human well-being*, Vol. 1. Washington, D.C.: Island Press, 77–122.

May, R.M. (2004) Tomorrow's taxonomy: Collecting new species in the field will remain the rate-limiting step. *Philosophical Transactions of the Royal Society of London, Series B*, 359, 733–734.

Pyle, R.L., Earle, J.L., and Greene, B.D. (2008) Five new species of the damselfish genus *Chromis* (Perciformes: Labroidei: Pomacentridae) from deep coral reefs in the tropical western Pacific. *Zootaxa*, 1671, 3–31.

Rodman, J.E. and Cody, J.H. (2003) The taxonomic impediment overcome: NSF's Partnerships for Enhancing Expertise in Taxonomy (PEET) as a model. *Systematic Biology*, 52(3), 428–435.

Wheeler, Q.D. (2007) Invertebrate systematics or spineless taxonomy? *Zootaxa*, 1668, 11–18.

Wheeler, Q.D. (2008) Introductory. Toward the new taxonomy. In Wheeler, Q.D. (Ed.) *The New Taxonomy. Systematics Association Species Volume*. Boca Raton, FL: CRC Press, 1–17.

Wilson, E.O. (2003) The encyclopedia of life. *Trends in Ecology and Evolution*, 18(2), 77–80.

Wilson, E.O. (2004) Taxonomy as a fundamental discipline. *Philosophical Transactions of the Royal Society of London B*, 359, 739.

Zhang, Z.-Q. (2006a) The first five years. *Zootaxa*, 1111, 68.

Zhang, Z.-Q. (2006b) The making of a mega-journal in taxonomy. *Zootaxa*, 1358, 67–68.

Zhang, Z.-Q. (2008) Contributing to the progress of descriptive taxonomy. *Zootaxa*, 1968, 65–68.

10 Provisional Nomenclature
The On-Ramp to Taxonomic Names

David E. Schindel and Scott E. Miller

CONTENTS

THE OTHER TAXONOMIC IMPEDIMENT

Many authors have discussed the "taxonomic impediment"—the lack of trained taxonomists and the technical infrastructure needed to halt and reverse the loss of biodiversity (Lyal and Weitzman, 2003; Rodman and Cody, 2003). The Convention on Biological Diversity (CBD) entered into force at the end of 1993, and taxonomists were quick to point out the critical role that taxonomy would play in reaching the CBD's objectives. In 1995, the CBD Second Conference of the Parties (COP 2) called for a study of the lack of taxonomists needed to achieve the CBD's objectives.* In 1998, COP 4 endorsed the need to increase research capacity in taxonomy and a recommendation† to create the Global Taxonomy Initiative (http://www.cbd.int/gti/).

Evenhuis (2007) reviewed the use of the term taxonomic impediment as the limitations based on lack of external resources (e.g., funding, people, facilities). He went on to describe eight steps in the taxonomic process (presented in compressed form below). Evenhuis asserted that most taxonomists were enthusiastic about conducting the first four steps, but the last four were obstacles ("the other taxonomic impediment") that severely limited the progress of taxonomy. The steps he described were:

1. Venturing into nature
2. Collecting specimens
3. Sorting specimens into species
4. Discovering new or rare species among the sorted specimens
5. Confirming new discoveries through comparison with publications and types

---

* Decision II/8.7 of COP 2
† Decision IV/1 of COP 4 endorsed the Recommendation II/2 of the Subsidiary Body on Scientific, Technical and Technological Advice (SBSTTA)

6. Describing and illustrating new species
7. Submitting manuscripts for publication
8. Educating others about the new species

In this chapter, we will focus on steps 5, 6, and 7. We will argue that creating a system of provisional nomenclature can fundamentally alter and accelerate the taxonomic process.

STEPS AND MISSTEPS IN THE TAXONOMIC PROCESS

"Nomenclatural events"—including the publication of formal, Code-compliant Linnaean taxonomic names—are the critical datapoints in taxonomy. Each nomenclatural event places one or more formal names into the literature and the corpus of claims for nomenclatural priority. Each proposed name becomes subject to scrutiny by the research community that will test its formal validity. For example, was the name proposed in an allowed form, and was it accompanied by the required documentation (e.g., designation of a holotype, if required by a nomenclatural code, or deposition of the required type in a repository)? If a name was formally correct, does it represent a new taxon or is it based on a holotype that bears a previously proposed name, and may therefore be an objective synonym? Beyond these objective concerns lie the longer-term subjective considerations. Does the proposed name represent a taxon concept that has already been proposed, and may therefore be a subjective synonym? Does the name represent a taxon concept that is too broad and may contain distinct taxa that deserve to be formally recognized under different names?

In the traditional practice of taxonomy, researchers are expected to anticipate these and other challenges to the validity of the names they propose. Prior to publication, taxonomists prepare careful descriptions of new taxa and can devote years, sometimes decades, to examining the publications and type specimens associated with similar taxa to ensure that theirs are truly new and distinct. In principle, taxonomists are being asked to prove a negative assertion when they propose a new taxon, e.g., "This species has never been described before." In practice, taxonomists make the practical decision to formally propose and publish a new name when it seems very unlikely (but not certain) that they have overlooked a previous nomenclatural event that will invalidate their claim for priority.

This process bears many similarities to the system of patent protection for inventions. An inventor creates an innovation that he or she thinks is novel and significant. To avoid the cost of filing a patent claim that may later be judged invalid, an inventor normally conducts a patent search to determine whether the idea has already been proposed. The longer inventors wait to file a patent claim, the greater is the likelihood that someone will either steal their idea or devise it independently. Once a patent application is made, it enters the public domain where anyone can examine it and decide whether the new patent infringes on a previously filed patent claim. Lawsuits are the recourse open to an inventor who thinks a new patent is, in essence, a junior synonym of his or her earlier invention, just as taxonomic revisions are the recourse used by taxonomists to lump, split, or place names in synonymy.

The parallel between taxonomy and patenting ends here. Patenting can lead to licensing, product development, commercialization, and financial gain for the inventor. None of these consequences or incentives is associated with naming a new species. Except in rare circumstances, taxonomists do not profit from the naming of new taxa. Nevertheless, taxonomists often conduct their research on potential new species as solitary individuals behind closed doors, restricting pre-publication access to their findings as if they were trade secrets (see Figure 10.1). Conducted in this way, it can take a taxonomist years to decades to generate tangible products in the form of the nomenclatural events.

For an inventor, the goal of financial gain can be attained only if the asset (the innovation) is kept in private hands until it is ready for the marketplace. For a taxonomist, the goal is the professional recognition and personal satisfaction of discovering, describing, and recording new forms of biodiversity. For historical reasons, the taxonomic process has defined nomenclatural events as the goal-

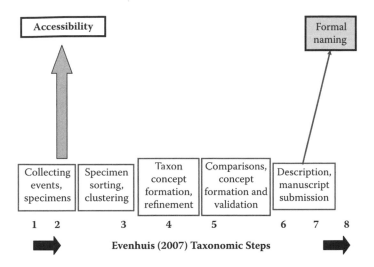

FIGURE 10.1   The results of each step in the taxonomic process, as traditionally practiced, are not shared until publication.

line over which taxonomists must cross if they are to receive formal recognition. Linnean names have become the only currency with which taxonomists can be rewarded. It is this system of incentives that is causing taxonomists to carry out their research as an individualistic, secretive pursuit.

In the following sections, we propose the introduction of standardized taxon labels as intermediate products in the process of producing new taxonomic names. As described, creating a standardized system of provisional nomenclature can create a new incentive and reward system for taxonomists and can greatly accelerate the documentation of biodiversity.

This is not a new problem, and others have suggested approaches in the past (e.g., the "interim taxonomy" of Erwin 1991, 1995), but a standardized system has never been widely adapted. Many taxonomists and ecologists have used various systems to refer in publication to species that lacked formal names. For example, Marks (1983: 534) explicitly formatted such references as "*Aedes* (*Finlaya*) 'Marks sp. no. 104' an undescribed species which has been studied in some detail by the author" versus named species and species for which identities were not resolved; and Holloway (1984) has made a practice of referring to undescribed lepidopteran species by the genitalia slide number of the critical voucher.

TAXON NAMES, CONCEPTS, AND LABELS

Taxon names provide the framework for information exchange and retrieval in taxonomy. The rules governing the formal naming of taxonomic units are overseen by several international commissions of nomenclature. Unlike patenting systems, taxonomy does not have centralized registries of patent applications or a regulated system for publishing patent applications. Polaszek et al. (2005) proposed the creation of ZooBank as a registry of taxonomic names for the International Commission of Zoological Nomenclature. Such a centralized registry would facilitate access to formal taxon names and their associated documentation.

Taxon concepts are biological hypotheses that are represented by taxon names. Concepts emerge in a taxonomist's thinking after examining specimens, sorting and separating them into groups of similar individuals, and seeking discontinuities in the variation among these groups. In most cases, these groups will correspond to previously described taxa, but some can emerge as new discoveries. The holotype designated for a new taxon is selected as a singular representation of the newly described taxon concept, and paratypes are often designated to represent the variation included by the author under that name. Descriptions that form a critical part of a nomenclatural act are meant to transmit the author's concept of the taxon and its limits in various biological dimensions (e.g., morphology, development, geographical distribution, ecological preferences, DNA sequence variation).

The term *taxon label* has no generally accepted meaning, but we propose one here. Taxon labels with these characteristics can be used to communicate taxon concepts that are not yet ready for formal description and publication. In principle, they can be posted on public Websites, submitted to taxonomic databases, and even published without compromising the priority or clarity of taxonomic names.

> A taxon label is a unique, stable, text-phrase applied to an unpublished taxon concept. The text-phrase of a taxon label should link the concept to reference specimens and/or a geographic locality, and should be in a form that makes it clearly unusable as a formal taxon name.

This definition has the following important components:

Uniqueness. A taxon label must clearly separate a taxon concept from all other concepts. Labels such as *Genus sp. A* are inadequate for distinguishing concepts.

Stability. Once a taxon label has been applied and made public through Websites, databases, presentations, or publications, it should remain stable in its linkage to associated specimens and localities. Any change in the underlying concept should trigger a change in the label.

Unsuitability as taxon names. As illustrated below, the text-phrase of a taxon label can include multiple words, numerals, punctuation marks, and mixtures of upper- and lowercase letters. The use of these character types will make it impossible to misinterpret a taxon label as a taxon name. In this way, publication of a taxon label will not compromise the later publication of a taxon name based on the same concept.

Examples of Taxon Labels

Taxonomists have routinely used provisional names, but their practices have not been standardized or consistent. For example, records in GenBank carry taxon identifiers such as *Ocyptamus* sp. MZH S143_2004, *Argentinomyia* sp. CR-12; *Eunotia* sp. EUN392T, and *Xenopus* (*Silurana*) sp. new tetraploid. These are clearly not meant to be formal Linnean names, but neither are they interpretable to the user. Other than contacting the researcher who submitted the record to GenBank, there is no information available about the taxon concepts associated with these provisional names.

Janzen et al. (2009) proposed a system for interim nomenclature that would include provisional taxon labels in a variety of formats such as "Patelloa xanthuraDHJ02," "Belvosia Woodley06" and "Astraptes LOHAMP." Their proposed system does not fully meet the criteria we propose above. Specifically, appending numerals to a person's name does not provide clear uniqueness and stability, and capitalization is not a secure way to distinguish taxon labels from taxon names. Is "Belvosia Woodley06" a provisional species in the genus Belvosia or a provisional label for a new genus? Could "Astraptes LOHAMP" be transcribed as "Astraptes lohamp," which can be mistaken for a formal name?

In contrast, a few communities of practice in taxonomy have developed standardized systems of provisional nomenclature. Barker (2005) described a decision by the Council of the Heads of Australian Herbariums (CHAH) to regularize the formation of "informal names" (termed "taxon labels" here, to distinguish them from taxon names). The proposed standard CHAH format for a taxon label was:

- *Genus-name sp. Phrasename (Voucher-specimen identifier)* Source
- where:
- Genus-name is a previously published generic name.
- "sp." is a standard delimiter that indicates species rank.

- Phrasename refers to a locality.
- (Voucher-specimen identifier) is a two-part field consisting of a collector's name and the voucher specimen number attached to the exemplar of the taxon concept, or its herbarium sheet number.
- Source refers to the name of the concept's proposer.

The CHAH constructed this standard to address two problems. First, state-level floras or censuses had been conducted independently and in some cases different taxon names had been published for a single species. "Informal names" provided a system of communication among regions that can aid the process of constructing a consensus taxonomic list for the country as a whole. Second, non-taxonomists responsible for the conservation of rare and endangered plant species needed a way to cite taxa in regulations. The CHAH standard provided an objective reference system without being mistaken for taxonomic judgments.

The CHAH standard offers a starting place for discussion of the preferred format for taxon labels. We suggest that TDWG (Biodiversity Information Standards, formerly known as the Taxonomic Databases Working Group) is an appropriate organization for developing a global standard for taxon labels.

Taxon Labels in the Taxonomic Process

Taxonomic research requires the skill, experience, and judgment of a craftsman, and the steps in the taxonomic process (especially Evenhuis's steps 5, 6, and 7) require precision and scholarship. However, nothing in the process requires it to be carried out as the proprietary work of an individual. Quite the contrary, enlisting collaborators would make the work of examining the literature and type specimen collections faster and easier. The widespread availability of information technology and the Internet are reducing distance as an obstacle to taxonomic research (see Godfray et al., 2007; Zhang, 2008).

The proposed system of taxon labels would enable researchers to make their interim results accessible on public databases and Websites. Articles could be published with references to taxon labels linked to the databases and Websites that provide the information associated with the taxon labels. Ecologists and other non-taxonomists could publish results using taxon labels, thereby avoiding the delay often associated with waiting for taxonomists to put formal names on specimens.

Putting interim results into the public domain would have two significant impacts. First, taxonomists could cite these interim products as publications and other deliverables. Second, authors could engage other taxonomists as collaborators in making critical comparisons with type specimens. Initiatives such as the European Distributed Institute of Taxonomy (EDIT; see http://www.e-taxonomy.eu/) are developing tools for "cybertaxonomy" that will enable Web-based collaboration and consensus-based taxonomy research conducted by teams of researchers, not solitary individuals. Figure 10.2 illustrates how taxon labels can be used for information exchange, collaboration, and publication of interim results.

Taxon Labels and DNA Barcoding

Hebert et al. (2003) proposed the use of a short, standardized DNA sequence as a diagnostic marker for species identification—a "DNA barcode" analogous to the Universal Product Code used to link products in stores to inventory records. Since that proposal, the Barcode of Life Initiative has mushroomed and has gathered DNA barcode records from almost 700,000 specimens representing more than 60,000 species. There are at least three circumstances in which taxon labels could be used:

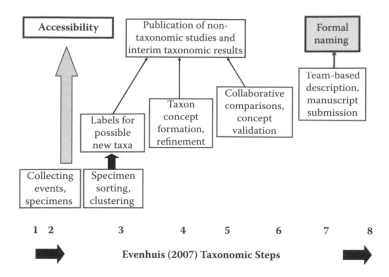

FIGURE 10.2   How taxon labels can be used for information exchange, collaboration, and publication of interim results.

1. Barcodes have uncovered hidden variation within species that may merit recognition as new species. These intraspecific barcode clusters need to be studied by taxonomists with experience in that group, and attaching a taxon label to each cluster would facilitate this research.
2. Field collections are also yielding specimens that do not appear to belong to any known species. Taxon labels attached to these specimens would facilitate communication between specialists as barcode data and digital images of specimens are exchanged for study.
3. Metagenomic analyses of environmental samples are producing enormous volumes of data on short gene fragments. The barcode gene sequences of the organisms in these environmental mixtures can be assembled to produce populations of barcodes. Clusters of these barcode records reflect discontinuities in genetic variation, but in the absence of discrete voucher specimens there is no way to associate taxonomic names to these clusters. Taxon labels provide placeholders for these barcode clusters until they can be associated with barcodes from identified specimens.

CONCLUSIONS

Formal taxonomic names that comply with nomenclatural codes are the mainstays of taxonomy and biodiversity research, but they are not the only medium for information exchange. Taxon labels can be very useful in conveying the results of non-taxonomic research and for accelerating the progress of taxonomic research. To achieve these ends, taxon labels will need to be unique, stable, and formatted to convey critical information without interfering with formal nomenclature.

ACKNOWLEDGMENTS

This work was supported by Alfred P. Sloan Foundation grant 2008-3-4 to the Smithsonian Institution for support of the Consortium for the Barcode of Life. The authors benefited from discussions with Dan Janzen, Arthur Chapman, James Croft and Terry Erwin.

REFERENCES

Barker, B. (2005). Standardising informal names in Australian publications, *Australian Systematic Botany Society Newsletter* 122:10–11.

Erwin, T.L. (1991). Establishing a Tropical Species Co-occurrence Database. Part 1. A plan for developing consistent biotic inventories in temperate and tropical habitats. *Memoria del Museo de Historia Natural, Universidad Nacional Mayor de San Marcos, Peru* 20:1–16.

Erwin, T.L. (1995). Measuring arthropod biodiversity in the tropical forest canopy. In M.D. Lowman, and N.M. Nadkarni, Eds. *Forest Canopies*. Academic Press. 109–127.

Evenhuis, N.L. (2007). Helping solve the "other" taxonomic impediment: Completing the eight steps to total enlightenment and taxonomic nirvana, *Zootaxa* 1407: 3–12.

Godfray, H.C.J., Clark, B.R., Kitching, I.J., Mayo, S.J., and Scoble, M.J. (2007). The Web and the structure of taxonomy, *Systematic Biology*, 56:6, 943—955

Hebert, P.D.N., Cywinska, A., Ball, S.L., and deWaard, J.R. (2003). Biological identifications through DNA barcodes. *Proceedings of the Royal Society London B* 270:313–321.

Holloway, J.D. (1984). The Moths of Borneo: Family *Notodontidae*. *Malayan Nature Journal* 37:1–107.

Janzen, D.H., W. Hallwachs, P. Blandin, J.M. Burns, J.-M. Cadiou, I. Chacon, et al. 2009. Integration of DNA barcoding into an ongoing inventory of complex tropical biodiversity. *Molecular Ecology Resources* 9 (supplement 1):1–26.

Lyal, C.H.C. and Weitzmann, A.L. (2004). Taxonomy: Exploring the impediment. *Science,* 305, 1106.

Marks, E.N. (1983). Mosquitoes of the Purari River lowlands. In T. Petr, Ed. *The Purari: Tropical environment of a high rainfall river basin*. Dordrecht, Netherlands: Dr. W. Junk Publishers. 531–550.

Polaszek, A., Agosti, D., Alonso-Zarazaga, M., Beccaloni, G., de Place Bjørn, P., Bouchet, P. et al. (2005). A universal register for animal names. *Nature* 437: 477.

Rodman, J.E. and Cody, J.H. (2003). The taxonomic impediment overcome: NSF's Partnerships for Enhancing Expertise in Taxonomy (PEET) as a model. *Systematic Biology,* 52(3), 428–435.

Zhang, Z.Q. (2008). Zoological taxonomy at 250: Showcasing species descriptions in the cyber era, *Zootaxa* 1671: 1–2.

11 Future Taxonomy

David J. Patterson

CONTENTS

INTRODUCTION

Taxonomy is at a critical point in its history. The remaining few thousand taxonomists are increasingly concerned about the prospects of the discipline (Agnarsson and Kuntner, 2007; de Carvalho et al., 2007a, b). At the same time, the impact of the growing human population on the biosphere has created a clear and urgent need for an understanding of biodiversity that can be integrated with sociological, historical, economic, geographical, and atmospheric sciences (inter alia) at a scale capable of predicting, and therefore preparing for, change. We look to biodiversity informatics to organize data about the biosphere. Biodiversity informatics may emerge as a discipline that will compete with taxonomy, or it may become a part of taxonomy. If it is adopted by taxonomists, it will bring a relevance to taxonomy that can revitalize the discipline. "Future Taxonomy" refers to a taxonomy that takes responsibility for shaping Biodiversity Informatics.

WHAT WAS TAXONOMY?

Biologists typically think of contemporary taxonomy as starting in the middle of the 18th century with Linnaeus's efforts to catalogue all known life (Linnaeus, 1753). This arbitrary point of departure minimizes many earlier synthetic efforts (such as those of Frederico Cesi and his colleagues through the Accademia dei Lincei (Freedberg, 2002) and Linnaeus's dismissal of the microbial world that was being documented by individuals such as van Leeuwenhoek, Joblot, and Müller. Nonetheless, Linnaeus established three key properties that continue to characterize taxonomy. They are: the intent to catalogue all known diversity, the application of binomial nomenclature for members of the catalogue, and the use of a formal hierarchy to provide an organizational framework for the catalogue.

To achieve its inventory of all diversity, taxonomy is a process that involves continuing discovery—whether through the investigations of new habitats, new technologies, or through a sceptical reappraisal of what is known. It is a discipline that requires awareness of its own history, nomenclatural practices, and phylogenetic insights. Taxonomists know the opinions of other workers about the entities and of their relationships as expressed in synonymy lists, classifications, or phylogenies. Through their heightened and taxonomically focused familiarity with the literature, taxonomists are often aware of all types of statements made about taxa within the scientific literature and through other outlets. As those statements have been made over 250 years, taxonomists appreciate and interpret older writings. Taxonomists are the experts to whom others go when they want to know something about organisms. That is, taxonomists are custodians and integrators of information about organisms. Their capacity pre-adapts taxonomists to be the architects of biodiversity informatics.

THE PROBLEMS WITH TAXONOMY

Taxonomy is imbued with a number of weaknesses that impede its extension into biodiversity informatics. A concerted effort is needed to overcome them (Chapter 10). The community of folk who regard themselves as taxonomists is small—estimated to be 6,000 or fewer (Hopkins and Freckleton, 2002). The size of this community, with its responsibility for the full spectrum of biodiversity, is dwarfed by the communities that focus on a small number of aspects of the biology of selected species. The active community responsible for some taxonomic areas is smaller and less vibrant than the community studying one enzyme in one species. Yet, the numbers aspect is not as dire as may first appear, because phylogeneticists, ecologists, and a very substantial community of naturalists complement taxonomists with their own awareness of biodiversity.

A second problem is that the discipline of taxonomy, despite the common features established by Linnaeus, is fragmented. Individual taxonomists rarely hold detailed knowledge on more than 1000 species. Nomenclatural practices are determined by an array of codes (botanical, zoological, prokaryotic, cultivated plants being the dominating ones). Especially within the domain of animals, conventions differ from one taxonomic area to another. There will be further division by the incoming Phylocode (Cantino and de Queiroz, 2008) which is philosophically very different from the traditional codes, because phylogenetic nomenclature does not have the benefit of being anchored on type specimens, but is linked to the considerably less solid opinions about relationships (see also Chapter 17). The consequence is that taxonomy is not integrated, but is fragmented into an archipelago of variously populated and sized islands of knowledge.

The consequences of the marginalization of taxonomy are widely recognized. The absence of good taxonomic practices in dependent disciplines, such as ecology, undermines the credibility of insights in those subjects (Bortulus, 2008). The gap between taxonomy and ecology has been widening. Students have less access to biodiversity training, and so lack awareness of the component players and the subtleties of their interactions within ecosystems. This weakness is particularly worrisome in the context of habitats and communities that exhibit declining resilience to change, because we collectively lack access to knowledge that would inform best management practices.

THE NARRATIVE CHARACTER WITHIN BIOLOGY

Biology is an unusual science (Mayr, 2004). Reductionism has a smaller role in extending the envelope of our knowledge about all species than in the "hard" sciences. Biology is the totality of cause and effect of the existing biota and their ancestors. All individuals are interconnected through ancestor–descendant relationships. The paths that the genealogies weave through space and time have a serendipitous character. Variation, divergence, abundance, and termination are all contingent on complex interactions and largely unpredictable events. The entities, such as the things we refer to as "species," cannot be unambiguously defined because we (collectively) observe a tiny subset of characteristics, of a tiny subsample of individuals, in a very small number of places, and for very short periods of time. We know so little about biology, and always will. Bigger insights come less from a disciplined analytical methodology, and more from well-informed insightful storytelling. At best, these stories can be regarded as models or hypotheses. The balance of fact and judgment in these narratives is rarely clear. Typically, some observations are discarded early in the game because they don't seem to fit a pattern, and usually most data are discarded after the story is told (in a publication). This approach has been criticized as "just so" stories by phylogeneticists (D.S. Wilson, 2003), or in ecology. While the criticisms have validity, rejection of less credible narratives through processes of reciprocal illumination remains the primary shaper of our understanding of biology. The greatest weakness of the narrative approach is not that it is unable to provide understanding, but that the data are not valued or retained, and so are not available for subsequent reanalysis. The strength of future taxonomy will depend on complementing the narrative perspective with greater datacentricity, and the products (stories) with processes that promote collective participation and accumulation of knowledge.

WHAT CAN WE DO?

Taxonomy can be revitalized. Quentin Wheeler, in Chapter 5, refers to taxonomy's need for an international collaborative paradigm. With such a paradigm driving the organization and behavior of taxonomy, activities that are currently parochial can become more integrated.

A communal pool of relevant information is a valuable component of such a paradigm. Traditionally, such a pool exists in the form of the libraries and in the minds of the experts. An emerging complement to the traditional pool is the virtual pool of information that is accessed through the Internet. This pool is currently growing as an ad hoc structure and is indexed and accessed via generic tools such as Google, but it can be intentionally shaped to improve its fitness for purpose. A pool that accumulates data for all of us to use would be an affirmation of the significance of data over the narrative, would speed access to relevant information, would expose data to higher scrutiny, and lead to improved data quality (Robertson, 2008). The model has been implemented with existing repositories of molecular information. Initiatives such as the Biodiversity Heritage Library are already moving along this path as the consortium converts literature records wholesale to an electronic format (http://www.biodiversitylibrary.org/). Key information, such as the literature that contains taxonomic and nomenclatural acts, will be increasingly available and will nourish taxonomic research in regions without access to good libraries. Taxonomy will be democratized and accelerated. Larger data pools will mean that we can aspire to asking grander questions than are currently feasible.

Assembling a pool and building the associated structures is not, obviously, without challenges. The architecture needs to overcome the idiosyncrasies that have grown up within the various subdisciplines. Distinctions that subordinate opinion to fact have to be clearly made. The agenda has to be both comprehensive and adaptable so that the structure will facilitate cooperative research, focus on relevance, and reward participation in the process rather than in products.

These challenges are not daunting, as a variety of projects are already moving in this direction. Various federated environments have emerged to bring together isolated initiatives. They include the Catalogue of Life Partnership for a list of species, Ocean Biogeographic Information System (OBIS) for marine biodiversity data, or the Global Biodiversity Information Facility (GBIF) for specimen data. The Biodiversity Heritage Library is mobilizing the literature. Central databases have emerged for certain areas of biology—such as the consortium of National Center for Biotechnology Information (NCBI)'s GenBank, the European Molecular Biology Laboratory (EMBL) Nucleotide Sequence database, and the DNA Database of Japan for nucleotide information.

The discipline of biodiversity informatics is emerging to manage information about organisms (Chapter 13). Biodiversity informatics has been said to deal with

capture, storage, provision, retrieval, and analysis, focused on individual organisms, populations, and taxa, and their interactions. It covers the information generated by the fields of systematics, evolutionary biology, population biology, and ecology, as well as more applied fields such as conservation biology and ecological management. (http://www.bgbm.org/BioDivInf/def-e.htm, http://www.jrsbdf.org/v2/home.asp).

This grey statement does not capture the potential of purpose, scale, or destiny of biodiversity informatics. The future is more successfully heralded by the efforts of a small community of champions who have knowledge both of information sciences and biology, and whose efforts, at least at the global level, are often captured by agencies such as the Taxonomic Database Working Group (TDWG) (or Biodiversity Information Standards). Their efforts toward metadata schemas, globally unique identifiers, and transfer protocols are facilitating the machine-to-machine dialog upon which future taxonomy must depend (see Chapter 16). This semantic component is essential to overcoming parochiality and achieving a cumulative knowledge environment.

UNIONIZATION

Unionization marks a shift from a fragmented endeavor to a unified discipline. The history of science is peppered with unionizing events such as Martin Behain's first use of the globe to unify maps in 1492. Linnaeus's *Systema Naturae* in the eighteenth century made a (fairly) comprehensive statement about known biodiversity and formed the foundation of subsequent work. Unfortunately, despite unifying principles, it failed to unionize the discipline, which evolved too fast and lacked a deeper truth upon which all that followed could be founded. It was Darwin's emphasis on the evolutionary process that provided us with that deeper truth and unified biology in a fashion similar to Mendeleev's periodic table contribution to the unification of the chemical sciences (Darwin, 1859; Mendeleev, 1869). Crick, Franklin, and Watson's recognition of the significance of DNA in inheritance and regulatory processes provided the mechanistic understanding that defined and interconnected much of the biology of the 20th century. Google Earth is another development around which disparate efforts can be linked together. The reasons, characteristics, and paths to unionization are not the same, but all contribute a framework around which an array of previously parochialized information became organized and unified. When viewed in retrospect, unionization is not so much distinguished by originality, but by the catalytic effect and timeliness of the common framework.

Some of what ails taxonomy could be addressed through unionization. That unionization, technically, is achievable now. The recent release of the Encyclopedia of Life (EOL, http://www.eol.org) has shown that biodiversity informatics need not be parochial. The project seeks to produce pages with access to information for all known species in compliance with the vision articulated by E.O. Wilson (2003), and within 10 years. The approach is not limited to more than two million named living species, but can be extended to extinct taxa, to the many taxa that we know of that have yet to be named, and to accommodate the species that have yet to be discovered using different taxonomic practices. EOL demonstrates that we have now moved to a position where biodiversity information can achieve the scale that is required for future taxonomy.

EOL is taxonomic inasmuch as it relies on a catalogue of species names against which to index information about those species. It is unlike previous exercises because it seeks to index any and all data objects for all taxa. EOL uses aggregation techniques to gather data objects from other environments and recombine those data objects to provide species pages that meet the needs of a diversity of users. To achieve this, it must be capable of participating in machine-to-machine dialog. The semantic dimensions that allow for machine-to-machine interactions are essential and necessary components of future taxonomy because the Web services, metadata, standards, and unique identifiers offer the key technical components of deparochializing the discipline.

When compared with the traditional approach of authoring species pages, EOL is whittling away at the narrative character of biology, and is promoting a shift toward shared data and their enhancement. This shift is made possible because the availability of data objects in an electronic form has enabled an e-taxonomy to emerge. The bandwidth of the Internet makes Web taxonomy feasible. But to unionize, Web taxonomy is not sufficient, and we need to lift to another level of scale, that of cybertaxonomy or semantic taxonomy. That will be a decentralized approach to a taxonomy that is designed to work with distributed knowledge. As discussed later, cybertaxonomy almost certainly must use a names-based indexing infrastructure and apply information management tools. The addition of participatory dimensions (social networking) would be a natural component of future taxonomy as it will help address the challenges of scale by engaging larger communities, and will promote improvements to quality through ongoing scrutiny of data.

Datacentricity, semanticization, participation, and an emphasis on process are essential because the narrative approach is not suited to integrative analyses or visualizations, nor can it serve the meta-analyses that are now called for. The solution will include the conversion of all data into semantically minimal elements. In this atomized form, and by adopting agreed standards, information can pass freely among data stores and analytical environments and be enriched as it is accessed and used through Web services and social networks, gathering both substance and value as it does so. The risk is that this conversion is done without sensitivity to unusual character of the discipline. This risk is heightened if the process of mobilizing electronic information about biology is led by information scientists. The risk is addressed if the course into semantic taxonomy is charted by the community of taxonomists.

THE NAMES-BASED INFRASTRUCTURE

EOL unifies independent insights, skills, and expression in significant part through its use of a taxonomically intelligent (Patterson et al., 2006) names-based infrastructure. This infrastructure is a key component of the technical aspects of unionization because names are almost invariably incorporated within biodiversity databases. That is, names can be adapted into a universal component of metadata. In a names-based infrastructure, names management is not coincidental to, but sits at the core of, handling data objects.

A names-based management system is assembled independently of the data objects that it serves to index. This separation or layering offers flexibility and a capacity to focus on optimizing the features of the indexing structure while minimizing compromises from other aspects. This separation is familiar to taxonomists who use different approaches for nomenclature and to make judgments about what are taxa.

The essential features of a names-based infrastructure are that: it can be applied to all life forms and so transcends the parochialism of the codes, it can index any data object for any organism, it will scale to handle any and all relevant data objects, it overcomes the "many names for one species" and the "one name for many species" problems that are discussed in the following paragraphs, it can discriminate among taxonomic concepts (Kennedy et al., 2005), it exploits the hierarchical structure (to disambiguate homonyms, browse content, drill toward more taxonomically precise groups, or expand searches or permit hierarchical aggregation of data), it benefits from improvements made in other online expert systems (such as nomenclators)—which will be part of a peer-to-peer federated information system; it allows the participation of experts and others for issues of quality control or to correct glitches that arise from algorithms that can at best only approximate to the complexity of biology, and through that participation captures the authority of the taxonomic community (Patterson et al., 2008).

The widespread adoption of Linnaeus's naming conventions has willed to us a very extensive system of metadata. The shift to the informatics use of names is not without challenges. Several problems confound an integrated names management system, especially when assisted through algorithms. The first is that many species have more than one name. A simple names-based data organization will not bring complementary data sets together because it is not aware that a piece of information labeled with the term "polar bear" is about the same species as data labeled with *Ursus maritimus* or *Ursus arctos*. Second, some names are used for more than one species, and simple names-based solutions, such as standard Internet search engines, will draw together information on *Aotus* the monkey with *Aotus* the legume. These are not new problems. Traditionally, taxonomists bring together data under different names, filter out information attached to homonyms, and then present appropriate data under a preferred name. A names-based infrastructure is designed to emulate these skills of taxonomists. Finally, a names-based infrastructure must embrace non-traditional names. High throughput molecular techniques are discovering new taxa, especially in the microbial realm, at a rate that can never be matched by the normal processes of taxonomic description. An increasing proportion of taxa will be known by their molecular signatures and not by Latin binomials. There is no reason that a names-based infrastructure cannot embrace such surrogates for names.

There are two solutions to the many names for one species problem. The first is for everyone to adopt the same unique standardized name for each taxon, with the correct name arising from an authoritative source. The online sources of correct names are nomenclators (such as Index Fungorum, ZooBank, or the International Plant Names Index) and aggregators (such as the Catalogue of Life). This approach is not feasible, sustainable, or scalable. It is not feasible because resolution of the "one-name-for-many-taxa" problem would lead to conflict with the codes of nomenclature. It is not sustainable because there is a substantial management cost to correcting the system to respect the 20,000 new or changed names (about 1% of all species) every year (Patterson et al., 2006; Pyle and Michel, 2008). Finally, in many cases, names cannot be corrected—such as the obsolete names embedded in older data objects. At the time of this writing, the Biodiversity Heritage Library has indexed about 30 million names instances, but only about a third of the names recovered occur in contemporary online sources of authoritative names. Very substantial resources are required if we are to incorporate the content of BHL into a well-indexed knowledge environment using the "standardized name" approach.

The alternative to using standardized names is to cross-map alternative names. This is the approach adopted by EOL, in which the names for the same taxon are linked by placing them in "reconciliation groups." A reconciliation group includes scientific names (whether code-compliant or not), lexical variants of names (including misspellings), out-of-date names, vernacular names, surrogates for names, and synonyms—both subjective (heterotypic) and objective (homotypic). Reconciliation groups allow data objects on the same species, but labeled with different names, to be brought together. A query using any name can be exploded to collect data under all names in the same reconciliation group. Names and their relationships can be flagged to explain why names are in the group, or to reveal what is currently regarded as the current and appropriate name for a taxon. The assembly of reconciliation groups is one of the most significant challenges for future taxonomy. About 20,000,000 scientific names have been compiled in GNI (www.globalnames.org) from repositories such as GBIF, nomenclators, uBio, NameBank, EOL, etc. To these have yet to be added a myriad of vernacular names. The challenges to place these into 2,000,000 or so reconciliation groups that represent the taxa will require innovative algorithms and community participation. Many of the names that we know about are minor lexical variants of each other—authors may or may not have dates, may or may not be abbreviated, species epithets have different endings, letters in Latin names have become transposed or distorted through transliteration or optical character recognition. The emergence of a suite of biological rules for language processing algorithms will help address these problems. The software must also be trained to understand synonymy lists, and be able to use them to assemble subjective synonyms into the same reconciliation groups. But, in the end, it will be the taxonomists and users of the names who will ensure accuracy, and they in turn will need suitably configured interfaces through which they can interact with the underlying names.

Reconciliation groups can be used to represent taxonomic concepts—that is, to provide the informatics foundation that allows experts to clarify what they believe a name represents. One expert can index data objects against "their" reconciliation group. A second expert is able to map a different subset of data objects to the same or different reconciliation group with the same or different preferred name. Through their indexing activities, experts express their concept of what a name refers to.

On the problem of homonyms, about 13% of the names of plant genera have homonyms in other areas (McNeill, 1997). Automated name recognition algorithms will encounter these and other identically spelled words (Virginia is the name of a genus of snakes, two U.S. states, a person's name, and is embedded in numerous vernacular names). Name discovery algorithms have to contend with legal and medical terms that look like species names (Anorexia nervosa, Habeas corpus). The incidence of similarly spelled words can be increased through misspellings, exclusion of authority information, abbreviations, or even the tendency of some environments, such as the image sharing environment Flickr, to modify the names (in this case, by removal of spaces between words in tags). Disambiguation is achieved through a variety of devices. Authority information added to names, the taxonomic location of the name, co-occurring terms in the data sources, or indeed the use of warning flags can all be used to help establish whether *Aotus* refers to the monkey or the legume.

NAMES ARCHITECTURE

The architecture being developed by EOL is summarized in Figure 11.1. Names are gathered from any source. The sources include nomenclators and aggregators, content partners, the Biodiversity Heritage Library, and databases. The process of gathering does not require any taxonomic scrutiny, and simply seeks to bring together every name that has been used to refer to a piece of information. Because of the scale of this task, the processes of gathering include the use of automated name discovery tools. Names are compiled within a name repository of all unique name-strings modeled on the uBio NameBank (www.ubio.org). This approach is complemented with an emerging Global Names Architecture (GNA) and its core element, the Global Names Index (GNI). GNI is a dynamic federated indexer of online name sources. Within this community of sources lie the nomenclators (such as ZooBank or Index Fungorum). GNI is designed as a semantically ready environment, allowing input from remote content partners through Web services that transmit key information using standards-compliant transfer protocols. Similarly, GNI is designed so that its content is accessible by other machines through appropriately configured APIs. That is, GNI has been designed as the first component of the semantic infrastructure for biology.

Once assigned to a repository, the names are to be added to a comprehensive hierarchical compilation referred to as "Union." Union holds names within reconciliation groups that are linked together by parent–child relationships to create the hierarchical component of the architecture. The parent–child structure allows for taxonomic concepts to be included. Homonyms and other

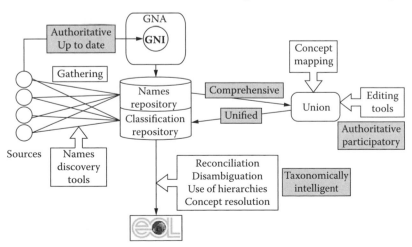

FIGURE 11.1 Outline architecture of names infrastructure.

spelled-alike names will occur in more than one reconciliation group. When coupled with thoughtful indexing, this achieves the solution to the "one name for many species" problem. Union is to be an editable environment that allows experts to correct, improve, add, or annotate the compilation.

Reconciliation groups are particularly important components of the architecture as they address the problem of "many names for one species." Names are assigned to reconciliation groups by algorithms or by participating experts. Reconciliation groups can be grouped in an unlimited array of taxonomic hierarchies such that the different opinions of taxonomists and users are accommodated. The architecture is designed to allow access to all classifications within a single navigational framework.

This approach is creating a system that is comprehensive, unified, participatory, and located in a distributed and federated names management environment. It is designed with the flexibility to express an unlimited number of taxonomic opinions, is authoritative, current, and overcomes the names problems.

BEYOND THE NARRATIVE

In addition to the names-based infrastructure, future taxonomy requires a framework for datacentricity. The solution must be capable of handling data for any species, draw information from any source, be easily translated from one format to another, and render the data in a semantically minimalistic form. Various metadata schemes do exist (such as the structured descriptive data (SDD), Species Profile Model, Plinian core etc.), but these are relatively flat approaches. Even with built in extensibility, flat schemas are not well suited to a landscape that is defined by the phylogenetic nature of characters and character states. They are also inherently constraining (Weinberger, 2007). Triples (Entity-Attribute-Value structures) offer a flexible alternative format that meets all of the requirements (Page, 2006). Within biology, triples would include a taxonomic identifier (whether for a species, a concept, or an individual), a second identifier for a character, and a value for the character (Figure 11.2).

The taxonomic identifiers of the triples can be provided through the names infrastructure, whereas a system of metadata for the characters has yet to emerge. The metadata terms for characters would have to be unambiguous within the context of machine-to-machine dialogue. Much of this information will be drawn from pre-existing databases, the literature, and other sources, each of which will

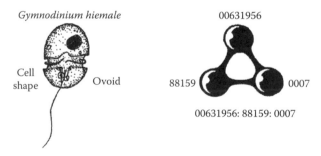

{3DF25O4E0-4F89-11D3-8A0C-0305E823307} 00631956:88159:0007

FIGURE 11.2 Example of tripletization of biodiversity information. A feature of an organism is converted into a globally unique numerical code. The "species" *Gymnodinium hiemale* is represented by a code (00631956) provided by a names or reconciliation-group registry system. The characteristic of cell shape is represented by a number (88159) drawn from an ontology. The final component represents the character state (0007 represents the state of ovoid for cell shape). The three codes when combined form the triple. 00631956:88159:0007 that carries the information in a semantic form that *Gymnodinium hiemale* has an ovoid body. The larger number {3DF25O4E0-4F89-11D3-8A0C-0305E823307} 00631956:88159:0007 includes a universally unique multidigit identifier that can be used to refer to registry and ontology systems, or attach attribution, and prevent duplication of the triple.

be idiosyncratic. A process of normalizing the identifiers is required. This could be done using the same options as with names: standardization through a controlled vocabulary, or taxonomically intelligent mapping of terms in a way that allows for reconciliation, disambiguation, and hierarchical relationships. By coupling Globally Unique Identifiers such as unique multi-hexadecimal numbers or Life Science Identifiers (LSIDs) to the triple (rendering them into quads—see Pyle and Michel 2008 for an example), more information can be linked to each triple (such as its derivation or comments that improve the quality of the content). Character and character states identifiers have phylogenetic relationships, and this property needs to be represented in the architecture. Building this system is no more challenging than building a system for names, and indeed can adopt similar devices to overcome the inevitable synonymies, homonymies, and hierarchical relationships. The bigger challenge is that of scale. Triple technology does not approach the capacity to handle the more than 10^{12} or 1 trillion triples that are expected to emerge from analyses of genomes and of traditional literature sources.

AND SO ...

The need for change within taxonomy was widely discussed a few years ago (Agosti and Johnson, 2002; Godfray, 2002; Saarenmaa, 2002; Stein, 2002). That message has been amplified because of the impact of the growing human population and its aspirations on the biosphere. Mechanisms to enhance taxonomic activities are now available, in part in the form of biodiversity informatics. Some taxonomists have demonized the situation, portraying the opportunities for which taxonomists are pre-adapted as competitive threats to their traditions (de Carvalho et al., 2008a, b). Taxonomists can and should add the new approaches to their traditions to improve the performance and relevance of taxonomy. We can and should train our descendants for an expanded role. If we do not, then the history of taxonomy is likely to report this as a period of missed opportunity.

REFERENCES

Agnarsson, I. and M. Kuntner. 2007. Taxonomy in a changing world: Seeking solutions for a science in crisis. *Systematic Biology*, 56, 531–539.

Agosti, D. and N. F. Johnson. 2002. Taxonomists need better access to published data. *Nature* 417, 222.

Bortulus, A. 2008. Error cascades in the biological sciences: The unwanted consequences of using bad taxonomy in ecology. *Ambio,* 37, 114–118.

Cantino, P. D. and de K. Queiroz. 2008. *PhyloCode: A Phylogenetic Code of Biological Nomenclature. Version 4b.* Available from: http://www.ohiou.edu/phylocode (accessed 20 October 2008).

Darwin, C. R. 1859. *On the origin of species by means of natural selection or the preservation of favoured races in the struggle for life.* London: Murray.

De Carvalho, M. R., F. A., Bockmann, D. S. Amorim, C. R. F. Brandão, M. de Vivo, J. L. and de Figueiredo, et al. 2008. Taxonomic impediment or impediment to taxonomy? A commentary on systematics and the cybertaxonomic-automation paradigm. *Evolutionary Biology*, 34, 140–143.

De Carvalho, M. R., F. A., Bockmann, D. S. Amorim, and C. R. F. Brandão. 2008. Systematics must embrace comparative biology and evolution, not speed and automation. *Evolutionary Biology*, 35, 150–157.

Freedberg, D. 2002. *The eye of the lynx.* Chicago: University of Chicago Press.

Godfray C. H. J. 2002. Challenges for taxonomy. *Nature* 417, 17–19.

Hopkins, G. W. and R. P. Freckleton. 2002. Declines in the numbers of amateur and professional taxonomists: Implications for conservation. *Animal Conservation*, 5: 245–249.

Kennedy, J. B., Kukla, R., and Paterson, T. 2005. Scientific names are ambiguous as identifiers for biological taxa: Their context and definition are required for accurate data integration. In Ludascher, B. and Rachsid, L. (eds). *Data Integration in the Life Sciences*, Springer. pp 80–95.

Leary, P. R., D. P. Remsen, C. Norton, D. J. Patterson, and I. N. Sarkar. 2007. uBioRSS: Tracking taxonomic literature using RSS. *Bioinformatics*, 23: 1434–1436.

Linnaeus, C. 1753. *Systema naturae, sive Regna tria naturae systematice proposita per classes, ordines, genera, and species.* Leiden, Netherlands: Theodorum Haak.

Mayr, E. 2004. *What makes biology unique?* Cambridge, U.K.: Cambridge University Press. 246 pp.

McNeill, J. 1997. Key issues to be addressed. In *The new bionomenclature: The BioCode debate.* Ed. D.L. Hawksworth. *Biology International* Special Issue 34:17–40.

Mendeleev, D. 1869. On the relationship of the properties of the elements to their atomic weights. *Zeitschrift für Chemie* 12, 405–406.

Page, R. D. M. 2006. Taxonomic names, metadata and the semantic web. *Biodiversity Informatics,* 3, 1–15.

Patterson, D. J., S. Faulwetter, and A. Shipunov. 2008. Principles for a names-based cyberinfrastructure to serve all of biology. *Zootaxa* 1950: 153–163.

Patterson. D. J., D. Remsen, C. Norton, and W. Marino. 2006. Taxonomic Indexing: Extending the role of taxonomy. *Systematic Biology,* 55: 367–373.

Polaszek, A., D. Agosti, M. Alonso-Zarazaga, G. Beccaloni, P. de Place Bjørn, and P. Bouchet, et al. 2005. A universal register for animal names. *Nature,* 437: 477.

Pyle, R. and Michel, E. 2008. Zoobank: Developing a nomenclatural tool for unifying 250 years of biological information, *Zootaxa* 1950: 39–50.

Remsen, D., C. Norton, and D. J. Patterson, 2006. Taxonomic informatics tools for the electronic *Nomenclator Zoologicus. Biological Bulletin* 210: 18–24.

Robertson, D. R. 2008. Global biogeographical data bases on marine fishes: Caveat emptor. *Diversity and Distributions* 14: 891–892.

Saarenmaa, H. 2002. Technological opportunities and challenges in building a global biological information infrastructure. In Saarenmaa H. and E. H. Nielsen (Eds). *Towards a global biological infrastructure.* Copenhagen: European Environmental Agency. 49–59.

Shrader-Frechette, K. S. and E. D. McCoy. 1992. Statistics, costs and rationality in ecological inference. *Trends in Ecology and Evolution* 7: 96–99.

Stein L. 2002. Creating a bioinformatics nation. *Nature* 417: 119–120.

Weinberger, D. 2007. *Everything is miscellaneous. The power of the new digital disorder.* New York: Times Books.

Wilson, D. S. 2003. *Darwin's cathedral: Evolution, religion, and the nature of society.* Chicago: University of Chicago Press.

Wilson, E. O. 2003. The encyclopedia of life. *Trends in Ecology and Evolution,* 18, 77–80.

12 The Encyclopedia of Life
A New Digital Resource for Taxonomy

James Hanken

CONTENTS

> Imagine an electronic page for each species of organism on Earth, available everywhere by single access on command. The page contains the scientific name of the species, a pictorial or genomic presentation of the primary type specimen on which its name is based, and a summary of its diagnostic traits. The page opens out directly or by linking to other databases, such as ARKive, Ecoport, GenBank and MorphBank. It comprises a summary of everything known about the species' genome, proteome, geographical distribution, phylogenetic position, habitat, ecological relationships and, not least, its perceived practical importance for humanity.... .
>
> The page is indefinitely expansible. Its contents are continuously peer reviewed and updated with new information. All the pages together form an encyclopedia, the content of which is the totality of comparative biology.
>
> **E. O. Wilson (2003)**

Human activities pose an ever-growing threat to biological diversity (Thuiller 2007). Population growth, global climate change, and other environmental perturbations heighten the urgency with which we must discover, formally describe, understand, and protect the world's species of living organisms. At the same time, the general public shows increasing interest in biodiversity and support for efforts to preserve it. In an attempt to respond to both of these imperatives, in May 2007 representatives of several of the world's leading natural history institutions, with initial financial support from two major private philanthropies, have joined together to create the Encyclopedia of Life (EOL).

The EOL will dynamically synthesize biodiversity knowledge about all known species, including their taxonomy, geographic distribution, specimens in collections, genetics, evolutionary history, morphology, behavior, ecological relationships, and importance for human well-being, and distribute this information freely through the Internet. It will serve as a primary resource for a wide audience that includes scientists, natural resource managers, conservationists, teachers, and students around the world. The EOL's broad scope and innovative tools will have a major global impact in facilitating biodiversity research, conservation, and education. Many countries, especially those in the developing world, regard biodiversity as an important element in their economic and societal development. The EOL will greatly facilitate access to biodiversity information in the countries where most of the planet's biodiversity is found and where it must be managed sustainably.

The work required to assemble and maintain the EOL is organized into a series of five subprojects, or components. Each subproject's activities are coordinated by a component group, which is hosted by one or more cornerstone institutions.* The overall effort, in turn, is project-managed by an administrative center, the EOL Secretariat. Specific goals and activities of each component group and the EOL Secretariat are summarized below. As these activities increase in scale and scope, more host institutions and other partner organizations, and possibly new component groups, will be added to make the EOL a truly international effort and to most effectively utilize the talents, knowledge, recommendations, and insights of the global biodiversity community.

SPECIES SITES GROUP

Online "species sites" form the core of the EOL (Figure 12.1). Individual sites are being assembled for each of the approximately 1.8 million named species of living organisms known at this time. Each site will consist of an entry-level species page designed for the general public, and more specialized resources for particular audiences. As new species are discovered and named, sites for each will be added. To assemble and manage these sites, the Species Sites Group:

- *Designs the "look and feel" and determines the recommended contents of entry-level species pages.* A standard configuration of each entry page specifies the placement of different kinds of information and the minimum content, and provides tools that allow users to optionally reconfigure a given page for specialized use.
- *Solicits content from potential data providers and users.*
- *Develops protocols for authenticating information on entry-level species pages.* Each entry-level species page will clearly differentiate authenticated from non-authenticated content. Scientific experts on particular taxonomic groups are being encouraged to serve as "curators" who will vet content, thereby helping to ensure that information on species sites is reliable and current.
- *Implements a robust intellectual property regime.* To the greatest extent possible, data provided through the Encyclopedia are intended to be freely available for all to use. Some data providers, however, choose to limit the reuse of information they provide. The Species Sites Group develops and implements mechanisms whereby use limitations are easily communicated to and understood by users. It also works to ensure that public credit is given to all data providers.
- *Develops specialized portals for different audiences.* For example, an educators' portal might provide access to special tools for classroom instruction or field trips. A birders' portal could showcase behavior of local species. Scientists and informatics specialists will be encouraged to develop new online tools for capturing and analyzing the enormous amount of basic biodiversity data that will be available via the Encyclopedia.

BIODIVERSITY INFORMATICS GROUP

To succeed, EOL requires novel informatics tools to capture, organize, and present knowledge about biodiversity. The underlying information technology (IT) infrastructure must be able to seamlessly

---

* Smithsonian Institution, The Field Museum of Natural History, Harvard University, Marine Biological Laboratory, Missouri Botanical Garden, and Biodiversity Heritage Library.

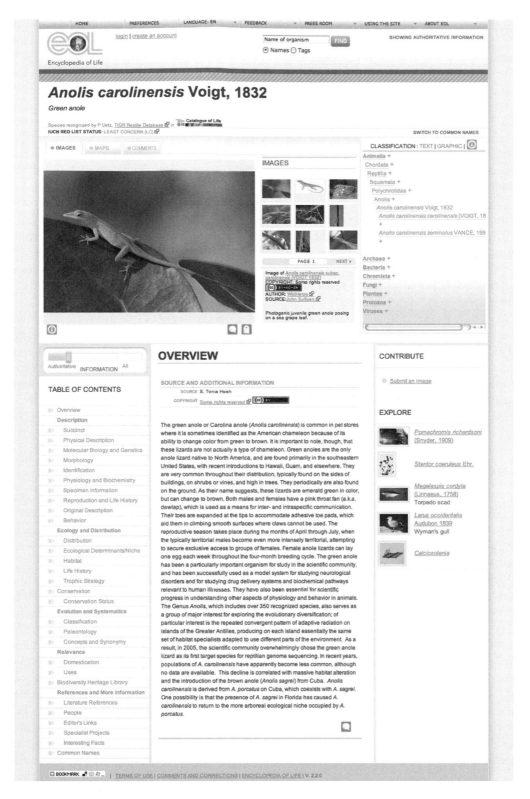

FIGURE 12.1 Entry-level EOL species page for the green anole, *Anolis carolinensis*, in February 2009. Links to additional content for this species are listed in the Table of Contents.

aggregate data from thousands of existing and future digital sources into individual species sites for expert and novice users alike. To achieve this functionality, the Biodiversity Informatics Group engages with the broader informatics community, collaborates with data providers, develops and employs software tools that exploit the full capacities of emerging Web technology, and places these tools in an open-source environment where they can be continually improved.

Aggregation ("mash-up") technology can assemble different data elements from remote sites and make these data available via a single Web portal (Butler 2006). Data aggregated in this manner on EOL, where it is organized by both taxonomy and subject matter, can be further manipulated through a range of tools and services now under development to index, organize, and associate data elements or create new elements. Such tools also include new methods of graphic analysis and visualization (e.g., Microsoft Live Labs Photosynth; www.photosynth.net/Default.aspx). The portal also provides services to both experts and enthusiasts upon whom the Encyclopedia depends for content and its authentication, and allows these and other users to reshape data for their own needs and for EOL Web pages. The software is modular and placed in an open content-management environment to guarantee that it continues to evolve. An attribution system acknowledges credit and shows data providers how their data are used. Information from EOL is delivered through a portal that is sufficiently flexible to allow users to tailor the display format and range of content to suit their needs, abilities, and personal preferences.

Working with partners such as the Species 2000 & ITIS Catalogue of Life (www.catalogueoflife.org/annual-checklist/2008/; Bisby et al. 2008), the Biodiversity Informatics Group is developing a global indexing system of scientific and common names for all organisms (Patterson 2003; Remsen et al. 2006). This list will be integrated within a Global Names Architecture, a shared virtual repository that links all major compilers of names to ensure that the EOL index is current, comprehensive, and authoritative. Finally, the Biodiversity Informatics Group actively supports efforts to develop standards for data interoperability and promotes their widespread adoption through TDWG (Taxonomic Database Working Group) and other standards bodies (Saarenmaa 2002).

SCANNING AND DIGITIZATION GROUP

The goal of the Scanning and Digitization Group is to make the published literature of biodiversity openly available online. Since most of this legacy literature presently exists only in print form, these works are being converted to digital form and made available to anyone via the Internet at no charge. These activities are being performed by an initial consortium of ten of the world's largest natural history libraries that together constitute the Biodiversity Heritage Library (BHL) (www.biodiversitylibrary.org).*

Benefits of the BHL are impressive and immediate. As of February 2009 more than 11 million pages, representing more than 28 thousand volumes, have been processed and are available for free and instantaneous download from the Library's Website (Figure 12.2). The resource is digitally linking the enormous biological diversity of tropical and developing countries to the corresponding scientific literature about biodiversity that is held primarily in a few major North American and European libraries. Such intellectual "repatriation" enables anyone who is not affiliated with major research and educational institutions to readily search, obtain, and fully utilize articles either previously unavailable to them or available with difficulty or at considerable cost. Educators are able to guide students' research projects with access to a wealth of examples incorporated in lesson plans and assignments. Scientists have immediate access to articles in the professional literature that are critical to their studies. Artists for the first time have routine access to inspiring illustrations in rare taxonomic works. Moreover, "taxonomic intelligence" software tools developed by the Biodiversity

---

* American Museum of Natural History; Field Museum of Natural History; Harvard University Botany Libraries; Harvard University, Ernst Mayr Library of the Museum of Comparative Zoology; Marine Biological Laboratory and Woods Hole Oceanographic Institution; Missouri Botanical Garden; Natural History Museum, London; The New York Botanical Garden; Royal Botanic Gardens, Kew; Smithsonian Institution Libraries.

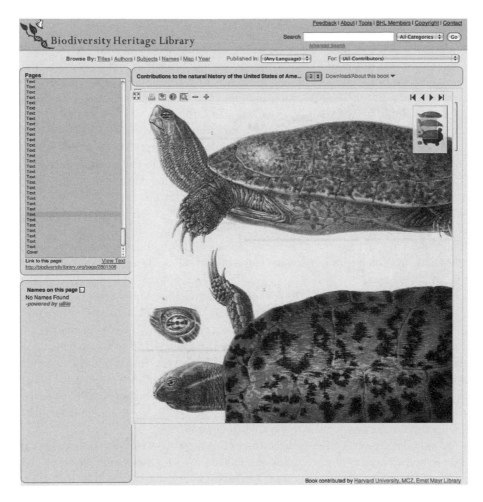

FIGURE 12.2 The Biodiversity Heritage Library provides immediate and unlimited access to high-quality digital reproductions of the legacy literature of biodiversity. Shown here is plate 27 from Louis Agassiz, *Contributions to the natural history of the United States of America, vol. II, pt. III, Embryology of the turtle,* published in 1857 by Little Brown and Company. This work is considered the acme of 19th-century zoological illustration (Blum 1993).

Informatics Group are being used to index each work in the BHL according to the scientific name(s) of species discussed within it, thus helping to unite users to their relevant literature, even if they don't know it exists (Patterson et al. 2006).

No single library within the BHL consortium holds the complete legacy literature of biodiversity, yet together their collections constitute a comprehensive assemblage. The BHL also provides a trusted body to negotiate permissions to legally reproduce works with the respective copyright holders. The Scanning and Digitization Group estimates that about 80 million pages of the core early literature may be freely scanned right now. The entire biodiversity literature—in all languages—likely totals 280–320 million pages. A single human-operated "Scribe" machine can scan and digitize about 70,000 pages per week. The Scanning and Digitization Group has already established EOL scanning centers in London, Boston, New York City, St. Louis, Urbana, and Washington, D.C. Additional centers are anticipated as scanning activity increases. By 2010, the Scanning and Digitization Group will offer approximately 25 million digitized pages of literature rich in taxonomic descriptions of biological species. For the first time, global audiences will have access to the core contents of our natural history museum and botanical garden libraries (Weissmann 2007).

EDUCATION AND OUTREACH GROUP

By putting a vast amount of information regarding all living species within easy reach of anyone with access to a computer, the EOL has enormous potential to enhance the teaching, understanding, and appreciation of biodiversity among the general public and to engender a greater sense of stewardship of life on Earth. The Education and Outreach Group seeks to realize this potential by exploring and promoting the use of EOL in educational settings.

Initially, the Education and Outreach Group is targeting three audiences where innovative uses of the Encyclopedia are especially likely. In formal education—ages 10–18 and university undergraduates—the EOL offers new access to data and novel ways to use them to answer questions about the distribution of species over space and time. In informal education settings, such as natural history museums and nature centers, the EOL will extend learning outside the classroom. Finally, citizen scientists will contribute content to the EOL by entering species information from activities such as community- or school-based "bioblitzes" and individual observations. In turn, by extracting relevant species pages to create personalized field guides, anyone who would like to do so will be able to enhance his or her knowledge of particular species and of biodiversity within the community.

Two pilot projects launched in fall 2008 suggest that the above approach is very promising. In the first, undergraduate mycology students at five U.S.-based universities were offered the opportunity to assemble entry-level pages for individual species of fungi as part of their regular class work (http://www.mushroomobserver.org/project/show_project/3). Fifty-two students accepted the assignment, and together they assembled accounts for 72 species. Drafts of each student's species page(s) were prepared on the Mushroom Observer Website, an online resource for both amateur and professional mycologists, where they were reviewed and the content authenticated by the student's professor. Later in 2009, these pages will be "harvested" by the EOL and made public on its Website, with appropriate credit given to the student curators.

In the second pilot project, EOL established an online mechanism for citizen-scientists anywhere in the world to upload images of any species of organism, which provides additional content for the corresponding EOL species site. Images are uploaded initially to the "Encyclopedia of Life Images" group on the Flickr Website (www.flickr.com/groups/encyclopedia_of_life), where they can be tagged by any Flickr member with various kinds of identifying information and other metadata. Those images that are tagged with a valid scientific name (viz., Latin binomial) are automatically harvested by EOL and posted as "not reviewed" content on the corresponding species site. Pending further review by the site's curator, individual images that are regarded as correctly identified to species are designated as such on the site. By February 2009, 599 members had uploaded 16,559 images.

BIODIVERSITY SYNTHESIS GROUP

The vast amount of data about living organisms that is being made routinely available by the EOL enables scientists to study biodiversity and the evolution of life in important new ways. These data also have the potential to facilitate work in applied fields, such as conservation biology, land-use planning, and environmental management. The Biodiversity Synthesis Group promotes such activities by organizing and assisting specialist working groups, convening cross-disciplinary conferences and workshops to explore integrative topics, and helping to develop relevant Web-based tools for integrating and analyzing species data. Its initial activities have addressed four primary themes:

- *Biodiversity in space and time.* Specialist working groups are developing new tools for mapping biogeographic data available via the EOL platform. There are, for example, unparalleled opportunities to integrate knowledge of biodiversity hot spots, collecting localities of museum specimens, geological history, and environmental variables in new ways.

- *The Tree of Life.* Phylogeny is the historical record of life's diversification and the conceptual framework for evolution. Additional working groups seek to integrate taxonomic and phylogenetic databases to both discover and illustrate trends in the evolution of biodiversity. For example, phylogenetic data provided by the already existing Tree of Life Web project (www.tolweb.org/tree/), when linked to the rich biological content on EOL species sites, offers a tremendous resource of interest to professional scientists, students, and lay people alike. One of the most exciting projects under way is the development of Web-based interactive visualization and navigational tools for large and complex evolutionary trees.
- *Discovering and describing biodiversity.* Vast numbers of unknown, unnamed species remain to be discovered and formally described. Working groups in this area integrate specimen data and taxonomic information for large, diverse, yet understudied or problematic taxa (e.g., decapod crustaceans, fungi, Diptera). An additional goal of this theme is to develop new digital tools to facilitate and quicken the recognition and naming of new species.
- *Conservation of biodiversity.* Synthesis of data sets on species distribution, evolutionary trees, genetics, and ecology can link biodiversity science and conservation. A deeper understanding of biodiversity should lead to improved management and conservation by governments, conservation organizations, and the general public.

SECRETARIAT

The role of the Secretariat is to coordinate and facilitate the entire EOL development process in consultation with the Encyclopedia's Steering Committee, Institutional Council, and Distinguished Advisory Board. The Secretariat oversees activities of the EOL's five component groups and, in doing so, helps to make E. O. Wilson's dream of the Encyclopedia of Life a reality. Specific activities include:

- Managing the entire EOL project and ensuring that core activities proceed effectively and synchronously.
- Providing administrative support to the Steering Committee, the Institutional Council, and the Distinguished Advisory Board.
- Recommending to the Steering Committee additional component groups or other mechanisms needed to achieve the EOL.
- Engaging additional funding and partnering organizations, and developing a long-term plan to sustain the EOL.
- Developing and implementing principles and policies of the EOL with regard to intellectual property, credit, and recognition.

CONCLUSION

Over the last several years there have been numerous calls for systematic biology, and taxonomy in particular, to embrace digital technologies and other computer-based protocols (e.g., Godfray 2002; Scoble 2008; Wheeler 2004, 2007). Such actions are seen as a means of increasing both the productivity and scientific impact of taxonomy—especially the rate of species discovery and description—through more effective data access and transfer as well as the application of new tools for biological discovery. Yet, despite the launch of several valuable and successful Web-based initiatives, such as the Global Biodiversity Information Facility (GBIF; Speers and Edwards 2008), and the almost routine application of various methods for digital measurement and imaging, taxonomy is far from realizing the dream of a "Cyberinfrastructure [that would] unite the world's museums and taxonomists into a seamless virtual biodiversity observatory ... and constitute the most powerful instrument ever conceived for taxonomic research and education" (Wheeler 2004: 579). Instead, taxonomy is frequently viewed, at least by those outside the professional community, as "one of the slowest disciplines to adopt computers" (Kelly 2008), a syndrome that reflects both practical and sociological constraints (Hine 2008). The Encyclopedia of Life is not intended exclusively, or even principally, to serve the needs of professional taxonomists. Yet, it can benefit this science enormously by providing immediate and convenient access to much of the information on which sound taxonomy is based, including the relevant scientific literature, collection data for specimens held in institutional repositories, gene sequences, and other essential data. The EOL, in turn, stands to benefit tremendously from the active involvement by and contributions from taxonomists. Indeed, the EOL requires such "buy-in" from this professional community if it is to attain its ambitious long-term goals.

Comprehensive, open-access data sets can facilitate scientific discovery in important ways that are not anticipated when the same data sets are proposed and initially assembled (e.g., Harris 2001; Mora et al. 2008; Raczkowski and Wenzel 2007; Schneider et al. 2007). In the same way, it is likely that taxonomy and the EOL, by working together, can generate new and synthetic knowledge about the world's biological diversity that neither approach could produce alone. Moreover, by empowering individuals around the globe with unparalleled access to knowledge and information, and by fostering international communication and collaboration, the EOL has the potential to transform the way all people think about, study, and appreciate biology and the world around us.

ACKNOWLEDGMENTS

James Edwards, Thomas Garnett, Graham Higley, Cynthia Parr, David Patterson, Marie Studer, and Mark Westneat contributed to the ideas expressed in this chapter. Breen Byrnes, Martin Kalfatovic, and Constance Rinaldo provided additional assistance and materials. The John D. and Catherine T. MacArthur Foundation and the Alfred P. Sloan Foundation have generously funded the startup costs of the EOL, which is being sustained with critical support from other donor institutions and individuals.

REFERENCES

Bisby, F. A., Y. R. Roskov, T. M. Orrell, D. Nicolson, L. E. Paglinawan, N. Bailly, P. M. Kirk, T. Bourgoin, and J. van Hertum, Eds. 2008. Species 2000 & ITIS Catalogue of Life: 2008 Annual Checklist Taxonomic Classification. CD-ROM; Reading UK: Species 2000.

Blum, A. S. 1993. *Picturing nature: American nineteenth-century zoological illustration.* Princeton, NJ: Princeton University Press.

Butler, D. 2006. Mashups mix data into global service. *Nature* 439:6–7.

Godfray, H. C. J. 2002. Towards taxonomy's "glorious revolution." *Nature* 420:461.

Harris, H. C., B. M. S. Hansen, J. Liebert, D. E. Vanden Berk, S. F. Anderson, G. R. Knapp, X. Fan, et al. 2001. A new very cool white dwarf discovered by the Sloan Digital Sky Survey. *Astrophys J* 549:L109–L113.

Hine, C. 2008. *Systematics as cyberscience: Computers, change, and continuity in science.* Cambridge, MA: The MIT Press.

Kelly, K. 2008. Technological twist on taxonomy. *Nature* 452:939.

Mora, C., D. P. Tittensor, and R. A. Myers. 2008. The completeness of taxonomic inventories for describing the global diversity and distribution of marine fishes. *Proc R Soc Lond B Biol Sci* 275:149–155.

Patterson D. J. 2003. Progressing towards a biological names register. *Nature* 422:661.

Patterson D. J., D. Remsen, C. Norton, and W. Marino. 2006. Taxonomic indexing—extending the role of taxonomy. *Syst Biol* 55:367–73.

Raczkowski, J. M. and J. W. Wenzel. 2007. Biodiversity studies and their foundation in taxonomic scholarship. *BioScience* 57:974–979.

Remsen D. P., C. Norton, and D. J. Patterson. 2006. Taxonomic informatics tools for the electronic *Nomenclator Zoologicus. Biol Bull* 210:18–24.

Saarenmaa, H. 2002. Technological opportunities and challenges in building a global biological information infrastructure. In *Towards a biological information infrastructure,* Eds. H. Saarenmaa and E. H. Nielsen, 60–72. Copenhagen: European Environmental Agency.

Schneider, D. P., P. B. Hall, G. T. Richards, M. A. Strauss, D. E.Vanden Berk, S. F. Anderson, et al. 2007. The Sloan Digital Sky Survey quasar catalog. IV. Fifth data release. *Astronom J* 134:102–117.

Scoble, M. J. 2008. Networks and their role in e-taxonomy. In *The new taxonomy,* Ed. Q. D. Wheeler, 19–31. Systematics Association Special Volume Series 76. Boca Raton: CRC Press.

Speers, L. and J. L. Edwards. 2008. International infrastructure for enabling the new taxonomy: The role of the Global Biodiversity Information Facility (GBIF). In *The new taxonomy,* Ed. Q. D. Wheeler, 87–94. Systematics Association Special Volume Series 76. Boca Raton: CRC Press.

Thuiller, W. 2007. Climate change and the ecologist. *Nature* 448:550–52.

Weissmann, G. 2007. Encyclopedias of life: From Diderot to the Yeti crab. *FASEB J* 21:2267–70.

Wheeler, Q. D. 2004. Taxonomic triage and the poverty of phylogeny. *Philos Trans R Soc Lond B Biol Sci* 359:571–583.

Wheeler, Q. D. 2007. Invertebrate systematics or spineless taxonomy? *Zootaxa* 1668:11–18.

Wilson, E. O. 2003. The encyclopaedia of life. *Trends Ecol Evol* 18:77–80.

13 Future Taxonomy Today
New Tools Applied to Accelerate the Taxonomic Process

Norman F. Johnson

CONTENTS

INTRODUCTION

One of the primary objectives of projects funded by the Planetary Biodiversity Inventory (PBI) Program of the U.S. National Science Foundation is the description of the species that compose a major taxonomic group. In groups such as insects, particularly those that occur outside of the north temperate zones, most such species are neither described nor even recognized among the holdings of collections. Furthermore, the characters that might be useful in distinguishing and characterizing these species are very imperfectly understood. An attempt to comprehensively describe the diversity of such taxa is a major scientific undertaking. Even though we have more than 250 years of experience in what is now called alpha-taxonomy, it remains a slow, labor-intensive undertaking. To compound this problem, the number of specialists willing and able to focus their research time on any single taxon is usually very limited. To aggressively pursue large-scale species discovery and description, the taxonomic community needs tools that accelerate the taxonomic process and maximize the efficiency of the time that researchers are able to invest while maintaining high standards of accuracy and completeness. This has been one of the objectives of the Platygastroidea PBI.

The superfamily Platygastroidea is a monophyletic group of parasitoid wasps (Murphy et al., 2007) comprising 5,065 species currently considered to be valid. My colleagues and I recently published a taxonomic revision of *Heptascelio* (Hymenoptera: Platygastridae) (Johnson et al., 2008), a genus of parasitoids of the eggs of acridoid grasshoppers. Prior to our study, *Heptascelio* was known from three species, one described in 1916 from the Philippines and two described in 1996 from southern India (Kieffer, 1916; Narendran and Ramesh Babu, 1996) (Figure 13.1).

FIGURE 13.1   The genus *Heptascelio*. 1, *H. lugens* Kieffer, holotype male; 2, *H. striatosternus* Narendran and Ramesh Babu, holotype female; 3, *H. punctisternus* Narendran and Ramesh Babu, holotype male; 4, Collecting localities of 684 specimens of *Heptascelio*. Scale bars in millimeters.

In fact, as far as could be determined from the published literature, these three species were known from only the three holotype specimens. Following our revision, we now believe that this group is widespread in the Old World, extending from West Africa in Cameroon and Gabon east to Irian Jaya.

The two species from India turn out to be synonyms, one the male and the other the female of the same species. In addition to the two previously described valid species, our revision added an additional 16 species new to science. Such a high ratio of new to "known" species, heavily skewed toward the new, is typical in this group of insects, repeated in nearly every genus in every part of the world. For example, in recent comprehensive revisions, the proportion of previously described valid species to the total number of valid species ranges from 6–33% (Early et al., 2007; Johnson and Masner, 2004, 2006; Taekul et al., 2008; Yoder et al., 2009). Given that more than 5,000 species of platygastroids have been described, it appears that the total diversity in this one relatively small corner of the insect world is on the same scale as even the largest classes of vertebrates.

The platygastroid PBI team is made up of 28 people from 13 countries and 6 continents (Figure 13.2). However, this number cannot be taken simply at face value. The taxonomic expertise is heterogeneous, and some have but limited experience with this particular group. Further, many of them have other responsibilities associated with their professions: teaching, advising students, collection curation, and so on. Thus, it is imperative to maximize the efficiency with which the limited amount of time available for this work is used. The goal of this chapter is to describe the tools and resources that we have developed as part of this project to facilitate the taxonomic process and to increase its efficiency and productivity.

TAXONOMIC WORKFLOW

Figure 13.3 illustrates a model of the taxonomic process. This is presented not as a definitive and comprehensive proposal, but as a tool to help identify the bottlenecks involved, points

FIGURE 13.2 The members of the Platygastroidea PBI team at the inaugural workshop held in Columbus, Ohio, 5–9 February, 2007. From left: E. Talamas, J. Cora, B. Coelho, M. Dowton, A. Aguiar, L. Musetti, Rajmohana K., N. Johnson, D.C. Darling, A. Polaszek, L. Masner, M. Yoder, J. Early, I. Mikó, A. Austin, J. Jennings, S. van Noort, H. Klebsch (not present in photo).

at which the development of appropriate tools could accelerate the process and improve its efficiency and accuracy.

The taxonomic revision of a group is built from two resources: samples of the organisms involved (either specimens or observations) and the legacy of taxonomic investigation found in the published literature. Specimens may already be present in the holdings of natural history collections around the world, or they may be acquired through new collecting activities. The literature provides hypotheses of taxa, characters to discriminate among taxa, and records of the distribution in time

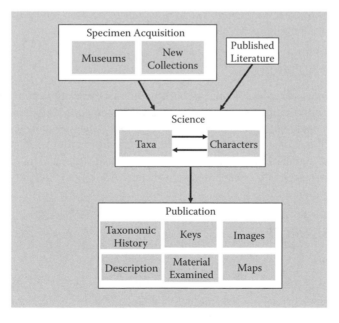

FIGURE 13.3 Model of taxonomic workflow.

and space. The process described here is focused on groups where such a preexisting taxonomic framework is minimal. The two resources, specimens and literature, provide the raw materials that feed into the scientific process of taxonomy.

The core of a revision is encapsulated in a feedback loop between hypotheses of characters and hypotheses of taxa. The investigator attempts to allocate specimens into taxa on the basis of empirical observations of the characters they exhibit. These taxonomic concepts and the degree to which specimens are congruent with them may suggest new characters and character states. The hypothesis of new characters may lead to modifications in the proposed taxa. The process is iterative and open-ended. Acquisition of new material may force the taxonomist to begin the whole process anew. At some point, the process reaches a stable equilibrium, and the results are formally presented to the scientific community.

Six elements are common components of taxonomic publications: a summary of the published literature; an identification key; formal descriptions (and diagnoses) of the taxa; a characterization of the material examined for the study; extensive imagery; and distribution maps. The literature review is often presented in the formalized structure of a synonymic list, and the individual entries may be briefly annotated. Identification keys first began to be commonly included in the entomological literature in the mid- to late-19th century, as the numbers of known taxa made some sort of mnemonic aid essential. Dichotomous keys are the traditional and still most widely used format (Walter and Winterton, 2007). The taxonomic description is a listing of the character states exhibited by the taxon for each character. A diagnosis is a condensed listing of character states, concentrating on those that "best" distinguish the taxon at hand from those others, usually explicitly cited, with which it is most likely to be confused. The section entitled "Material examined" serves to document, in more or less detail, the specimens upon which the study is based. For newly described taxa, these specimens make up in whole or in part the so-called type series. Images are extremely valuable adjuncts to textual descriptions. Finally, illustrations of the geographic distributions translate the data on collecting localities from the material examined into a visualization that should be easily understood.

The quality of a taxonomic publication is a function of the quality of each of these steps in the taxonomic process. Our objective has been to develop the tools and resources to both establish and maintain high standards for each of these in both accuracy and completeness. Additionally, though, time is the critical limiting resource, and we have sought to develop methods that will reduce the investment in time while maintaining quality.

MATERIALS AND METHODS

The tools and resources below are primarily built upon an Oracle implementation of our revised version of the ASC information model (Association of Systematics Collections, 1993). Although the database schema could be implemented in any relational database management system, Oracle includes a procedural language, PL/SQL, that enhances the programming capabilities beyond that available through the standard SQL (structured query language) alone. The code is compiled and stored within the database itself, enhancing performance. Additional functionality is built on Perl and JavaScript programming.

TOOLS AND RESOURCES

PUBLISHED LITERATURE

The published record of taxonomic studies is unusual in two respects. First, because of the importance of the Principle of Priority in the International Code of Zoological Nomenclature, the entire corpus of literature is relevant, or at least that beginning with the publication of the tenth edition of *Systema Naturae* (Linnaeus, 1758). Second, the literature is highly dispersed. For the first century or so following Linnaeus, the primary mode of publication was an individual book. Many of these early publications are now rare and difficult to access in even the best libraries of the world. Since then, the journals of learned societies and institutions have proliferated widely. Taxonomic publications on the Platygastroidea are dispersed among 402 serials published in at least 20 languages. Very few individual libraries have all of these resources. Therefore, it is a significant task for taxonomists to locate and obtain the literature that documents the taxonomic study of their group.

For the Platygastroidea, the catalogues of taxonomic literature were last published in the early 1990s (Johnson, 1992; Vlug, 1995). The content of these publications has been recorded in the database, with the fundamental elements including the taxonomic name as used in the publication, the complete citation data, and annotation concerning the use of the name in an individual publication. These data have been updated and maintained, and are accessible through the Web portal Hymenoptera On-Line, http://purl.oclc.org/NET/hymenoptera/hol). Currently, a total of 10,026 names of platygastroid wasps are recorded, annotated, and linked to citations in the systematic literature. Assembling this body of literature, working with the not inconsiderable assets of The Ohio State University Library system, required more than five years of work. All of these documents are now stored as .pdf files (portable document format). As a result, no future workers need to make the same investment of time, money, and effort in gathering the published taxonomic literature.

For the Platygastroidea alone, the corpus of systematic literature is composed of 1,281 articles, books, and chapters, comprising a total of 56,110 pages. Of that total, 93% of the documents have been scanned, converted into .pdfs, and linked to citations within the database. To facilitate online access and reduce bandwidth use, each document is stored both as a single, sometimes fairly large, file and as individual pages. A simple application allows a user to view an individual page in a Web browser and to navigate from one page to another. Currently, the .pdf files present simple images of the pages. In collaboration with plazi.org, we are working to convert these images into text files using optical character recognition software, and then to annotate the internal structural elements of the document using the TaxonX schema (http://research.amnh.org/informatics/taxlit/schemas). With a very small number of exceptions, copyright holders have agreed to make these documents freely available.

SPECIMEN DOCUMENTATION

The traditional standard protocols for specimen documentation in collections differ among taxonomic groups. In studies of insects, the researcher typically has to process hundreds or thousands of specimens. With very few exceptions, the institutions from which this material is borrowed have not formally acquisitioned and recorded the data associated with individual specimens. In some cases, the data associated with specimen lots—that is, specimens in the aggregate—are recorded and occasionally such acquisition numbers are indicated on specimen labels. But in the vast majority of cases, there is no record, digital or otherwise, of the data associated with individual specimens.

Therefore, in our taxonomic work, we routinely add a label to each pinned specimen that individually identifies it. The label conforms to the recommendation of the Entomological Collections Network (Thompson, 1994) and consists simply of a coden (usually, OSUC, standing for Ohio State University Collection) and a number, presented both in human-readable form and as a machine-readable barcode (Code 49 or DataMatrix). The coden does not indicate ownership of either the specimen or the data, but rather points to the institution where the data were initially recorded and are available. The OSUC+number serves as the primary key for specimen data within the database.

The label data are typically recorded verbatim in a spreadsheet. In the case of languages that do not use a Latin alphabet, we photograph the labels for later transcription or transliteration. If geographic coordinates are not included on the label, then the collecting locality is georeferenced, making use of online gazeteers. The data in the spreadsheet is imported into a Perl application, the Data Entry Assistant, that enables the user to check the data against the database and enter the record into the database. Individual data records can then be accessed through a Web portal by searching for the code+number combination.

This protocol is a significant investment in time and effort, but we view it as a critical element in the taxonomic process. It provides a ready mechanism for keeping track of specimens that may have some peculiarities in characters. Once georeferenced, the distribution can be easily visualized and compared with other hypothesized taxa. Correlations between characters and with temporal or spatial distribution can be examined. Finally, at a practical level, tracking of individual specimens facilitates the management of loans from multiple institutions.

Documentation of specimens through images is important in both the process of research and of publication. For specimens in the size range of platygastroids, generally 0.5–5 mm in total length, light photography of pinned specimens usually has insufficient depth of field. Traditionally, that leaves two general means of illustration: drawing and scanning electron micrographs. Both have their strengths, but both are time-consuming and either expensive (SEM, professional artists) or dependent upon the artistic abilities of, often, the scientist. Over the past decade, the combination of increased and affordable computing capacity, high-quality digital cameras, and extended-focus software have made imaging of even these small specimens fast (< 1 minute) and relatively inexpensive. Portable imaging systems allow the scientist to take photographs of specimens when visiting other collections.

It has become so easy to take high-quality photographs that managing and archiving the files is a significant issue. One set of images from our project, especially those that appear in the finished publications, are archived at Morphbank (http://www.morphbank.net). For local use however, particularly with large numbers of images, we have developed a simple image management system called Specimage (Figure 13.4).

This is a small Perl/Javascript application that is integrated with the taxonomic and specimen information in the primary database. Upon submission, an image is automatically converted and stored in three formats: the full-resolution image as submitted (usually in TIFF or BMP format); a compressed .jpg file suitable for viewing in a Web browser, and a thumbnail version. Metadata on the body parts in the picture and the view angle, together with the taxonomic name (or working name for provisional taxa) and specimen bar-code identifier, can be used to search for pictures of interest. Images may be organized by specimen, arrayed in a matrix of thumbnails, or pictures of two taxa may be viewed side-by-side. This last option enhances the ability to compare the characters of one specimen with another.

TAXON-CHARACTER LOOP

This is the scientific core of a taxonomic revision. Taxonomic concepts and characters are both sets of hypotheses, and the underlying assumptions are that these two sets are mutually compatible and that there exists a set of character states that is unique to each taxon. Generating such hypotheses is a creative process, the intuitive recognition of patterns among the observations at hand. As earlier noted, this is an open-ended process, for there are no criteria by which to judge whether all taxa or all characters have been found. As a result, this is probably the most significant rate-limiting step in the taxonomic process. Coming to the decision to exit this process, to conclude that the results are sufficiently close to the "truth," is often difficult and painful. Authors who publish prematurely may be roundly and appropriately criticized for superficial character analysis or insufficient study material. In these cases, at least, the results are available for criticism, improvement, or rejection. Failure to break out of the taxon-character loop can result in inordinate delays in publication or hoarding of specimens. Somewhere between these two extremes is a happy medium.

The goal, then, becomes one of increasing the efficiency at which we exit the taxon-character loop in a reasonable amount of time at an appropriate level of analysis. Our approach has been to do this by increasing the number of participants in each individual taxonomic project. This may sound counterintuitive in that more opinions can lead to more disagreement, but more input can also reduce

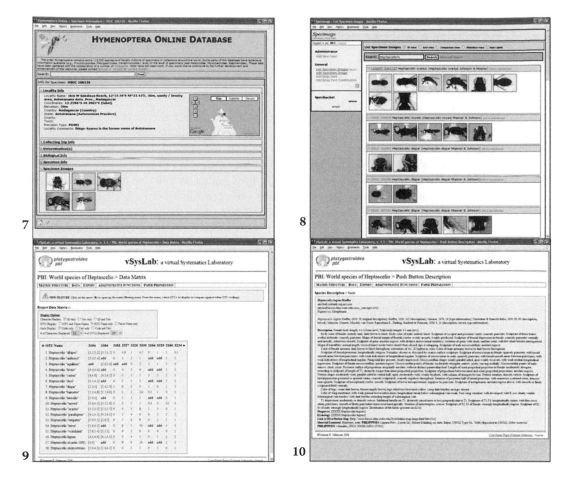

FIGURE 13.4 Screen shots of Web-accessible taxonomic database applications. 7: Hymenoptera On-Line, showing data for selected individual specimen; 8: Specimage, image database displaying thumbnails of available images of species of *Heptascelio*; 9: vSysLab, data matrix view; 10: vSysLab, displaying output of automated species description.

the amount of time spent on unproductive lines of inquiry and more creative minds can apply their acumen to resolve problems. The traditional approach to taxonomy, particularly in groups in which the number of researchers is quite small relative to the number of taxa, is for an individual to accumulate a body of research material and to work in isolation on these specimens. In the 250 years of taxonomic effort since Linnaeus (1758) this methodology has resulted in the recognition and characterization of hundreds of thousands of species. Yet our experience in Platygastroidea demonstrates that this effort of ten or more generations has not completed even one half of the task of discovery and description of the world's biodiversity. We can at least afford to try alternative methods.

To bring together researchers around the world to collaborate on individual taxonomic projects, we have deployed a Wiki site, that is, a Web site at which registered persons can easily contribute text, files, and images for other users to see. One important aspect of such a site is the history of changes to each page, so that when changes are made it remains possible to retrieve all previous versions. The site is secured by user names and passwords, appropriate for an environment in which ideas are to be freely proposed, criticized, and improved. Each taxonomic project is set aside as its own set of pages. Within them, taxonomic hypotheses can be proposed, described, and freely illustrated, particularly by embedding images and annotating those images. Thus, users on opposite sides of the world can do the equivalent of sharing a microscope and specimen to propose, discuss, hone, and improve ideas. The key issue is ease of use for the contributor and a collaborative ethos among all participants.

CHARACTER MATRIX DEVELOPMENT

A number of computer applications have been developed to facilitate the recording of characters exhibited by taxa and the production of natural-language taxonomic descriptions. In our project we opted to independently develop such a facility as a database application so as to integrate those functions with the existing specimen, taxonomic, and literature data. The application, vSysLab, has a Web interface in which the user can add or edit taxa, either as existing taxonomic concepts or as working taxa, or add or edit characters, including both continuous data or characters divided into discrete states. Images can be associated with character states to clarify the meaning of text descriptions. With the Web interface, any user with the proper credentials can add or edit data from anywhere in the world. Again, this facilitates collaboration among workers who are physically separated.

The data can be viewed either as lists of taxa or characters, or as a two-dimensional taxon character matrix. In the latter view, missing data are highlighted with links to functions to score that taxon for that character. Ordering of characters and taxa may be specified, and characters may be grouped into clusters (e.g., head, thorax, abdomen). The resulting matrix can be exported in several forms: as natural language text, a Nexus file, in Lucid Interchange Format (for the multi-entry key program Lucid), or as an .xml file using the Structure of Descriptive Data schema (SDD). The natural language option produces output in the form of "Character: Character state(s)." The degree to which this text actually is natural language depends on the form in which the textual descriptions of each character and state were originally defined.

This "Character: Character state" format can be viewed as repetitive or overly verbose. It has two important advantages though. First, it makes it clear what the hypothesized character is and what the character states are. For example, character 1 may be the color of the first abdominal segment and character 2 the color of the second segment. If the color of the two segments is identical, it might be more succinct to state "segments 1 and 2 red" rather than "color of segment 1: red"; "color of segment 2: red." However, in the first formulation, the reader may be uncertain whether the character identified by the author as important is the individual colors of the two segments, or the fact that the two of them are the same color. In some circumstances, the possible misinterpretation may be important. Second, documenting the characters in a verbose, but explicit format substantially increases the feasibility with which such texts may be automatically analyzed and mined for information. A consistent and predictable structure will facilitate this option.

Accurate recordkeeping is not the only advantage to this approach to character coding. Since missing data are easily seen in the matrix, this makes it easier to be certain that, insofar as possible, all taxa have been scored for all characters. A complete matrix is most useful for multi-entry identification key development. Additional functions enable the user to specify a subset of taxa and generate a list of the characters in which they differ. One can then critically compare the differences between very similar species and assess whether the hypothesis that they are different species is actually well supported by the data at hand.

TAXONOMIC TREATMENTS FOR PUBLICATION

From the export of natural-language textual descriptions from the database, it is but a short step to the automatic production of most of the elements of a taxonomic treatment. A basic set of such elements includes (1) the name of the taxon under consideration; (2) a synonymic list; (3) the description; (4) a diagnosis; (5) the etymology of a newly proposed name; (6) a distribution map; (7) a listing of the material examined; and (8) commentary.

One further export function from vSysLab is a "push-button description." Using this, most of these elements are generated directly from the database. The name of the taxon is straightforward, and the database generates a unique Hymenoptera Namer Server life-sciences identifier (LSID) for that taxonomic concept. A synonymic list is generated from the taxonomic portion of the database, with catalog-style abbreviated citations and the annotations of citation content. The description is produced as described earlier. The diagnosis is left blank as a placeholder. The characters that distinguish the taxon at hand from any other are easily extracted using vSysLab. Similarly, the etymology and comments are left for the author to craft. A Web link to a distribution map is automatically generated. The coding behind this link extracts the georeferenced coordinate data and passes them on to Google Maps to produce a display of the distribution of collecting localities for that taxon. Finally, the material examined section is automatically produced. Where relevant, the data for the holotype are reported first, followed by a summarized presentation of all other specimens. These specimens are reported by their barcode identifiers, sorted by country and depository. The full data for each of the specimens or for the entire data set can then be retrieved by querying the Hymenoptera On-Line Web service to which the data are linked.

Automating this process contributes more than speed enhancements. Consistency is vastly improved, as characters are always reported in the same predefined order for each description within a larger taxonomic work. "Material-examined" sections are notoriously prone to error, particularly when an author attempts to condense text so as to conserve space. The details in these data are usually not critical to the taxonomic work at hand, so removing the data from the text and storing them remotely, essentially as an appendix, maintains access, significantly conserves space, and eliminates the type of clerical errors that plague such listings. This is consistent with practices in other scientific disciplines in which the raw data themselves are rarely included in the publication; rather, a summary is provided. Importantly, though, providing an explicit listing and link to the specimen data, including the unique identifier for the specimen and the depository in which it is found, maintains a critical audit trail, documenting the factual basis upon which scientific conclusions are based.

Although several of the elements of our database applications have been described as important in the production of the finished publications, their utility is not limited to recordkeeping. Visualizing geographic distributions is very helpful in the development and testing of taxon hypotheses. The Specimage software allows the scientist to quickly compare specimens side-by-side on a computer screen. The comparison function in vSysLab contrasts the entire set of characters among designated taxa, permitting a quick analysis of the level and nature of support for separating potential species. All these tools were primarily designed to facilitate taxonomic research, enhance the level of accuracy and completeness, and encourage collaboration among researchers.

SOFTWARE ACCESSIBILITY

Each of the components of the software tools is either publicly accessible or guest user accounts with limited privileges are available for potential users to explore the system. The components can be viewed at the following persistent URLs by appending the name of the component (in the right-hand column) to the base URL http://purl.oclc.org/NET/hymenoptera/

Hymenoptera On-Line (access to specimen, taxonomic, and literature data)	hol
vSysLab (taxon character matrix development and manipulation, descriptions). Username: ASPGUEST; password: ASPGUEST	vsyslab
Specimage (image management and archiving)	specimage
Hymenoptera Name Server (taxonomic data)	hns

Both the database schema and code are available on request. The fundamental components needed to deploy the system are the Oracle relational database management system, a Web server, and the Perl scripting language.

THE FUTURE OF TAXONOMIC RESEARCH

Even as we have developed tools and resources to facilitate production of taxonomic papers, it has become clear that the time is ripe to enhance both the content and function of such documents. As taxonomy becomes a more collaborative endeavor, the number of people who contribute in a meaningful way to the results of the research increases, and the number of authors should increase in turn. In our initial project publications, we have taken an intentionally expansive approach to authorship, a practice that brings us more in line with the standards in other fields of science. To recognize and address the variation in the roles and contributions of these authors, we have adopted the practice of more explicitly stating the parts of the project in which each author participated. Similarly, the question of authorship of taxa arises. We have not shied away from assigning multiple individuals as such authors and of changing both the names and the order of authors' names for different species. Insofar as taxon authorship is meant to reflect the intellectual contributions of individuals, this seems like the most appropriate approach. There are no clear criteria by which to make such determinations however, and perhaps a more inclusive approach would be to give authorship to a named group of individuals (e.g., the Platygastroidea PBI team). Whether this is consistent with the ICZN remains to be tested.

The mode of publication of taxonomic work is strongly influenced by the requirements of the Code, and these have predominantly been implemented as traditional ink-on-paper publications. The advantages to such methodology include the relative permanence of the medium over the long term, the need to have a definitive initial version of the proposal of new taxa, and its accessibility to readers without the requirement of special equipment. Traditional publications, however, have disadvantages. The production of color illustrations is relatively expensive and kept to a minimum. Publications quickly go out of date as new discoveries are made, often even during the time period required for printing and distribution. Multi-entry keys cannot be implemented. Navigation through documents can be difficult. Libraries are always pressed for funds, and with journals becoming more expensive, subscriptions are decreasing. Therefore, availability becomes an issue. Under the current version of the Code, some publishers are working around the requirements by simultaneously publishing traditional hard-copy and .pdf versions that are identical in terms of the printed text. The .pdfs, however, can include links to external supplemental resources that can be activated by users connected to a computer network.

The cyberinfrastructure we have developed to support our work demonstrates that many of the components of taxonomic publications can be implemented as computerized database applications. From this it is but a short conceptual step to entirely electronic, dynamic publications. The mode in which such "papers" would be consumed includes not only reading of text by a person, but automated extraction of content by data-mining software. Such extraction can be greatly facilitated by the implementation of developing biodiversity informatics data standards. Schemas implemented by our Platygastroidea PBI database applications encompass most of the types of data in taxonomic publications, including the Darwin Core (DwC) for specimen collecting data; Distributed Generic Information Retrieval (DiGIR) for querying databases for specimen data; Structure of Descriptive Data (SDD) for characters of taxa; Metadata Object Description Schema (MODS) for literature and other bibliographic data; the Taxonomic Concept Schema (TCS) for taxa. Life sciences identifiers (LSIDs) serve as the mechanism for defining globally unique identifiers for all of these elements. Finally, schemas such as TaxonX can be used to identify the major components of taxonomic papers and thus help to accurately locate information relevant to a query.

Discussions have now begun on developing a new edition of the Code, a version that can take advantage of the capabilities offered by electronic forms of publication while retaining the stability and authoritativeness of hard-copy publication. If done with care, the taxonomic endeavour can be significantly accelerated toward achieving its long-term goals of discovering, delimiting, and describing the entirety of our planet's biota.

ACKNOWLEDGMENTS

Thanks to the entire Platygastroidea PBI team: A. D. Austin, A. P. Aguiar, F. Bin, S.- P. Chen, H. Clebsch, B. Coelho, J. Cora, D. C. Darling, M. Dowton, J. Early, A. Guidotti, S. Hemly, J. Hoke, J. Jennings, H. Klompen, L. Masner, I. Mikó, L. Musetti, A. Polaszek, Rajmohana K., R. Romani, C. Taekul, E. Talamas, A. Valerio, S. van Noort, and M. Yoder. I wish to express particular appreciation to L. Musetti for essential input and advice on all aspects of the project, to A. D. Austin, co-principal investigator on the grant, J. Cora for taking over the computer end of the project, and A. Polaszek for the invitation to participate in the symposium. This material is based on work supported by the National Science Foundation under grant Nos. DEB-0344034 to N. F. Johnson and DEB-0614764 to N. F. Johnson and A. D. Austin.

REFERENCES

Association of Systematics Collections. 1993. Committee on Computerization and Networking. An information model for biological collections. http://www.nscalliance.org/bioinformatics/asc%20model/Ascmodrpt.pdf.

Early, J. W., L. Masner and N. F. Johnson. 2007. Revision of *Archaeoteleia* Masner (Hymenoptera: Platygastroidea, Scelionidae). *Zootaxa* 1655:1–48.

Johnson, N. F. 1992. Catalog of world Proctotrupoidea excluding Platygastridae. *Memoirs of the American Entomological Institute* 51:1–825.

Johnson, N. F. and L. Masner. 2004. The genus *Thoron* Haliday (Hymenoptera: Scelionidae), egg-parasitoids of waterscorpions (Hemiptera: Nepidae), with key to world species. *American Museum Novitates* 3452:1–16.

Johnson, N. F. and L. Masner. 2006. Revision of world species of the genus *Nixonia* Masner (Hymenoptera: Platygastroidea, Scelionidae). *American Museum Novitates* 3518:1–32.

Johnson, N. F., L. Masner, L. Musetti, S. van Noort, Rajmohana K., D. C. Darling, A. E. Guidotti, and A. Polaszek. 2008. Revision of world species of the genus *Heptascelio* Kieffer (Hymenoptera: Platygastroidea, Platygastridae). *Zootaxa* 1776:1–51.

Kieffer, J. J. 1916. Neue Scelioniden aus den Philippinen-Inseln. *Brotéria* 14:58–64, 172–187.

Linnaeus, C. 1758. *Systema naturae. Regnum Animale.* 10th ed. W. Engelmann, Lipsiae.

Murphy, N. P., D. Carey, L. R. Castro, M. Dowton, and A. D. Austin. 2007. Phylogeny of the platygastroid wasps (Hymenoptera) based on sequences from the 18S rRNA, 28S rRNA and cytochrome oxidase I genes: Implications for evolution of the ovipositor system and host relationships. *Biological Journal of the Linnean Society* 91:653–669.

Narendran, T. C. and M. G. Ramesh Babu. 1996. On the systematics of *Heptascelio* Kieffer (Hymenoptera: Scelionidae). *Uttar Pradesh Journal of Zoology* 16(2):89–93.

Taekul, C., N. F. Johnson, L. Masner, Rajmohana K., and Chen S.-P. 2008. Revision of the world species of the genus *Fusicornia* Risbec (Hymenoptera: Platygastridae, Scelioninae). *Zootaxa* 1966:1–52.

Thompson, F. C. 1994. Bar codes for specimen data management. *Insect Collection News* 9:2–4.

Vlug, H. J. 1995. Catalogue of the Platygastridae (Platygastroidea) of the world (Insecta: Hymenoptera). *Hymenopterorum Catalogus* 19:1–168.

Walter, D. E. and S. Winterton. 2007. Keys and the crisis in taxonomy: Extinction or reinvention. *Annual Review of Entomology* 52:193–208.

Yoder, M. J., A. A. Valerio, L. Masner, and N. F. Johnson. 2009. Identity and synonymy of *Dicroscelio* Kieffer and description of *Axea*, a new genus from tropical Africa and Asia (Hymenoptera: Platygastroidea: Platygastridae). *Zootaxa* 2003:1–45.

14 The All Genera Index
Strategies for Managing the BIG Index of All Scientific Names

David Remsen

CONTENTS

In their book *Evolution of the Insects* authors David Grimaldi and Michael Engel (2005) acknowledge that

> All accumulated information of a species is tied to a scientific name, a name that serves as a link between what has been learned in the past and what we today add to the body of knowledge.

In the two and half centuries since Linnaeus established a formalized process for naming and classifying species, scientific names have become the primary means for referencing a taxon whenever a piece of information is intended to refer unambiguously to a particular type of organism or group of organisms. The creation and management of these names is governed by formalized codes of nomenclature that seek to ensure conformity, mediate conflict, and ensure persistence and accuracy (ICZN, 1999). As a result, some of the names originally published in the 10th edition of *Systema Naturae* over 250 years ago (Linnaeus, 1758) continue to be used to refer to the same species today.

A recent biodiversity informatics symposium in Stockholm (SBIS, 2008) focused on two broad areas around a core theme. It compared the collective knowledge of all species to an enormous virtual book, a volume spanning 250 years of history back to the publication of *Systema Naturae* as the first organized catalogue of living organisms linked within a single systematic system. And it posed the question "Can we regain the lost book of all living species?" Presenters then provided a look at some parts of the book that have become available online via a range of initiatives. Also presented were some of the technical practices being directed at managing it.

The "accumulated information of a species" referred to by Grimaldi and Engel constitutes the content of the pages of this virtual book of all living species, and this content is increasingly becoming available online. Entering a species name as a keyword in any of a number of Internet search engines will likely reveal thousands of "hits." Such results may span a broad range of content ranging from scholarly sources such as journals and scientific databases to discussion lists, blogs, and a range of other sources.

One scholarly example is the network of hundreds of museums and observational networks that contribute to the Global Biodiversity Information Facility (GBIF). The current GBIF network provides access to over 163 million "occurrence records" (GBIF, 2008), references to museum specimens, and observations of living species. This collective index continues to increase at the rate of 5% per month with billions of such records in existence in the world's museums and observation networks. All of these, spanning the full range of post-Linnaean biology, are tied to a scientific name.

The National Center for Biotechnology Information hosts many biomedical and biological databases such as GenBank and PubMed. The GenBank database provides access to nearly 100 million sequence records, each of which is derived from an organism that is identified by a name. Since 1982, the number of base pairs in the Genbank database has doubled approximately every six months (Genbank, 2008). Each GenBank sequence is tied to a scientific name.

The Biodiversity Heritage Library, a consortium of life science libraries, is digitizing the full corpus of our heritage biodiversity literature going back more than 200 years. Over 26,000 volumes and 10 million pages are already available online, with a target of more than 100 million pages to add in the coming years (Biodiversity Heritage Library, 2008). Among these pages are millions of scientific names that provide the context that transforms literature into biology.

Every month, thousands of journals publish many thousands of new articles that reference taxa by scientific names. Articles originate in a wide range of disciplines from oil exploration to oceanography, medicine, chemistry, and natural history. Universal Biological Indexer and Organizer (uBio) has created an application (Learyet al., 2007) that identifies scientific names in regularly updated RSS[1] feeds from approximately 600 sources. In just over two years, more than 108,000 articles have been identified having names contained in the Catalogue of Life.

Numerous databases of taxonomic expertise provide authoritative access to information on a wide range of species. Fishbase[2] is a widely known and comprehensive resource for historical and contemporary information on fishes and is complemented by the Catalogue of Fishes at the California Academy of Sciences. Many other groups of plants and animals ranging from Amphibia[3] to Solanaceae[4] to Algae[5] are equally represented and add to the available body of online knowledge.

These are just a few examples of an enormous range of digitization activities that intentionally or incidentally reference organisms and are transforming the conceptual notion of all collective knowledge of all species into something tangible, although still enormously vast and diffuse. While they may vary widely in scope, intent, source media, subject, and date of publication, they are unified through their links to living organisms made by reference using scientific names. They are, in a very real sense, reconstituting the components for the "lost book of all living species" that can be defined as the set of all knowledge or data that is tied to a scientific name. It might be premature to characterize the raw totality of all uncollated information tied to a species as a "book." Instead, we might view regaining the conceptual book as a process: a process that begins with first transforming the inaccessible to the accessible and then proceeds from discovering the accessible to a subsequent synthesis. A real world analog might be to characterize the totality of all information regarding all species as an enormous virtual "library of biology" from which the book of all species, or for that matter, different books, might be composed. Libraries contain the raw materials from which such syntheses can proceed, and a primary function of libraries is to make information discoverable. Discovery precedes synthesis, and libraries are instruments of discovery from which new knowledge can be synthesized.

Libraries and books make information discoverable via indices. Indices are familiar as lists of words and associated page numbers generally found in the backs of books or via card catalogs. They serve as a key means for information discovery. If the book of all living things is composed of all information tied to scientific names, then we might visualize the index of this book as a list of all existing scientific names. Instead of page numbers, one could imagine an array of Uniform

Resource Identifiers[6] (URIs), Globally Unique Identifiers[7] (GUIDs), and other methods for referencing online content. Each scientific name would provide pointers to occurrences of that name within the virtual book, providing the baseline for discovery and renewed synthesis and knowledge.

The remainder of this discussion is focused on these two concepts: a vast, virtual and global collection of information on all living species, and an associated index of scientific names within these data that provide a potential key to its discovery and enable its ultimate re-synthesis into new knowledge of living species. This activity is already under way and the relationship between this collection of information and the index of names is beginning to emerge. Utilizing it as a means for discovery presents challenges to all indexers of biodiversity content and addressing these challenges has required us to look at this index as an entity in itself, with its own properties and scope. For convenience we sought to give this collective index of all scientific names a name itself. We coined the term "the BIG Index" to refer to this superset of all scientific names. The term "BIG" is an acronym for a collective "biodiversity indexing group" representing all who publish, manage, digitize, or otherwise mobilize species information and face common challenges in enabling discovery of these data via scientific names and from which the BIG Index is emerging.

The ubiquity of scientific names annotating all information about species, combined with their grounding in formalized rules of nomenclature, would appear to position the BIG Index as having a central role as the core metadata from which a collective index of all species information could be derived (Patterson, 2006). Realizing this vision has been frustrated because scientific names are not fixed, stable, or unique, and because continuing evolutionary and taxonomic studies lead to changes in the potential metadata structure. As a result, 5–10% of scientific names become taxonomically invalidated each decade (Froese, 2000). Historical names, however, remain in books and on museum jars even as taxonomic opinion has moved on. It may seem surprising to those outside of biology that, given the central role scientific names play regarding information about a species, no collective list of scientific names, nor even a reasonable estimate of their number, exists (Stork, 1997). Yet, as a virtual concept, it does exist, and to a large group of indexers of biodiversity content, it is emerging as a reality and bringing both opportunities and challenges in facilitating access to information.

QUALITIES OF THE BIG INDEX

Given 250 years of post-Linnaean biology and an estimated 1.8 million (Chapman, 2005) currently described species, there are least 1.8 million currently accepted scientific names of species.

There are many individual lists of species names, some quite comprehensive within a given taxonomic domain or period of time. Sherborne's *Index Animalium* (Sherborn, 1933) attempted to catalogue all animal names from the publication of the tenth edition of *Systema Naturae* to 1850. Thomson Reuters inherited the Zoological Record, which has been indexing the scientific literature for animals' names since 1864, and the Index of Organism Names, which contains over 1,806,748 names as of this writing (Thomson Reuters). The International Plant Names Index (IPNI, 2008) combines three overlapping plant names datasets into one. Index Fungorum (Index Fungorum, 2008) has amassed a large collection of over 300,000 fungal names. The Catalogue of Life provides access to more than 1.9 million names representing 1.1 million species (COL, 2008). Some of these initiatives attempt to identify new names published in the literature. Registration of new names is mandatory for all prokaryotes (IJSEM, 2008) and the non-mandatory registration of new animal names is under current development under the auspices of ZooBank (Polaszek et al., 2005). Collectively, one might assume, these form a relatively comprehensive component of the BIG Index; however, resolving just how many names exist and quantifying the completeness of the BIG Index is more complicated than it may initially appear.

One complication is due to different degrees of cardinality inherent to different definitions of a distinct name. If one were to compile a giant list or spreadsheet, one row or line for each name, these different definitions would result in very different numbers of rows. Which of these is most appropriate depends on different needs that might be ascribed to different potential uses and user groups. The differences appear to be tied to three basic views or properties that can be ascribed to scientific names as:

1. Labels for taxa
2. Units of nomenclature
3. Strings of characters

Each of these will be briefly explored and their bearing on the BIG Index considered.

SCIENTIFIC NAMES ARE LABELS FOR TAXA

Names are used as labels for currently recognized taxa. If names can be described as the "words" of taxonomy, then taxonomic descriptions define these words. Changes in taxonomic opinion lump and split previously distinct taxa, with the result that some names no longer refer to recognized taxa and are rendered unaccepted or, in the lexicon of zoology, invalid (International Commission of Zoological Nomenclature, 1999).

The BIG Index is clearly composed of both currently accepted and unaccepted names. For many biologists, conservationists, and managers of both information and physical organisms, the current and correct name for a taxon is the priority, and given that a comprehensive list of correct names for all species does not currently exist, justifies a perspective that places a priority on assembling it. This definition forms the basic priority for the development of the Catalogue of Life, which seeks to provide an authoritative name for all 1.8 million species (About Species 2000, 2008). This worthy effort can be realized only through the work of taxonomic experts who must laboriously review all relevant material relating to species in order to define its boundaries. This approach constitutes the most visible efforts in assembling and organizing the BIG index. It also represents the most labor-intensive and subjective aspect of cataloguing it. Defining the BIG Index solely in the context of such efforts is an approach that ultimately results in the most refined list of names, but it also imposes a significant bottleneck in an effort to catalogue all names, which regardless of their ultimate treatment from a taxonomic perspective, remain as persistent nomenclatural facts—facts that can be inventoried and catalogued via separate and inherently faster processes than the cataloguing of species.

A challenge for biodiversity indexers presented by names as labels for taxa is that taxa are prone to polysemy (Berendsohn, 1995; Koskela and Murphy, 2006), where the same taxon may have quite different taxonomic circumscriptions or senses. Polysemy differs from homonymy in that a taxonomic polyseme shares the same name-bearing type but differs in circumscription, resulting in different taxon concepts. Since taxon names are derived from types, they do not confer circumscription information. The implications of polysemy in the inferred synthesis of knowledge about a species based on the use of scientific names in information retrieval must be considered, as it may provide results that are incongruent with an intended taxon definition. This provides the rationale in distinguishing discovery from synthesis in the potential application of the BIG Index for providing collective information on species.

SCIENTIFIC NAMES ARE UNITS OF NOMENCLATURE

Names form the basic "words" of taxonomy. Scientific names are tied to nomenclatural acts that govern their creation. Species names are based on type specimens and, while these specimens were originally typified based on an assumption of uniqueness, subsequent opinion may result in their being deemed conspecific to other types. Based on rules of priority, one name becomes the accepted name for the taxon and all other names become unaccepted names or synonyms (ICZN, 1999). The 2008 Catalogue of Life Annual Checklist is composed of discrete taxonomic sectors or

Global Species Datasets (GSDs). Some of these GSDs provide both a list of accepted taxon names and unaccepted names that are associated with the listed taxa. The GSD for fishes is particularly well treated. It lists 31,000 accepted names and more than 50,000 unaccepted names.

A list of validly published scientific names, therefore, is not the same as a list of species. Names may be unaccepted for a number of reasons. Some of these names may be tied to types (heterotypic synonyms) other than the one bearing the name. Some of these names may be linked to the same type as the accepted name (homotypic synonyms) only one of which may serve as the accepted name.

I can find no firm estimate on the number of distinct types that have been described nor the number of distinct binomial or trinomial combinations that are based on them, but it clearly exceeds the number of distinct species, perhaps by several times (Patterson, D. J., 2006). Stork (1997) estimated the level of synonymy in the insects is about 20%, and gives a figure of over 9,000 mammalian names for mammal types within the Natural History Museum in London. Dubois (1977) reported a rate of 64% of valid names for the amphibia in 1970. The number of distinct binomials arising from these figures is hard to assess. One confounding factor in assessing this number is due to differences in the botanical and zoological codes of nomenclature. These codes define nomenclatural acts differently, with the result that botanists and zoologists have different underlying definitions of what constitutes a "name" (de Jong et al., 2004).

In zoology, a binomial name is not a recognized code-governed nomenclatural unit (ICZN, 1999). When a new species name is created, the species epithet forms the basis of the name. That name is tied to another code-governed act that created the genus. If the species is subsequently assigned to a new genus, creating a new, distinct binomial, this event is not governed by the code. Indeed, it is not formally considered a new name at all, even though the species epithet may itself change if the gender of the genus has changed, rendering the name quite distinct.

This is not the case in botany, where re-combinations are governed by the code (ICBN, 2000). As a result, botanical names may be rendered more complex as the author of the new combination is concatenated to that of the original author. This complexity provides additional details regarding the provenance of the name but also increases the number of different ways the name may be expressed.

This disparity among the codes as to what constitutes a name can mean that a botanist may see a multitude of names in a case where a zoologist may only see one. Table 14.1 illustrates two sets of names, each containing five distinct combinations tied to two types. Botany recognizes five code-governed acts for species and recognizes five distinct names. In zoology, there are two code-governed acts for species and two names. For this reason, the cardinality of a row of names in a botanical list would differ from a zoological listing. Since zoological combinations are not code-governed, they may not be as comprehensively treated as binomials within published synonymies.

TABLE 14.1
Five Different Name Combinations Are Treated Differently among the Zoological and Botanical Codes

Botanical Code	Zoological Code
5 code governed acts/2 types	2 code-governed/2 types
5 names	2 names
Agalinus paupercula var. borealis Pennell	*Gasterosteus saltatrix* Linnaeus 1766
Gerardia paupercula var. borealis (Pennell) Deam	*Temnodon saltator* (Linnaeus 1766)
Gerardia paupercula ssp. borealis (Pennell) Pennell	*Pomatomus saltatrix* (Linnaeus 1766)
Gerardia purpurea var. parviflora Benth	*Tetrao afer Müller 1776*
Agalinis purpurea var. parviflora (Benth.) Boivin	*Francolinus afer* (Müller 1776)
	Pternistes afer (Müller 1776)

The disparity of definitions and the overall lack of formal treatment of binomial combinations as code-governed units of nomenclature in zoology, where most binomial combinations exist, present challenges to indexers of biodiversity content. The BIG Index is composed of the sum of all of these combinations and a logical de-aliasing of all combinations to their original types would provide a significant boost to overall discovery. Such an exercise is fundamentally different from developing taxonomic catalogs as it entails fact checking, not expert taxonomic judgment. This implies it can be undertaken in parallel with global taxonomic efforts and progress governed by different rate-limiting steps.

Scientific Names Are Strings of Characters

At their most fundamental, names are a series of alphanumeric characters, and it is in this form that indexers of biodiversity content are presented with them. Computer processors manage digitized text as strings of characters. Information retrieval based on keywords relies on matching strings of characters. To a computer, the names "*Loligo pealeii*" and "*Loligo pealei*" and "*Loligo pealii*" represent three distinct sets of characters. Likewise "*Pomatomus saltatrix* Linnaeus 1768" and "*Pomatomus saltatrix* L. 1768" are two distinct strings. Lacking a comprehensive authority list of binomial names often requires indexers to see all combinations as potentially distinct.

Scientific names may vary in spelling or format for a number of reasons, only one of which is outright misspellings. Taxon names have a variety of component parts, many of which may be selectively omitted in a particular usage. These include authorship, infra-specific rank markers, subgenera, year of publication, and other annotations (Table 14.2). Many of these are prone to further differentiation due to different formats, sequences, and abbreviations resulting in considerable latitude in the composition of the same nomenclatural unit. The net effect is that a single correctly spelled name can vary in length by 100% or more. Such variation presents considerable challenges in reconciling such vastly different formats for the same name. Additionally, indexers of biodiversity content may not possess sufficient familiarity with nomenclatural structure to accurately map these forms.

The genus and species components of taxon names may also demonstrate spelling differences that result in multiple representations for the same name. Reasons range from ordinary misspellings at the time of recording, human and machine-mediated transcription errors, to different formatting conventions of Latin.

The bacterium *Actinobacillus actimycetemcomitans* is misspelled 26 different ways within PubMed, an index of more than 16,000,000 titles and abstracts in the biomedical sciences (Table 14.3). Each of these names is tied to at least one, and often many, publications. In some cases, misspellings appear to outnumber the correct form of the name. A search using the online search

TABLE 14.2
Variation in Formatting Conventions for a Single Taxon as Recorded from Different Institutional and Government Sources[10]

Unparsed Taxon Name	# Characters
Agalinus paupercula var. *borealis*	34
Agalinus paupercula var. *borealis* Pennell	42
Agalinus paupercula Britton var. *borealis* Pennell	50
Agalinus paupercula (Gray) Britt. var. *borealis* Pennell	56
Agalinis paupercula (A.Gray) Britton var. *borealis* Pennell	59
Agalinus paupercula (Gray) Britton var. *borealis* (Pennell) Zenkert 1934	72

TABLE 14.3
Variation in Spelling of the Bacterium *Actinobacillus Actimycetemcomitans* as It Has Been Recorded in *The Pubmed Citation Index*

Actinobacillus actimycetemcomitans	*Actinobacillus actinomycetemcomitants*
Actinobacillus actinmycetemcomitans	*Actinobacillus actinomycetemcommitans*
Actinobacillus actinomicetemcomitans	*Actinobacillus actinomycetemocimitans*
Actinobacillus actinomy	*Actinobacillus actinomycetencomitans*
Actinobacillus actinomyce	*Actinobacillus actinomycetum*
Actinobacillus actinomycemcomitans	*Actinobacillus actinomyctemcomitans*
Actinobacillus actinomyceremcomitans	*Actinobacillus actinomyectomcomitans*
Actinobacillus actinomycetam	*Actinobacillus actinomycetemcomintans*
Actinobacillus actinomycetamcomitans	*Actinobacillus actinomycetemcomitance*
Actinobacillus actinomycetecomitans	*Actinobacillus actinomycetemcmitans*
Actinobacillus actinomyetemcomitans	*Actinobacillus actynomicetemcomitans*
Actinobacillus actinonmycetemcomitans	*Actinobacillus actinomycetemcomitans*
Actinobacillus actionomycetemcomitans	*Actinobacillus antinomycetemcomitans*

engine Google, for the three variations of *Loligo pealeii* cited above, returns many times the number of results for the two spelling variants than for the correct form of the name (Sp2000 Loligo, 2009).

Accounting for these types of variations in names is critical for computerized systems to make effective linkages among content and between users who may spell a name one way and need to match content labeled in another. Reconciling orthography can be facilitated through the use of software algorithms such as TaxaMatch (Rees, 2008). An index of orthographies tied to a distinct lexicon of code-compliant names would serve a strategic function for managing the BIG Index. Until then, it is difficult for indexers to a priori determine whether a name is a variant spelling of an existing name or a distinct name itself.

In sum, it is important to recognize that the BIG Index of all scientific names tied to the collective knowledge of all species possesses all of these three properties. Digitization initiatives such as GBIF, the Biodiversity Heritage Library (BHL), the The National Center for Biotechnology Information (NCBI) and indexes such as Google reveal the sheer volume of existing information that is linked to scientific names. These billions of units of information are tied to many millions, possibly 10 or 20 million, distinct sets of character strings. These different strings ultimately will resolve to some number of distinct binomial and trinomial nomenclatural combinations, perhaps 5–7 million, that themselves are derived from some 2–4 million named types. Nearly all of these resolve to approximately 1.8 million currently identified species. Given taxonomic polysemy, the exact number of species and the overall distribution of the BIG index will always be a consequence of the specific collective taxonomic views that form a particular global treatment.

As the BIG Index moves from a concept to a virtual reality as a consequence of increasing digitization of biodiversity information, we are provided with an opportunity for enhanced discovery fraught with numerous challenges characterized by the three properties of names outlined here. Their resolution is necessary if we are to enable discovery, and ultimately synthesis, of all the information tied to the BIG Index in a way that makes taxonomic sense. This requires the application of methodologies that holistically account for all three properties and do not rely solely on one. The GBIF network provides a useful exemplar where taxonomy, nomenclature, and the character string properties of names present these challenges.

THE BIG INDEX AND GBIF

The most recent index of the GBIF network at the time of this writing is composed of over 163 million records, each of which represents an occurrence of a taxon either as a specimen reference or as a discrete observation of a living organism. This represents approximately 100 gigabytes of structured text formatted according to some version of the DarwinCore (TDWG Darwin Core, 2008) or Access to Biological Collections Data (TDWG ABCD, 2008). Every month this text is processed into a series of index files that can be visualized as a giant table or spreadsheet where each column is a distinct data type (collector, location, data, taxon, etc.). One of those columns contains the terminal scientific names to which each record refers. This column might be thought of as the GBIF portion of the BIG index.

GBIF uses the Annual Checklist of the Catalogue of Life to provide an authoritative "taxonomic backbone" for the GBIF index to enhance management and retrieval. GBIF uses the Catalogue of Life classification of species to organize data records into a hierarchical set of nested folders, with each folder corresponding to a higher taxon ultimately linked to species. This enables GBIF-mobilized data to be browsed and retrieved in a manner akin to browsing files in a computer file system. Synonyms can be conflated so that a search for information in the GBIF index using one name can include synonymous terms that can increase access to potentially relevant records. Successfully using the Catalogue of Life or any authoritative taxonomic data for this purpose is proportional to the degree of matches between the index of names in the GBIF data and the index derived from the Catalogue of Life. Note also that given taxonomic polysemy and the undocumented taxonomic context of sources, these methods may return results that do not conform to a specific taxon concept view.

In 2007, GBIF performed an analysis of an index derived from approximately 120 million records. 4.86 million distinct character strings composed the Scientific Name column of the data. This figure excluded obvious strings that didn't conform to some basic structural conventions for species names. The author component was temporarily excluded to quantify the degree to which it contributed to the overall total. When authorship was removed, the number was reduced to 3.4 million distinct names. When the scientific names index was matched to the 2007 Annual Checklist of the Catalogue of Life, 328,301 names intersected. These numbers do account for a high proportion of occurrence records in the GBIF index due to the uneven distribution of names relative to records. Nonetheless, the high number of distinct names was of concern enough to merit further evaluation

One reason for the low overall match may be due to a high degree in variation in how names are recorded; in other words, it may be that the variation is at the level of character strings. Another explanation may be that the Global Species Datasets that form the basic constituent of taxonomic contributions to the Catalogue of Life, are often not nomenclaturally complete for the taxonomic sector they address. An example is the GSD for birds (Figure 14.1). This sector is based on a 2005 version of an avian taxonomy called Zoonomen (Zoonomen). This taxonomy provides a global list of valid names for species and subspecies for birds. Some taxa include synonyms. When comparing indexes from other global catalogs of birds as well as other well-known bird catalogues, it becomes clear that a collective catalogue of bird nomenclature far outnumbers the names represented in any one of a number of global and regional taxonomic catalogues of birds.[8]

Within the GBIF index itself there are additional names purporting to be birds that are not found in the Zoonomen taxonomy that may occur within these or other resources. The Darwin Core and ABCD formats allow data providers to include a taxonomic classification for any record. Records designated as birds (Aves) therefore may be inferred to be included within the Aves, either by matching against an authority list such as the Catalogue of Life or via an explicit record designation.

Collectively, this presents challenges to providers of biodiversity information who may now have multiple options as to what to return to a query for all records relating to birds (Aves). Using the Zoonomen bird taxonomy retrieves one set of records. Using the Howard and Moore 3rd edition retrieves a different set of records. Employing any classification as a retrieval structure omits records of birds whose names are inclusive only of the superset of all avian classifications. This

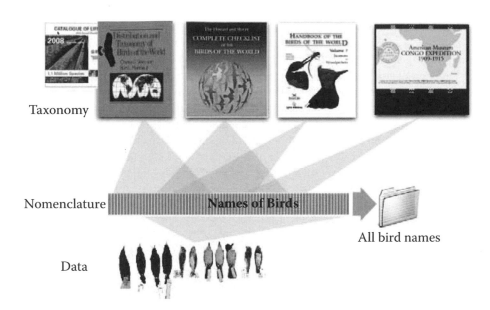

Taxonomy

Nomenclature — Names of Birds — All bird names

Data

FIGURE 14.1 Incomplete and overlapping nomenclatural coverage of different taxonomic sources relative to the superset of all nomenclature for a particular taxonomic sector, in this case the Aves or birds.

leads to a natural separation of nomenclature from taxonomy in terms of potential informatics application in the management and discovery of biodiversity and in the implementation of a supporting infrastructure. All avian taxonomies could derive their constituent nomenclature from a separate referent and more complete nomenclatural index, yet still remain distinct in their specific taxonomic circumscriptions. From the perspective of information discovery as distinguished from subsequent synthesis of knowledge, it may be more useful to draw upon a reliable nomenclatural lexicon than to be restricted to a nomenclaturally incomplete taxonomic view. Such a view will remain incomplete until a particular taxonomic sector has accounted for all nomenclatural combinations of all the types within that sector.

Such a separation also makes procedural sense in terms of developing methodologies to enhance information discovery in biodiversity information since the cataloging of nomenclature and nomenclatural relationships among species names is based on factual, not subjective, criteria. Validation of nomenclatural origins and conformance to the relevant code remains a time consuming process but nonetheless requires no new science to proceed. What is required is enhanced methods of discovery of sources to enable validation. Thus, GBIF and others seek to develop a coordinated infrastructure that integrates taxonomy and nomenclature in a manner that can be mutually informed while at the same time separating logical cataloging processes and enabling clear boundaries between different domains of nomenclatural and taxonomic quality.

A COMPREHENSIVE INDEX OF GENERA

All species binomials contain a parent genus. The 2008 Catalogue of Life Annual checklist lists 105,771 genera for 1,105,589 species, so an order of magnitude different between the two seems a reasonable estimate. In 2002, as a component of the uBio project, Patterson surmised that a comprehensive nomenclatural catalogue of genera (Patterson, 2003), placed within a provisional management classification would, in principle, provide a structure for tying all binomial combinations into a single nomenclatural framework. While such a structure would not provide taxonomic status for species names, it would provide the basis for improved organization of any names that could not be more effectively treated within accessible taxonomic catalogs.

A comprehensive and provisionally classified index of all genera would provide benefits to biodiversity indexers. A comprehensive list would improve the capacity to quantify the degree of homonymy among the genera. Identically spelled genera are statistically more likely than homonymous binomials, and initial attempts to catalogue the genera indicate as many as 10% are homonyms. An absolute count of homonyms among the genera can substantially improve the means by which GBIF and other biodiversity indexers provide access to information. Currently, GBIF relies on taxonomic context, provided in source data, to attempt to differentiate homonyms. Due to changes in taxonomy over time, this can result in a conflation of potential homonyms or even create phantom homonyms.

Table 14.4 illustrates a real-world example of homonym conflation taken from data in the GBIF network. The genus *Oenanthe* is presented in a number of records that provide different taxonomic contexts to the name. Without a consistent, comprehensive, and accessible nomenclatural foundation to these names, it is difficult to determine how many *Oenanthe* are being referenced. In this case, the different taxonomic contexts, while perhaps accurate according to a particular time or point of view, confound the resolution of the nomenclature. A comprehensive catalogue of genera would reveal that there are only two published *Oenanthe* (*Oenanthe* Linnaeus 1754 and *Oenanthe* Pallas 1771) and that these must ultimately be resolved to one of them in terms of intent (i.e., the name may or may not be accurately applied to the referenced specimen). Once the absolute nomenclatural count of *Oenanthe* is known, the different taxonomic contexts within the data transform from liabilities to assets in the sense that they can be used to provide clues on disambiguating subsequent usages of the name.

A comprehensive catalogue of all genera will provide the complement to a list of homonyms: the 90% or so of all genera that are lexically distinct. Barring homographic misspellings (inadvertent misspellings that match existing genera) any species combination with this genus name can be assumed to be assigned to it. If the genus is placed in a provisional taxonomic hierarchy, we have a structure for managing names that can serve as a sort of second tier scaffolding for names not yet treated in accessible taxonomies, and naturally organize these names in a manner that serves efforts such as the Catalogue of Life, the European Distributed Institute of Taxonomy (EDIT), Pan European Species Integration (PESI), and other initiatives that seek to develop comprehensive taxonomic catalogues.

GBIF, the BHL, and others form the front line of biodiversity digitization initiatives that can not only benefit from, but may also inform, taxonomic initiatives compiling species checklists. For example, as these initiatives uncover new and uncatalogued specimens and publications, it would be beneficial to have mechanisms in place to make such information discoverable to those experts who might best treat them.

Cross-referencing a taxonomically organized index of nomenclature against a similar index of taxonomic databases provides the means to identify content linked to names within a taxonomic sector that are not included in the referent taxonomic catalogue. If this content has some scholarly interest to the curator of the catalogue, for example if it references specimens (e.g., GBIF) or literature

TABLE 14.4
Sample Data Taken from Darwin Core Formatted Specimen Records in the GBIF Index

Kingdom	Phylum	Class	Order	Family	Genus
Plantae	Magnoliophyta	Magnoliopsida	Apiales	Umbelliferae	Oenanthe
Plantae	--	--	--	--	Oenanthe
Plantae	Magnoliophyta	Magnoliopsida	Apiales	Apiaceae	Oenanthe
Plantae	--	--	--	Orchidaceae	Oenanthe
Animalia	Chordata	Aves	Passeriformes	Muscicapidae	Oenanthe
Animalia	Chordata	Aves	Passeriformes	Turdidae	Oenanthe

(e.g., the BHL), we can enable the means for biodiversity indexers to segregate such knowledge into taxonomic domains that match those of experts and create the seeming incongruent possibility of informing experts of potential nomenclatural omissions from their catalogues. This might enable GBIF to inform the curators of the GSD for fishes of new fish names tied to specimens.

Initiatives like the BHL reference content that may not contain higher taxonomic references such as provided for in the data exchange standards utilized by GBIF. If the digitization of publications uncovers a novel species combination in the genus *Pternistes*, for example, a comprehensive list of genera can provisionally place that name within the Aves. This will enable the BHL to automatically include this item within the Aves by inference. Furthermore, services could be enabled that cross-reference external names indexes and report missing entries to initiatives seeking to build global species datasets such as the Catalogue of Life or in the case of birds, the Howard and Moore checklist group (Dickenson).

A real example that concerns a taxon mentioned in an article published in 1930 in the *American Journal of Tropical Hygiene and Medicine* entitled The Flight of *Stegomyia aegypti* (Figure 14.2). A review of the abstract clearly identifies this as the mosquito better known as *Culex aegypti* or *Aedes aegypti*. This binomial nomenclatural combination does not, however, appear within the Catalogue of Life or the source Diptera Global Species Dataset, the Biosystematic Database of World Diptera.[9] Assignment of *aegypti* to the genus *Stegomyia* appears to be the topic of some taxonomic controversy (Polaszek, 2006). Yet the placement of *aegypti* within *Stegomyia*, a genus or subgenus of mosquito, is a nomenclatural fact. A catalogue of genera would place this combination naturally within the Diptera as a consequence of its nomenclatural structure and serve as a provisional structure while the taxonomic work proceeds at its natural pace. As a nomenclatural management structure, it will enable indexers to better manage names in a manner that enables coordination with initiatives working to resolve taxonomy.

Am. J. Trop. Med., s1-10(2), 1930, pp. 151-156

The Flight of Stegomyia Aegypti (L.)[1]

R. C. Shannon AND N. C. Davis
Yellow Fever Laboratory of the International Health Division, Rockefeller Foundation, Bahia Brazil

During the course of three experiments undertaken to test the flight of *Stegomyia aegypti*, in which over 20,000 stained specimens were released in small villages, two specimens were recovered in houses at a distance of more than 300 meters from the point of release, while 95 were taken at intermediate points.

Another lot of 12,000 mosquitoes was released on a boat 900 meters distant from one shore and 300 meters from the opposite shore. Of these, eight were recovered on shore at an approximate distance of one kilometer from the boat.

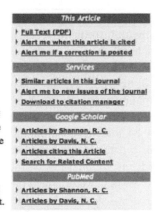

Less than 0.4 per cent of the entire number of released mosquitoes was recovered. The latest capture made was thirteen days after the date of release (experiment I). After a week's time all stained specimens became exceedingly scarce. This indicates either that dispersion is very thorough, widespread, and rapid, or that mortality is very high.

From the foregoing, it would appear that a flight of more than 300 meters is not exceptional among the stegomyiae of a representative community; and a sustained flight of one kilometer over water is well within the range of possibility.

FIGURE 14.2   Historical publication titled with a cryptic species name.

GBIF is developing a comprehensive list of genera through partnerships with existing nomenclatural initiatives and databases such as the Index Fungorum, the International Plant Names Index, ZooBank, the Thomson Reuters Index of Organism Names, and others. Rather than create a new database, GBIF seeks to build dynamic indices of these and other resources to create a virtual and collective knowledge base. This enables a resource to be derived from a federation of components instead of creating a new duplicate record. New techniques are required to resolve overlaps and, as indicated in this chapter, these are not just simple exact matches but reflect the sort of orthographic variation in format and authorship that characterize all uses of names, be it in specimen datasets or authoritative taxonomic publications.

A current version of an All Genera Index, composed of large collections of zoological, botanical, and fungal genera, contains approximately 400,000 distinct genera, of which nearly 30,000 are provisionally identified as homonyms. Additional sources, and the infrastructure to enable consistent and open access to them, are the target of current activities.

SUMMARY

The collective knowledge of information about all species will be a component of the larger efforts worldwide to mobilize physical information resources and make them available in some digitized format. Such efforts are likely to be executed and managed by institutions and initiatives that lack formal taxonomic training or even basic sensitivities regarding the format and importance of scientific names in biodiversity. This, combined with the great latitude in which scientific names are expressed, requires special attention to ensure that the data that are tied to these names become visible within a context that makes sense to biologists. It is our view that the ubiquity of names within all information tied to a species enables a transformative opportunity to the discipline of taxonomy as this information emerges into the digital world.

To effectively administer this knowledge, however, requires more than just the tools of taxonomy, particularly given the exacting nature of the practice and the amount of work that appears to be ahead. It requires a coordination of validated nomenclature and taxonomy with appropriately applied informatics techniques, appropriately segregated to distinguish boundaries and aspects of quality and fitness for particular uses. Logical boundaries exist between resolution of nomenclatural properties of names, which are generally objectively derived, from those of taxonomy, which require expertise and judgment. Informatics techniques can be applied, in combination with validated nomenclatural practices, to resolve and de-alias the wide latitude of orthography represented in the actual application of names within the collective "book of living things." The BIG Index of all names of all living things will be complete, and taxonomy will be recognized as the critical framework for organizing all information about all species, when there is a clear and simple mechanism for resolving any name to the highest available degree of nomenclatural validation and taxonomic scrutiny.

It may be some time before all names have received the fullest degree of taxonomic treatment. Therefore, we need to develop intermediate and provisional steps that coordinate with such efforts but provide resolution pathways that are inclusive for all names. The net result will be the capacity to regain the book of knowledge of all species and a framework for moving taxonomy into the future.

ENDNOTES

1. RSS or Real Simple Syndication. See Wikipedia (http://en.wikipedia.org/wiki/RSS_(file_format))
2. Fishbase (http://www.fishbase.org/)
3. Amphibian Species of the World (http://research.amnh.org/herpetology/amphibia/index.php)
4. Solanaceae Source (http://www.nhm.ac.uk/research-curation/research/projects/solanaceaesource/)
5. AlgaeBase (http://www.algaebase.org/)
6. For a definition of URI's see Wikipedia (http://en.wikipedia.org/wiki/Uniform_Resource_Identifier)
7. For a definition of GUIDs see Wikipedia (http://en.wikipedia.org/wiki/Globally_Unique_Identifier)

8. Additional catalogs of birds purporting to be global in scope include the Howard and Moore Checklist 3rd edition, the Morony, Bock and Farand 2005, Nomina Avium Mundi 2002, and Sibley and Munroe 1992. Additionally, historical bird catalogs contain referent bird nomenclature such as Sclater, Systema Avium Ethiopicarium 1930, Sharpes Catalogue of Birds of the British Museum 1893 and portions of Peters Checklist of the Birds of the World 1931.
9. The Biosystematic Database of World Diptera (http://www.sel.barc.usda.gov/Diptera/biosys.htm)
10. Sources for these name orthographies include The International Plant Names Index, the Ohio Department of Natural Resources Website, the Tropicos database of the Missouri Botanical Garden, and the Wisconsin Botanical Information Server.

BIBLIOGRAPHY

About Species 2000. (08/05/08). Retrieved 12/21/08, from http://www.sp2000.org/index.php?option=com_content&task=view&id=12&Itemid=26.

Berendsohn, W. G. (1995). The concept of "potential taxa" in databases. *Taxon* 44 (2), 207–212.

Biodiversity Heritage Library. (11/24/08). *News and updates from the Biodiversity Heritage Library.* (C. Freeland, Ed.) Retrieved 12/21/08, from Biodiversity Heritage Library: http://biodiversitylibrary.blogspot.com/2008/11/100000000-pages.html.

Chapman, A. D. (2005). *Numbers of living species in Australia and the world.* Toowoomba: Australian Biodiversity Information Services.

COL. (2008). Catalogue of life: 2008 annual checklist. (F. A. Bisby, Y. Roskov, T. Orrell, D. Nicholson, L. Paglinawan, N. Bailly, et al., Eds.) Retrieved 12/21/08, from http://www.catalogueoflife.org/annual-checklist/2008/info_2008_checklist.php.

de Jong, Y., Verbeek, M., Michelsen, V., Steeman, F., Bailly, N., & Basire, C. (2004). Overview of nomenclatural practices in zoology and botany with respect to the consequences of integrating zoological and botanical checklists data into a single virtual environment. *Taxonomic Databases Working Group 2004 Annual Meeting.* http://www.nhm.ac.uk/hosted_sites/tdwg/2004meet/paperabstracts/TDWG_2004_Papers_Jong_1.htm

Dickinson, E. (Ed.). (n.d.). Retrieved from The Howard and Moore Checklist: Birds of the World. http://birds-howardandmoore.org/.

Dubois, A. (1977). Les problèmes de l'espèce chez les Amphibiens Anoures. *Mémoires de la Société zoologique de France* 39, 161–284.

Froese, R. E. (2000). Challenges to taxonomic information management: How to deal with changes in scientific names. In H. Shimizu (Ed.), *Global environmental researches on biological and ecological aspects.* (pp. 3–10). Tsukuba, Japan.

GBIF. (2008, 12 1). GBIF data portal. (G. B. Facility, Producer) Retrieved 12/21/08, from GBIF data portal: http://data.gbif.org.

Genbank. (2008, 12 15). NCBI-GenBank Flat File Release 169.0. (N. C. Information, Producer) Retrieved 12/21/08, from GenBank: ftp://ftp.ncbi.nih.gov/genbank/gbrel.txt.

Grimaldi, D. and Engel, M. (2005). *Evolution of the insects.* Cambridge University Press.

ICBN. (2000). *International code of botanical nomenclature* (St. Louis Code ed.). (D. L. Hawksworth, Ed.) St. Louis, MO.

ICZN. (1999). *International code of zoological nomenclature* (4th ed.). (W. Ride, H. G. Cogger, C. Dupuis, O. Kraus, A. Minelli, F. C. Thomson, et al., Eds.) The International Trust for Zoological Nomenclature.

IJSEM. (04/01/08). *Instructions for authors. International Journal of Systematic and Evolutionary Microbiology*: http://ijs.sgmjournals.org/misc/ifora.shtml#06m.

Index Fungorum. (12/01/08). *Index fungorum.* (P. Kirk, Ed.) Retrieved 12/21/08, from http://www.indexfungorum.org/Names/Names.asp.

IPNI. (12/01/08). *The international plant names index.* (I. Partnership, Producer) Retrieved 12/21/08 from http://www.ipni.org.

Koskela, A. and Murphy, M. (2006). Polysemy and homonymy. In K. Brown (Ed.), *Encyclopedia of language and linguistics* (2nd ed). Elsevier. 742–744.

Leary, P. R., Remsen, D. P., Norton, C., Patterson, D. J., and Sarkar, I. N. (2007). uBIoRSS: Tracking taxonomic literature using RSS. *Bioinformatics* 23, 1434–1436.

Linnaeus. (1758). *Systema naturæ per regna tria naturæ, secundum classes, ordines, genera, species, cum characteribus, differentiis, synonymis, locis.* (Editio decima). Holmiæ.

Patterson, D. J. (2003). Progressing towards a biological names register. *Nature* 422, 661.

Patterson, D. J., Remsen, D. (2006). Taxonomic indexing—extending the role of taxonomy. *Systematic Biology* 55, 367–373.

Polaszek, A. (2006). Two words colliding: Resistance to changes in the scientific names of animals: Aedes vs. Stegomyia. *Trends in Parasitology* 22, 8–9.

Polaszek, A., Agosti, D., Alonso-Zarazaga, M., Beccaloni, G., Bjørn, P. d.P., Bouchet, P., et al. (2005). A universal register for animal names. *Nature* 437, 477.

Pyle, R. (Ed.). (2008). *ZooBank.* (I. T. Nomenclature, Producer) Retrieved 12/21/08, from ZooBank: http://www.zoobank.org/.

Rees, T. (2008). TAXAMATCH, a "fuzzy" matching algorithm for taxon names, and potential applications in taxonomic databases. *Taxonomic Databases Working Group 2008.* TDWG.

SBIS. (12/01/08). *The Book of Moments.* (S. M. History, Producer) Retrieved 01/02/09. Stockholm Biodiversity Informatics Symposium 2008: http://artedi.nrm.se/sbis2008/TheBookOfMoments2.pdf.

Sherborn, C. D. (Ed.). (1933). *Index Animalium.* London, England: Cambridge University Press.

Sp2000 Loligo. (2009). *Spelling of valid name: Loligo pealeii.* Retrieved from http://www.catalogueoflife.org/annual-checklist/2008/show_species_details.php?record_id=859107.

Stork, N. E. (1997). Measuring Global biodiversity and its decline. In M. Reaka-Kudla, D. E. Wilson, and E. O. Wilson (Eds.), *Biodiversity II* (pp. 41–68). Washington, D.C.: Joseph Henry Press.

TDWG ABCD. (2008). Retrieved from http://wiki.tdwg.org/ABCD/.

TDWG Darwin Core. (2008). Data standard definitions. Retrieved from http://wiki.tdwg.org/twiki/bin/view/DarwinCore/WebHome.

Thomson Reuters. (n.d.). Index of organism names metrics. (T. Reuters, Producer) Retrieved 12/29/08: http://www.organismnames.com/metrics.htm.

Zoonomen. (n.d.). Zoonomen.net. (A. Peterson, Editor) Retrieved 12/21/08: http://www.zoonomen.net/.

15 Linnaeus–Sherborn–ZooBank

Andrew Polaszek and Ellinor Michel

CONTENTS

INTRODUCTION

Taxonomy is the scientific discipline we use to identify, describe, and classify living organisms. Linnaeus, who made the first effective attempts to both describe and catalogue Earth's biodiversity, is rightly referred to as the "Father of Taxonomy." Linnaean zootaxonomy has the official starting date of January 1, 1758 (Linnaeus, 1758), and a quarter of a millennium later, with an arsenal of new tools for delimiting living diversity, we find ourselves entering a potential taxonomic Golden Age. E.O. Wilson has spoken and written many times of our priority task of "completing the Linnaean Enterprise" (e.g., Wilson, 2000), and the biodiversity informatics tools that are now available to us make the completion of the Linnaean Enterprise an imminent possibility. Given this current potential, for those outside the field of zootaxonomy it must come as a surprise and a disappointment to discover just how far we still are from completing the identification and reliable cataloguing of Earth's animal biodiversity.

More than twenty years ago Prof. Robert May expressed this clearly when he said:

Setting all the reservations and biases aside, the total number of living organisms that have received Latin binomial names is currently around 1.5 million or so. Amazingly, there is as yet no centralized computer index of these recorded species. It says a lot about intellectual fashions, and about our values, that we have a computerized catalogue entry, along with many details, for each of several million books in the Library of Congress but no such catalogue for the living species we share our world with. Such a catalogue, with appropriately coded information about the habitat, geographical distribution, and characteristic abundance of the species in question (no matter how rough or impressionistic), would cost orders of magnitude less money than sequencing the human genome; I do not believe such a project is orders of magnitude less important. Without such a factual catalogue, it is hard to unravel the patterns and processes that determine the biotic diversity of our planet. (May, 1988).

Clearly, the development of such a "catalogue" is precisely what a succession of taxonomists (effectively starting with Linnaeus), scientists, librarians, and, more recently, ICT specialists have been trying to achieve. We attempt here to provide an overview of some of these major initiatives since 1758, and then examine one particular initiative, ZooBank, which has the potential to greatly facilitate the successful completion of the Linnaean Enterprise.

CAROLUS LINNAEUS AND THE FIRST "COMPLETE" CATALOGUE OF KNOWN ANIMALS: *SYSTEMA NATURAE* 10TH ED. (1758)

Linnaeus's *Systema Naturae* project grew from a few pages in the first edition (Linnaeus, 1735) to 1,384 pages by the tenth (Linnaeus, 1758), which included 313 genera and 4,397 species of animals. Although owing a large debt to the earlier works of John Ray and Joseph Pitton de Tournefort, *Systema Naturae* was the first attempt to develop anything like a comprehensive catalogue of the Earth's species. It was described recently as "the first and last one-man encyclopaedia of life" (Alessandro Minelli, personal communication). Laying the foundation for binominal (genus+species) nomenclature in animals, the system still in general use for all animal and plant scientific nomenclature today, made the 10th edition a milestone in the history of zoology. "Nature's system" is what scientists are still gradually but successfully unraveling. This is a challenging task, requiring a unique degree of global collaboration among these scientists, and in particular among taxonomists. We have several excellent examples of such successful collaboration (e.g., for Diptera—Evenhuis et al., Chapter 7 this volume, as well as Antbase [http://antbase.org], Fishbase [http://www.fishbase.org], and many other resources), and we may be close to a universal system for cataloguing the names of animals described both retrospectively and in the future—ZooBank (Polaszek et al., 2005a).

While having provided the stable system of binominal nomenclature for the last 250 years, there is no doubt that the Linnaean system is problematic at other levels of classification, in particular at and above the generic level. Fixed, hierarchical relationships between "family-group" taxa do not exist in the real world. Indeed, it can be well-argued that such relationships hardly exist outside the concept of species, and perhaps not even there. Yet the Linnaean hierarchy does provide an extremely useful reference for working with taxa. However flawed our current picture of the "tree of life" may be, there is no doubt that many of the broad groupings and relationships that appear in it are accurate reflections of evolutionary history. The question is whether the accuracy of Linnaean nomenclature is adequate for the task of communicating about biodiversity. We are just discovering the degree to which such processes as horizontal gene transfer, gene duplication, and hybridization are complicating our image of the hitherto almost scientifically sacred tree of life. These complications notwithstanding, the constant modifications to that tree, based on increasingly accurate data and analysis of those data, are enriching but not replacing the traditional model, apart from at its deepest roots and nodes. While current advances in our knowledge of molecular evolution often challenge traditional models of hierarchical organism classification, we need to be pragmatic and conserve those aspects of Linnaean classification that serve the wider needs of society (Polaszek and Wilson, 2006). If a new system arises that works better than the old one, and changing systems doesn't derail the primary objective of communication about living diversity, the probability is that we will adopt it.

FROM THE STRICKLANDIAN CODE (1842) TO THE FIRST INTERNATIONAL CODE: THE "RÈGLES" (1905)

Several workers grasped and ran with Linnaeus's innovation: notably his foremost student Johan Christian Fabricius, who made considerable contributions to zoological nomenclature (Fabricius, 1778), especially in the field of entomology. But the process leading to the first zoological codes of nomenclature began in 1842, when the palaeontologist, malacologist, and ornithologist Hugh

Strickland formed a committee to "consider of the rules by which the nomenclature of zoology may be established" (Melville, 1995). This was the first attempt to go further than Linnaeus or Fabricius had done toward the establishment of a uniform and permanent set of rules (Strickland, 1843). Among the members of Strickland's committee were Charles Darwin and Richard Owen. Later Kiesenwetter (1858), Lewis (1872, 1875), Sharp (1873) and Blanchard (1889) introduced various alternative codes or amendments and improvements to the Stricklandian code. Without exception, their aims were to increase, or introduce, stability and an improved system for the naming of animals, though often in sharp contrast to previous workers' recommendations. Such widespread disagreement regarding some of the foremost principles of the codes, e.g., priority, continues to the present time.

The achievements of these successors to Linnaeus lacked authority; their codes were based on best practice and recommendations, and hence enforceable only by gentlemen's agreements. This situation was able to persist with some success, however, until greater numbers of taxonomists entered the arena, and by the late 19th century the situation with describing and naming new animal taxa was becoming unmanageable and even chaotic. The formation of the International Commission on Zoological Nomenclature (ICZN) in 1895 changed all of this. The establishment of an international, democratically appointed body of expert taxonomists to legislate and rule on matters of zoological nomenclature was a hugely important advance in zoological systematics. The Commission was founded at the International Congress of Zoology in Leiden in 1895, with Raphael Blanchard as the first president, continuing in this role until 1919. Despite much divided opinion within the Commission, in particular with respect to priority, the first international code, the "Règles Internationales de la Nomenclature Zoologique" (known subsequently as the "Règles") was published in 1905 (ICZN, 1905). Two years later, the Congress of Zoology in Boston announced the intention of applying this newly published code to the scientific names of several thousand of the more commonly used animal names in an attempt to effectively render them "set in stone." This has been regarded as the precursor of the ICZN's *Official Lists of Names and Works in Zoology* (Hemming, 1958; Melville and Smith, 1987; Melville, 1995; Smith, 2001) itself a precursor of the ZooBank register of animal names and nomenclatural acts (Polaszek et al., 2005a; Pyle and Michel, 2008).

This dual ambition, on the one hand of trying to set in place a set of rules to stabilize past and future animal nomenclature, and on the other to provide an authoritative and lasting list of animal names, has been a constant feature of the efforts of generations of zoologists attempting to address this area. While Strickland, Blanchard, and their contemporaries were addressing the rules, others were also working toward a complete catalogue of animal names.

INTEGRATING ZOOLOGY AND NOMENCLATURE. ALBRECHT GÜNTHER AND THE ZOOLOGICAL RECORD (1864–PRESENT)

A group of scientists from the Zoological Society of London and the British Museum founded *The Record of Zoological Literature* in 1864, which aimed to comprehensively catalogue publications on animal biology with a focus on biodiversity and systematics. Albrecht Günther (1830–1914) was the principal member and editor of the first six volumes. He has been described as its "founding father and leading spirit." Albrecht Karl Ludwig Gotthilf Günther FRS (later anglicized as Albert Charles Lewis Gotthilf Gunther), was a German-born, later British, zoologist. Günther studied theology in Bonn and Berlin, and then medicine in Tübingen. He came to the British Museum in 1857, where he studied ichthyology. Günther was Keeper of Zoology at the British Museum from 1875 to 1895. The major work of his life was the eight-volume *Catalogue of Fishes* (Günther, 1859–1870).

The Zoological Society took over publication of what is now called the *Zoological Record* in the 1880s, and supported it for a century until 1980. At this point, they entered into partnership with BIOSIS, a Philadelphia-based group of scientific publishers. In 2003 BIOSIS bought the rights to *Zoological Record* from the Zoological Society, introducing computerized technology for its production. A year later the Thomson publishing group (now Thomson Reuters) acquired BIOSIS and assumed full responsibility for its production. In 2005 a group of zoologists, led by ICZN, published a proposal to establish ZooBank—a universal registration system for the scientific names of animals (Polaszek et al., 2005a). At that time, *Zoological Record* was a critical element in the establishment of ZooBank. To set in place a global database and registration system for the scientific names of animals, huge resources were clearly needed. *Zoological Record*, with a dedicated staff of more than 30 zoology graduates daily scanning and databasing the zoological literature, was the ideal partner with ICZN for kickstarting ZooBank. In fact, ZooBank started in mid 2006 with 1.6 million scientific animal names, drawn from *Zoological Record*'s Index of Organism Names. The ZooBank initiative immediately attracted very positive global media attention, with plaudits in periodicals ranging from the *New York Times* to *The Economist* (Yoon, 2005; Anon, 2006). However, within a year, several organizations with vested interests in zoological biodiversity informatics decided unilaterally that *Zoological Record* was an inappropriate partner for ZooBank.

Nonetheless, by founding *Zoological Record*, Günther and his colleagues initiated something apparently unstoppable. One hundred and fifty years later, *Zoological Record* is still going strong, and currently indexes over 70,000 items yearly published in over 5,000 serials and non-serial sources, in many languages. Günther's legacy, the *Catalogue of Fishes*, has also led directly to one of the currently most popular and complete online zoological resources—FishBase. It is remarkable to note that of the current Commissioners of the International Commission on Zoological Nomenclature more than 20% are ichthyologists. Clearly a passion for fish systematics and nomenclature goes hand in hand with an interest in the development of zoological biodiversity informatics.

FIRST ATTEMPTS AT A COMPLETE INDEX OF ANIMAL GENERA: SCHULZE'S *NOMENCLATOR ANIMALIUM GENERUM ET SUBGENERUM* (1926–1954) AND NEAVE'S *NOMENCLATOR ZOOLOGICUS* (1939–2004)

Franz Eilhard Schulze (1840–1921) was a German anatomist specializing in sponges, in particular the Hexactinellida, and was a professor at the universities of Graz and Berlin. Sheffield Airey Neave (1879–1961) was a naturalist and collector, Secretary of the Zoological Society of London, and Deputy Director of the Imperial Institute of Entomology (Baker and Bayliss, 2009). Both scientists were convinced of the importance of developing a complete list of the generic names of animals, or a genus-level nomenclator, as a critical tool in organizing, and therefore comprehending, biological information. Both had the honor of their influential efforts' being continued after their deaths; Schultze's work covered names from 1758–1940, with a last publication date in 1954, and Neave's work was updated until 2004. Neave's work is still in regular use today despite recognized problems (http://uio.mbl.edu/NomenclatorZoologicus), and will be incorporated and improved as part of the foundation for new biodiversity bioinformatics tools such as ZooBank.

Unsurprisingly, given the state of Anglo-German relations during much of the first half of the 20th century, Schulze and Neave worked independently of each other on projects that aimed to provide precisely the same information. This familiar "reinventing the wheel" is something that our community should avoid in the future, maximizing the synergy of international collaboration instead of duplicating effort and scarce resources. On this note, and despite the comment above, it is worth reflecting on the fact that throughout the first half of the 20th century, spanning both world wars, ICZN (with a secretariat based successively in the United States or the United Kingdom) included Austrian, German, and Japanese Commissioners from 1913–1937 (Apstein), 1913–1930 (Kolbe), 1913–1945 (Handlirsch), 1930–1948 (Richter), 1935–1957 (Esaki), 1937–1944 (Arndt) (Melville, 1995). Typically, biologists were able to set aside political and nationalistic differences in the name of scientific progress and cooperation.

FIGURE 15.1 Charles Davies Sherborn aged 61.

THE FOUNDER OF BIODIVERSITY BIOINFORMATICS: SHERBORN AND *INDEX ANIMALIUM* (1902 AND 1922–1933)

With Charles Davies Sherborn (1861–1942; Figure 15.1) we come to a monumental figure, a titan in our historical succession of those documenting zoological biodiversity. *Index Animalium*, when completed in 1933, was the result of more than 30 years of ceaseless endeavor –11 volumes (including the indexes) and more than 9,000 closely printed pages in small typeface, including all animal names from 1758–1850. Sherborn personally researched and checked information for *Index Animalium*, almost exclusively from the original sources. Working daily in the Natural History Museum's library, a short walk from his Chelsea homes (Figure 15.2, Figure 15.3), he constantly sought out rare publications that the museum did not have, and after extracting the necessary information donated the volumes permanently to the museum. He also left a large legacy in his will for the purchase of further books for the museum libraries, a sum that, 67 years after his death, is still not exhausted. As Linnaeus can be rightly credited with the title "Father of Taxonomy," so Sherborn is for many of us the "Father of Biodiversity Informatics." Sherborn's biography (Norman, 1944) documents his wealth of scientific publications and other achievements besides *Index Animalium*.

Sherborn was born in 1861 in Chelsea, close to the present site of the Natural History Museum, although the museum itself was not completed and open to the public until 1881 when he was already 20. At least as early as 1883, Sherborn began to use the library of the Natural History Museum (or British Museum (Natural History) as it was then called), and continued to do so until the 1940s, being

FIGURE 15.2   The childhood home of Charles Davies Sherborn: 10 Gunter Grove, Chelsea, London SW10. Ellinor Michel, current Executive Secretary of ICZN, is standing in front.

a semi-permanent occupant of a particular corner desk. It was this persistent diligence that enabled him to successfully complete *Index Animalium*, the foundation for modern biodiversity informatics.

In August 2008, to honor Sherborn's achievements, the International Trust for Zoological Nomenclature (the organization responsible for the financial management of the ICZN Secretariat in London) initiated the Sherborn Award. At the 20th International Congress of Zoology, the award, a sculpture of the fish *Howella sherborni* (Norman), was presented to ICZN Past President Professor Alessandro Minelli. Sherborn's biographer, John Roxborough Norman, was himself an accomplished ichthyologist, and described *Rhectogramma sherborni* (later transferred to *Howella*) in 1930 after his friend and colleague. Professor Minelli is the first recipient of the Sherborn Award, largely due to his pioneering of the online Fauna Italia project (http://faunaitalia.it/index.htm), a huge achievement in its own right that also led directly to the highly successful Fauna Europaea project (http://www.faunaeur.org).

THE 21ST CENTURY—ZOOBANK

Building upon earlier zoological cataloguing and databasing initiatives such as the ICZN *Official Lists and Indexes*, Thorne (2003) proposed a system for registering the scientific names of animals via *Zoological*

FIGURE 15.3   240 King's Road Chelsea London SW10.

Record. Perhaps the most important aspect of Thorne's proposal was the suggestion that an amendment to the Code would require the mandatory registration of new names. This revolutionary proposal was later taken up by an international group of taxonomists and other zoologists, resulting in the ZooBank proposal (Polaszek et al., 2005a, 2005b; Pyle and Michel, 2008, 2009, Chapter 16 this volume).

ZooBank is an initiative that aims to harness the emerging power of biodiversity informatics. The term "biodiversity informatics" originated in 1993 in response to needs identified in the 1992 Rio Convention on Biodiversity (Behrendson, W. http://www.bgbm.org/BioDivInf/TheTerm.htm) and has been defined broadly as the application of information and communication technology (ICT) to biodiversity data, linking those data with other biological and non-biological data, and especially relevant at the species level. Johnson (2007) recently provided a working definition that is more applicable, from our viewpoint, to the task in hand: "Biodiversity informatics is an emerging field that applies information management tools to the management and analysis of species occurrence, taxonomic character, and image data." The EDIT initiative (European Distributed Institute of Taxonomy: www.e-taxonomy.eu) is an EU-funded Network of Excellence program with the goal of reducing the fragmentation of biological taxonomic research and coordinating an effort to facilitate taxonomic research using the World Wide Web (Smith et al., 2007). The development of the "scratchpad" as a taxonomist's content management system (http://scratchpads.eu/) has great potential to deliver "gold-standard" data on zootaxonomic descriptions and other nomenclatural acts in an open-access form, such as those required by ZooBank.

Registration in ZooBank of nomenclatural acts is both prospective (i.e., for new descriptions and acts) and retrospective, back to 1758. The critical mandatory aspect of registration would ensure completeness of the ZooBank registry, i.e., unless prospective names are ZooBank registered, they would be simply unavailable (Krell and Wheeler, 2007). Clearly, this is achievable only with an amendment to the current Code, and this process is already under way, with a published draft amendment (ICZN, 2008). Another crucial aspect of ZooBank is the acceptance of electronic-only published descriptions and acts as valid forms of publication under the Code (Knapp et al., 2007; Zhang, 2008).

The task of populating ZooBank requires a strategic approach. A minimum of 16,000–24,000 new taxon descriptions or other nomenclatural acts are added annually (N. Robinson, *Zoological Record*, pers. comm., with estimates of 30,000 possible, Philippe Bouchet, pers. comm.) to the 1.7–1.8 million already described animal species (Bouchet, 2006). Because an important function of ZooBank is to check these names and acts for Code-compliance, the time required to verify all names has been estimated at a minimum of 12,500 person days, or approximately 60 years (Pyle and Michel, 2008). This strategic approach would adopt in the first instance two lists of names, a "black" list of unavailable names and a "white" list of available names. The former would draw initially from the already available online *Official Indexes of Names and Works in Zoology* (Melville and Smith, 1987; Smith, 2001). The "white" list would be made of two main elements, prospective and retrospective names and acts, respectively. The former will require taxonomists to register names and acts with ZooBank online, similar to the way that gene sequences are registered with Genbank (http://www.ncbi.nlm.nih.gov/Genbank). Most authors and publishers are likely to support this process, partly because the LSIDs (Life Science IDentifiers) provide increased exposure to publications.

The Encyclopedia of Life (Hanken, Chapter 12 this volume) is a major biodiversity informatics initiative that will benefit greatly by linking to ZooBank. All species data in EoL are retrieved via the species names, each of which will be verifiable with ZooBank, thus conferring authority on these names. There appears to be great willingness among animal taxonomists to register their names and acts with ZooBank, although there is some indication that mandatory registration should wait until the project has had enough exposure and momentum. However, with the increasing likelihood of taxonomic data being published in electronic-only format, which is currently inadmissible according to the Code, a registration system may become an imminent necessity.

Retrospective registration of names and acts has begun relatively straightforwardly with the content of Linnaeus's *Systema Naturae* 10th edition, as well as a number of recent selected publications. The first species ever to be registered in ZooBank was the fish *Chromis abyssus* (Pyle et al., 2008). The timing of this act of registration was deliberately set for January 1st, 2008 to coincide exactly with the 250th anniversary of the start of zoological nomenclature. The subsequent incorporation of published names can draw from a variety of sources including *Index Animalium*, but also many up to date authoritative sources. However, the quality of these largely secondary source data needs to be clearly indicated. For example, a gold flag would indicate that the name has been checked to its original published source, a silver flag that it has been checked to a reliable secondary source (e.g., an authoritative checklist), a bronze flag indicating it comes from an unvetted source (Pyle and Michel, 2008).

We have demonstrated that progress in developing robust taxonomic checklists and nomenclators has been largely dependent on individuals who have the vision and determination to deliver products. There have been several instances where disillusionment with global initiatives has spurred individuals or small teams to go it alone, with the positive result that often these independent efforts have led to products' meeting the highest standards possible. Unfortunately, one of the potential negative consequences of this method of deriving robust nomenclatural information is occasional later reluctance by those authors to openly and freely share the products of their considerable labors.

We see a solution to this in encouraging the current move to open-access literature while working to develop appropriate individual recognition in all biodiversity information products.

SUMMARY

There is no doubt about the central role technology must have in our project to complete the Linnaean Enterprise, with ZooBank as a major driving force. Other contemporary initiatives should also benefit from the development of ZooBank, for example the Encyclopedia of Life and Catalogue of Life (Species2000). A sustainable process of benefit-sharing among data providers (usually taxonomists), data distributors (publishers, non-governmental organizations) and end-users therefore needs to be established. This will require something of a sea-change in the way we produce, process, and publish our zootaxonomic data. The products of taxonomic research belong squarely in the public domain. It has even been suggested that there is no room for publisher's copyright or other restrictions affecting the availability, use, and reuse of taxonomic data. Furthermore, the clear advantages of publishing taxonomy in a way that can be updated continuously, accommodate multimedia data at negligible cost, and is completely open-access are undeniable (Polaszek et al., 2008). ICZN is in a position to help effect this change by providing the immediately accessible ZooBank registry with authoritative information on names, allowing taxonomy to rage and flow around a stable, archived, verifiable core of nomenclature. Furthermore, ICZN is in a position to facilitate increased circulation of taxonomic information by recognizing alternative publication modes while maintaining a requirement for permanent archives for zootaxonomic data, as is appropriate for the legal guardians of original biodiversity information. Access to complete, reliable biodiversity information is too important to be made inaccessible for commercial reasons, and the completion of the Linnaean Enterprise depends on unconstrained information access.

REFERENCES

Anon, 2006. Today we have naming of parts. *The Economist* 11 February 2006.

Baker, R.A. and Bayliss, R.A. 2009. Two naturalists in Africa: Sheffield Airey Neave (1879–1961) and James Jenkins Simpson (1881–1937) with particular reference to their work on insects and ticks from 1910 to 1915. *The Linnean* 25(1): 20–28.

Blanchard, R. 1889. De la nomenclature des êtres organisés. *Compte-Rendu des Séances du Congrès International de Zoologie.* Paris, 1889: 333–404.

Bouchet, P. 2006. The magnitude of marine biodiversity. In: Duarte, C. (Ed.) (2006). *The exploration of marine biodiversity: Scientific and technological challenges.* Fundacin BBVA, Bilbao, pp. 31–62.

Fabricius, J.C. 1778. *Philosophia entomologica sistens scientiae fundamenta adiectis definitionibus exemplis, observationibus, adumbrationibus.*1–178. E. Carol, Bohnii, Hamburgi et Kilonii.

Günther, A.C.L.G. 1859–70. *Catalogue of the fishes in the British Museum.* London: British Museum (Natural History). Department of Zoology, 1859.

Hemming, A.F. 1958. (Ed.) *Official lists of family-group, generic and specific names in zoology and of works approved as available for zoological nomenclature. Official indexes of rejected and invalid family-group, generic and specific names in zoology and of rejected and invalid works in zoological nomenclature.* 1–712. International Trust for Zoological Nomenclature, London.

ICZN (International Commission on Zoological Nomenclature). 1905. Règles internationales de la nomenclature zoologique. (International rules of zoological nomenclature.) Internationale Regeln der zoologischen Nomenklatur. Rudeval, Paris, 1–57.

ICZN (International Commission on Zoological Nomenclature). 2008. Proposed amendment of the *International Code of Zoological Nomenclature* to expand and refine methods of publication. *Zootaxa*, 1908, 57–67.

Johnson, N.F. 2007. Biodiversity informatics. *Annual Review of Entomology* 52: 421–438.

Knapp, S., Polaszek, A., and Watson, M. 2007. Spreading the word. *Nature*, 446, 261–262.

Krell, F.T. and Wheeler, Q.D. 2007. Codes must be updated so that names are known to all. *Nature* 447:142.

Lewis, W.A. 1872. *A discussion of the law of priority in entomological nomenclature; with strictures on its modern application; and a proposal for the rejection of all disused names.* 1–68. Williams and Norgate, London.

Lewis, W.A. 1875. On entomological nomenclature, and the rule of priority. *Transactions of the entomological society of London* 8 (appendix): i-xlii.

Linnacus, C. 1735. *Systema naturae.* Ed. 1. Lugduni Batavorum.

Linnaeus, C. 1758. *Systema naturae.* Ed. 10. 1–824. Salviae, Holmiae.

May, R.M. 1988. How many species are there on Earth? *Science* 241: 1441–1449. doi: 10.1126/science.241.4872.1441.

Melville, R.V. 1995. Towards stability in the names of animals. A history of the International Commission on Zoological Nomenclature 1895–1995. 1–92. ICZN, London, UK.

Melville, R.V. and Smith, J.D.D. 1987. (Eds). *Official lists and indexes of names and works in zoology.* 366 pp. International Trust for Zoological Nomenclature, London, UK.

Norman, J.R. 1944. *Squire. Memories of Charles Davies Sherborn.* 1–202. Harrap, London.

Polaszek, A., Agosti, D., Alonso-Zarazaga, M., Beccaloni, G., Bjørn, P. de Place, Bouchet, P. et al. 2005a. A universal register for animal names. *Nature* 437: 477.

Polaszek, A., Alonso-Zarazaga, M., Bouchet, P., Brothers, D.J., Evenhuis, N., Krell, F.-T., Lyal, C.H.C., et al. 2005b. ZooBank: The open-access register for zoological taxonomy: Technical Discussion Paper. *Bulletin of Zoological Nomenclature* 62(4): 210–220

Polaszek, A., Pyle, R. and Yanega, D. 2008. Animal names for all—ICZN, ZooBank and the New Taxonomy. In: Wheeler, Q.D. (Ed.). *The new taxonomy.* Systematics Association Special Volume Series 76. CRC Press Boca Raton, London, New York. Chapter 8, 129–141.

Polaszek, A. and Wilson, E.O. 2006. Sense and stability in animal names. *Trends in Ecology and Evolution* 20: 421–422.

Pyle, R.L., Earle, J.L., and Greene, B.D. 2008. Five new species of the damselfish genus *Chromis* (Perciformes: Labroidei: Pomacentridae) from deep coral reefs in the tropical western Pacific. *Zootaxa* 1671: 3–31.

Pyle, R.L and Michel, E. 2008. ZooBank: Developing a nomenclatural tool for unifying 250 years of biological information. In: Minelli, A., Bonato, L. and Fusco, G. (Eds) *Updating the Linnaean heritage: Names as tools for thinking about animals and plants.* Zootaxa, 1950, 39–50.

Sharp, D. 1873. *The object and method of zoological nomenclature.* 1–39, London.

Sherborn, C.D. 1902–1933. *Index Animalium; sive, Index nominum quae ab A.D. MDCCLVIII generibus et speciebus animalium imposita sunt.* Trustees of the British Museum. London (available at http://www.sil.si.edu/digitalcollections/indexanimalium).

Smith, J.D.D. 2001. *Official lists and indexes of names and works in zoology supplement 1986–2000.* 1–139 pp. International Trust for Zoological Nomenclature, London, UK.

Smith, V S., Roberts, D., Gonzalez, M., and Rycroft, S. 2007. Scratchpads—a space for your data on the web. http://www.slideshare.net/vsmithuk/2007smith-scratchpads-edit.

Strickland, H.E. 1843. Report of a committee appointed "to consider of the rules by which the nomenclature of zoology may be established on a uniform and permanent basis." *Report of the 12th meeting of the British Association for the Advancement of Science*, 105–121.

Thorne, J. 2003. Zoological record and registration of new names in zoology. *Bulletin of Zoological Nomenclature* 60: 7–11.

von Kiesenwetter, H. 1858. Gesetze der entomologischen Nomenclatur. *Berliner entomologische Zeitschrift* 2: 11–22.

Wheeler, Q.D. and Krell, F.T. (2007) Codes must be updated so that names are known to all. *Nature* 447, 142.

Wilson, E.O. 2000. A global biodiversity map. *Science* 289: 2279.

Yoon, C.K. 2005. In the classification kingdom, only the fittest survive. *New York Times* 11 October 2005.

Zhang, Z.-Q. 2008. Opening the door to electronic publication of zoological nomenclature and taxonomy. *Zootaxa* 1908: 68. www.mapress.com/zootaxa.

16 ZooBank
Reviewing the First Year and Preparing for the Next 250

Richard L. Pyle and Ellinor Michel

CONTENTS

INTRODUCTION

COMPLETING THE LINNAEAN ENTERPRISE

With the publication of *Systema Naturae* in 1758, Carolus Linnaeus launched a system for naming and classifying animals that would endure for the next two and a half centuries, right up until today. Throughout this amazingly long (in the context of scientific methodology) span of time, the "Linnaean Enterprise" of establishing "sense and stability" to animal diversity has expanded far more in content than it has in its basic goals, methods, and vision. This is not in any way a failure by the taxonomic community to evolve over the years so much as it stands as testament to the deep understanding and appreciation Linnaeus had for the natural living world. But Linnaeus was fortunate, in some sense, to have lived at a time when the entirety of zoological diversity could be represented in a single paper-printed volume. The names he gave the species he recognized were unique and unambiguous. He could not have known that, 250 years hence, despite a continued and accelerating process of species discovery, humanity would still not know the true extent of animal diversity even within an order of magnitude. He would have been irritated to note that his system of ordering names, which worked so well for a few thousand names in a single work, began to fray at the edges when information became dispersed, when zoologists continued to add to the system in different languages from distant corners of the earth, with access to different subsets of the existing system of names. As the literature on species descriptions grew, the challenge of keeping it accessible and regularized was met by a few visionary biologists.

The Linnaean Enterprise was greatly expanded in the first half of the 19th century when Albrecht Günther founded the *Zoological Record* in 1834 as an effort to catalogue all publications in zoology. Since its inception, the *Zoological Record* has not only archived the past accomplishments of zoological research, it has provided a model for biodiversity informatics, in that it integrates disparate fields of biology and bridges them through the "common denominator" of all biological sciences: scientific names. The sweeping ambition of *Zoological Record* has been a remarkable success story, providing a critical tool that has improved biologists' ability to synthesize information of all kinds. It continues today, engaging a team of biologists to catalogue publications in over 5,000 serials and other publications, entering 75,000 records each year. Over 3.8 million records can be found through *Zoological Record*. Although this gives access to the majority of zoological publications, it is estimated that there are still a significant proportion that escape this catalogue as they are published in obscure sources.

By the turn of the 19th century it was clear that there was a need to bring order to this deluge of biological information by providing a backbone of a complete catalogue of scientific names, a nomenclator, to act as an authoritative source for the published origin of the names. Early attempts at nomenclators had been made by Agassiz (1848), Marschall (1873), and Scudder (1882), among others; however, they were insufficient and not maintained. In 1902 Charles Davies Sherborn published the first installment of what has become the most ambitious listing of all animal names to date, both fossil and recent, entitled *Index Animalium*. He was the right man for this job, as he was an exacting cataloguer, based at an institution with a superlative library of natural history (the then-named British Museum (Natural History), London) where he could check each reference personally. He had three unambiguous aims: (1) to provide a complete list of all generic and specific animal names, (2) to provide a reference for each description, and (3) to give an exact date for each page quotation (Sherborn, 1902 p. vi). His first volume of 1,200 pages treated 60,000 species that he had directly checked in 1,300 references that dated from 1758–1800; this took him twelve years, including a period of ill health (which he carefully documents as totalling three years' break from the intensive work). He worked on this catalogue for 31 more years, producing eleven volumes, commenting that

> … for accurate work it is necessary for the student to verify every reference he may find; it is not enough to copy from a previous author; he must verify each reference itself from the original. Bad work, for which there is little excuse, is only too common. (Sherborn, 1932)

Index Animalium is still seen as an essential resource today, with its online version receiving approximately 1,150 unique visitor hits per month. This monumental work has justifiably earned Sherborn the title of "The Father of Biodiversity Informatics" (see Polaszek and Michel, Chapter 15 this volume).

Sherborn's masterwork was arranged by species. This, for example, is the listing for our own taxon:

sapiens   Homo, Linnaeus, Syst. Nat., ed. 10, 1758, 20 ; ed. 12, 1766, 28 ; varr. ferus, americanus, europaeus, asiaticus, afer, monstrosus.

His reasons for this arrangement have at their core the understanding that his work was a resource for the bibliographic origins of scientific names, which should remain independent of decisions on taxonomy. A nomenclator provides access to the data needed to determine priority and avoid homonymy, or duplication, of the same name for different taxa. Sherborn's clearly stated thoughts on this (1921, p. ix) were that:

1. No synonymy of species is attempted: that depends on the idiosyncrasy of the systematist.
2. Any attempt at specific synonymy would be opposed to progress, as experience shows that vast changes may take place in a single year.

3. An arrangement under species permits of a generic synonymy, for by running the eye down the second column of the printed work, it will be possible to ascertain the various generic names with which a particular species name has been connected.

However, a reliable nomenclator was also needed for generic (and sub-generic) names; homonymy is an equally insidious problem for these names, and the generic level is often the point of entry for taxonomic questions. Two independent efforts to redress this were published in the early part of the 20th century. The first of these, *Nomenclator Animalium generum et subgenerum*, was published in 1926 by widely respected invertebrate zoologist Franz Eilhard von Schulze and included generic animal names from Linnaeus. Work on *Nomenclator Animalium* continued through 1954 by W. Kükenthal and K. Heider. Another approach to the same goal was made by Sheffield Airey Neave with his *Nomenclator Zoologicus*, which now catalogues the bibliographic origin of the 340,000 genera and sub-genera described from Linneus 1758 to 1994 (Remsen et al. 2006). This work is available online and heavily used, drawing approximately 2,300 searches per month from 19,000 individual users per year. Neave (1939, p. v) explained in the Foreword to what was to become an 11-volume work, that in 1934

… systematists found themselves in great difficulties for lack of up-to-date information relating to generic names, largely owing to the fact that no index to new generic names had been published since 1910, and that the *Nomenclator Animalium generum et subgenerum* published by the Prussian Academy of Science was neither complete nor up-to-date, besides being very costly.

Thus, this second effort at a generic nomenclator was undertaken for scientific, economic and, we might guess, a sliver of nationalistic reasons.

With the advent of computers and the Internet came much more powerful ways to organize and access information. The Linnaean Enterprise expanded into the electronic domain early on in the history of the personal-computer revolution, with scientists and amateurs immediately building taxonomic databases of varying degrees of scope, completeness, and reliability. However, these efforts remained disconnected and inaccessible. The advent of Internet connectivity and expanding vision sowed the seeds for ambitious Linnaean projects of an altogether grander scale, such as the Catalogue of Life (www.catalogueoflife.org), and the Encyclopedia of Life (www.eol.org). These initiatives aim to bring together taxonomic and other relevant biological information, on recognized (or valid) taxa of living metazoans with the ambitious aim to eventually construct largely complete information sources. However the accompanying nomenclatural information is included largely as a by-product of the taxonomy, without a mandate to ensure that names are International Commission on Zoological Nomenclature (ICZN) Code-compliant, and their authority does not include a legislative body on nomenclature. As so succinctly expressed by Sherborn, taxonomy needs a robust source for nomenclatural information that is independent of taxonomic judgments. Names come in and out of valid use depending on the perspective of the taxonomist making judgments on synonymy; however, only names that have been published according to the rules of the ICZN Code are available.

LINKING BIODIVERSITY INFORMATION THROUGH NAMES

Names are the natural organizing linkage for biological information, as language is the natural structure for thought and communication. Robust names with unambiguous meanings are important not only for taxonomists to recognize which units they are describing, but also for all other users of biological information, including people working in agriculture, human and veterinary medicine, conservation, ecology, molecular biology, law, and policy. In fact, any human concern that involves the living world should ideally have an unambiguous nomenclature; however, work involving science and its products may have a more explicit justification for unambiguous names than daily concerns. Linnaeus recognized this as a particular need for communication in science, as it is more international, more nuanced in detail, and with greater consequences for misinterpretation than many other kinds of communication. This stimulated him to develop his binominal system. In the first edition of the ICZN Code (1961, p. iv), J. Chester Bradley said, "Ordinary languages grow spontaneously in innumerable directions; but biological nomenclature has to be an exact tool that will convey a precise meaning for persons in all generations."

Having the tool of an (ideally) exact, unambiguous nomenclature with 250 years of scientific legacy linked to the names gives us powerful leverage on information. The power of information technologies allows us to expand the vision of *Zoological Record,* which worked well even as print on paper, to make an order of magnitude more information available to all users, and make it easier to organize once it is in hand. This can only improve scientific practice, which builds incrementally on past knowledge.

THE ROLE OF ICZN AND ZOOBANK

If names are the logical links for scientific information about animals, we need to ensure that those links are as robust and unambiguous as possible. As early as the mid-19th century it was realized that without guiding rules and the cooperation of the taxonomic community, zoological nomenclature would falter, with consequent serious loss of access to information. A group of concerned luminary zoologists headed by Hugh Strickland and including Charles Darwin (originator of the theory of evolution), Richard Owen (renowned anatomist and founder of the Natural History Museum, London) and John Westwood (first chair in Entomology at Oxford), set out a code of rules for the scientific naming of animals. Their body was the precursor for what is now the ICZN, an international organization of distinguished zoologists whose mandate is to continue to formulate and update rules for naming animals, and to adjudicate in situations of confusion or dispute. These rules are published in the ICZN Code of Nomenclature ("the Code," currently in its 4th edition, 1999, and online www.iczn.org) with supplemental amendments and declarations. By ensuring that nomenclature is applied in a globally consistent way, the ICZN provides continuity both for new species discoveries and for the correction of errors and inconsistencies in past works. The current ICZN mandate also includes developing tools for making nomenclature accessible through modern technology.

Having a complete listing of existing available names, exposing and preventing further homonymy were the central aim of the authors of the great nomenclators of the past; used appropriately, bioinformatics will make this task much easier to accomplish. In 2005, the ICZN Secretariat and Commissioners initiated the process of achieving this goal by proposing ZooBank as a Web-based registry of zoological names and nomenclatural acts (Polaszek et al., 2005a). A series of meetings among the Commissioners, and presentations to the taxonomic community allowed input in the initial model for ZooBank. ZooBank is not intended to replace existing nomenclatural catalogue databases, and it makes no assessment or judgment of the taxonomic content of any published work.

IMPLEMENTING ZOOBANK: THE DEVIL HIDES IN THE DETAILS

There are two aspects to implementing ZooBank: technical and policy. For the most part, the technical implementation is relatively straightforward. Policy decisions, however, are much more subtle, and have proven to require more careful consideration. ZooBank derives its legitimacy through the ICZN, the widely recognized international body charged with maintaining the rules of zoological nomenclature. As such, policy decisions related to ZooBank are at the collective discretion of the ICZN Commissioners. In August 2008, in connection with the Symposium from which this chapter was taken, the ICZN held a meeting in Paris that was attended by 13 Commissioners, with eight external professional participants. A full day of the three-day meeting was devoted to presentations and issues related to ZooBank, and during that meeting, a ZooBank Committee was formed to move forward with discussions and recommendations for ZooBank development and design (Pyle, 2008). In the days that followed, a series of related meetings addressed the role of ZooBank in the broader context of emerging global taxonomic data infrastructure, as well as for potential data content providers. One of the clear themes that emerged from those meetings was the sense that, while the grand vision of ZooBank is easy to articulate, the specific implementation details are what require the most careful thought and consideration. The remainder of this chapter provides an overview and brief discussion of some of those considerations.

INITIAL LAUNCH AND EARLY DEVELOPMENT

The first public implementation of ZooBank was launched at midnight, GMT, on January 1st, 2008—exactly 250 years to the day after the official start of zoological nomenclature (i.e., the date officially fixed for the publication of *Systema Naturae*). At that time, the content of ZooBank (zoobank.org) included 4,819 names established in *Systema Naturae*, as well as five new species of damselfish published concurrently with the launch of ZooBank (Pyle et al., 2008).

From its launch through to the time of this writing, the public ZooBank Web site allows only searching of and read-only access to existing ZooBank entries. Active development of a Web-based registration interface over the past year has resulted in nearly a thousand additional names from over 3,500 publications and over 4,000 authors. Two grants from GBIF (Global Biodiversity Information Facility) and one from TDWG (the biodiversity informatics standards body) have supported the implementation and content expansion process, and at least two journals (*Zootaxa* and *ZooKeys*) are proactively including ZooBank registration as part of their ongoing publication process.

Development and testing of Web-based interfaces to allow addition of content and editing of existing content continues, in preparation for a wider launch to allow more active and direct participation by a broader set of practicing taxonomists.

GLOBALLY UNIQUE IDENTIFIERS

One of the core aspects of registration is the assignment of persistent Globally Unique Identifiers (GUIDs). In a sense, the names themselves have served the function of identifiers from the perspective of humans communicating with humans; however, due to changing combinations, alternate spellings both mandated by the Code (gender agreement and emendations) and unintentional (misspellings and typographical errors), as well as homonymy (both within zoology, and among names governed by other codes of nomenclature), these names are neither persistent nor unique enough to serve the function of a GUID in the context of electronic information management.

ZooBank is following the lead of GBIF and TDWG in implementing Life Science Identifiers (LSIDs) as the GUID for ZooBank registration entries. LSIDs conform to a standard format consisting of minimally five components, each separated by a colon (:). The first two components, "urn:lsid," are common to all LSIDs, and simply establish the LSID as a "uniform resource name" (urn), and an LSID. The third component represents the "authority identifier," which usually takes the form of a Web domain name. As such, all ZooBank LSIDs incorporate the text, "zoobank.org" as the authority identifier. The fourth component of an LSID is the "namespace identifier." In the case of ZooBank, this corresponds with the "domains" of data that will be registered in ZooBank. These are discussed in more detail in the following section, but the four namespace identifiers already implemented in ZooBank are "act" (for nomenclatural acts), "pub" (for publications), "author" (for zoological authors), and "specimen" (for primary, or name-bearing, type specimens).

The fifth component of an LSID is the "object identifier," which is simply a number or other identifier that is unique within a particular namespace of a particular authority. According to the LSID specification, the combination of authority identifier, namespace identifier, and object identifier must collectively be unique on a global scale. However, the object identifier component of ZooBank LSIDs is itself a globally unique "Universally Unique Identifier" (UUID). UUIDs are 16-byte (128-bit) numbers, usually represented by hexadecimal characters (the numbers 0–9 and the letters A-F) in the format of 8 characters, a hyphen, three sets of four characters, each separated by a hyphen, followed by another hyphen and an additional 12 characters. Some unappealing aspects of using UUIDs in this context are that they are unpleasant to look at, nearly impossible for most mortals to memorize, and take a lot of space when typed out in their entirety. However, GUIDs in general, including ZooBank LSIDs in particular, are not intended to be optimized for human consumption and interpretation. Indeed, the taxon names themselves serve this function rather effectively (as they have for the past 250 years)—given the ability of human brains to resolve any ambiguity from homonymy or alternate spellings when provided sufficient context. Rather, ZooBank LSIDs (like all GUIDs) are optimized for consumption and interpretation by computers. Moreover, by using the globally unique UUIDs as the object identifiers within LSID, ZooBank is effectively safeguarding against the possibility that LSIDs will not persist as an adopted style of GUIDs in biodiversity informatics. With an object identifier that is itself globally unique, ZooBank can strip the LSID "wrapper" (i.e., the first four components of the LSID) and still maintain a true GUID for its data objects. This GUID (the UUID) can then be "wrapped" in other existing (or not yet invented) resolution protocols and syntaxes, and thus stand the greatest chance of persistence over the long-term.

Defining "Registration"

Somewhat surprisingly, one aspect of ZooBank that has not yet been well defined is exactly what constitutes "registration." Until the ICZN adopts procedures for electronic publication of new names and nomenclatural acts, the information that will ultimately end up in ZooBank begins (in virtually all cases) as ink on paper. Before this information can be entered into ZooBank, it must first be converted to electronic textual form. Whether this is accomplished by manual keystrokes or by optical character recognition (OCR) of electronically scanned pages, the information itself usually ends up in a computer database. Many, many such databases exist and predate the existence of ZooBank, and thus the majority of existing electronic content that will ultimately reside within ZooBank is not yet "registered."

The first step in getting this existing electronic information into ZooBank is to import it into the ZooBank database. Because of the way the ZooBank database is set up, at the time of data import every entry is automatically assigned its permanent UUID. However, this UUID is not converted into a formal ZooBank LSID until it has undergone some "prescreening" process. Exactly what this process entails is one of the many policy details of ZooBank implementation that has not yet been formalized. However, some of the basic criteria for assigning a formal ZooBank LSID would likely include:

- Assurance that the same data object has not already been assigned a ZooBank LSID
- Assurance that the data object is indeed among the domains of objects included within ZooBank (i.e., a nomenclatural act, a publication, an author, or a type specimen)
- Assurance that the data object falls within the scope of ZooBank (e.g., that it is, in fact, relating to zoological nomenclature as opposed to botanical nomenclature)
- Some minimum level of information content (i.e., minimum required data fields, which has yet to be defined).

Other, as yet undetermined criteria may also need to be fulfilled before the assignment of a ZooBank LSID.

The assignment of a ZooBank LSID is only the first formal step in the registration process. At least one (but possibly more than one) additional step is required to ensure that the information is complete and accurate. "Complete" can be defined in several ways, such as minimum required data fields, as well as minimum standards for what must actually be entered into those fields. "Accurate" is a somewhat nebulous term in this context, but is generally considered to mean that a qualified person (i.e., a person with substantial familiarity with the ICZN Code) has examined a copy of an original publication (or, perhaps, a facsimile of an original publication, such as high-resolution digital scans of the pages), and has verified that the information was accurately transcribed from the printed page to the ZooBank database, and that the original printed page meets the criteria for availability as prescribed by the ICZN Code. In this, we aim to adhere to the standards promoted by Sherborn.

Still unresolved is where, exactly, within this workflow of information capture, the word "registered" is appropriately applied. Furthermore, the specific steps of this workflow, who should be fulfilling those steps, and the exact process by which those steps are fulfilled, all remain among the details of ZooBank yet to be established through formal policy. Nevertheless, it seems reasonably safe to presume that the ZooBank data workflow will involve some sort of "staging" area for content prior to assignment of LSIDs, followed by a "validation" process whereby LSID-assigned data objects are scrutinized for completeness and accuracy, and finally some sort of indication that the record has been validated. In fact, we expect that by virtue of ZooBank's being updatable by experts, the effect of having "many eyes" examining the data will quickly bring it to a higher standard than Sherborn could achieve as a single individual scanning tens of thousands of entries over many years.

What Data Objects Should Be Registered?

There are two dimensions of how the scope of ZooBank data content needs to be defined. The first is in terms of what "domains" of objects should receive ZooBank LSIDs, and the second concerns the scope of instances within each domain.

Currently, four "domains" of data objects have been identified as falling within ZooBank's purview: nomenclatural acts, publications, authors, and type specimens. Each of these has some direct bearing on Articles of the ICZN Code, and therefore represent logical objects to register in ZooBank.

A nomenclatural act exists in the form of a taxonomic treatment appearing within a publication deemed available by the ICZN Code. In the context of ZooBank, "nomenclatural acts" are those particular taxonomic treatments (or "usage instances") that are themselves governed by the ICZN Code. Published treatments of new taxon names in the species group, genus group, and family group all constitute Nomenclatural Acts. Other kinds of nomenclatural acts include lectotypifications and neotypifications, emendations, and actions by first revisers. Other kinds of nomenclatural scts may or may not fall within the scope of ZooBank. For example, the botanical Code governs acts that form novel combinations of preexisting species group names within preexisting or newly created genus group names. The zoological Code does not govern such acts per se, but they do have some relevance to the ICZN Code in that they may create secondary homonyms. Also, the Code does include provisions that apply to names above the family group rank, but it is not immediately clear whether such names (and the nomenclatural acts that established them) should be included within ZooBank.

While it is clear that names and other nomenclatural acts that adhere to the provisions of the Code should be registered in ZooBank, less clear is whether names and acts that fail to meet such provisions should also be registered. Including them would open a Pandora's box of issues relating to the near-infinite scope of names for animals that are not governed by the Code. However, exclusion of ungoverned names could limit the practical utility of ZooBank, in that quick access to a list of names and acts unavailable under the Code would clearly provide a valuable function to the taxonomic community (Pyle and Michel, 2008). Balancing the practicality of initiating a useful list of known unavailable names with the Sisyphean task of assembling them needs to be weighed up in the development of ZooBank.

Regardless of the final scope of taxonomic treatments that are defined as nomenclatural acts within ZooBank, all such Acts come to exist through publications. Article 8 of the ICZN Code defines published works in the context of zoological nomenclature, and establishes criteria for determining whether such works are available under the Code. Certainly, all publications that contain registered nomenclatural acts (however they are defined) should be included within the scope of registered ZooBank publications. But there may be other publications worthy of registration, such as works that have been explicitly rejected—if for no other reason than to make such explicit rejection clear to all ZooBank users (Pyle and Michel, 2008).

Although not explicitly governed by the ICZN Code as nomenclatural acts or publications, authors of taxon names and acts are integral to zoological nomenclature, and are addressed by several Articles in the Code. Again, it is clear that authors of registered published works would themselves be registered, but there are other kinds of authors who may fall within the scope of ZooBank. For example, contributors to the ZooBank registry may also be registered as "authors," even if they have not authored registered publications.

The fourth domain within ZooBank—the "primary" or "name-bearing" type specimens (such as holotypes, syntypes, lectotypes, and neotypes)—obviously has important relevance to Code-governed zoological nomenclature. However, specimen data traditionally fall within the purview of the institutions that house natural history collections. The consensus among Commissioners at the Paris ICZN meeting in August 2008 was that type specimens represent a legitimate domain within ZooBank, but it remains unresolved whether "secondary" (non-name-bearing) types may also be entered into the ZooBank registry.

A fifth domain of data object considered for inclusion within ZooBank by the ICZN Commissioners is the type-specimen repository. There are certain Articles in the Code that address the collections in which type specimens may be deposited, so there are grounds for including this domain as part of the ZooBank initiative. However, there are many unresolved details about how such entities would be defined, and to what extent their registration would or would not have any bearing on the availability of names established based on type specimens housed therein. Moreover, at least two other initiatives (the Biodiversity Collections Index, and the Registry of Biological Repositories) already exist to serve this function, and it is not yet clear how ZooBank would interact with these other two registries.

Prospective Registration

As originally conceived, the primary function of ZooBank is to serve as a registry for new publications and nomenclatural acts, such that registration occurs before or immediately after the publication of a work containing nomenclatural acts (Polaszek et al., 2005a, 2005b). This aspect of ZooBank serves the community by allowing the rapid dissemination of new information, and establishes a foundation upon which a future version of the ICZN Code would require that all new publications and acts be registered in order to be regarded as available under the Code.

The major obstacle to implementing "prospective" registration in ZooBank is in getting taxonomists to willingly contribute their time to enter the content into the ZooBank database. Compulsory registration, as enforced by the Code, would be successful only with a willing taxonomic community, so requiring registration will not, by itself, guarantee the success of ZooBank (i.e., it may instead lead to large-scale non-compliance with that requirement of the Code). The first step in garnering willing support from the taxonomic community is to make the processes of entering and editing records in ZooBank very simple and intuitive. Efforts in this regard are currently in development. But beyond merely lowering the "cost" of entering content by simple user interfaces, ZooBank needs to provide taxonomists with value-added features that help them get their jobs done. Some of these features could be implemented within ZooBank—such as free "subscription" services that notify members of the taxonomic community of new publications with nomenclatural content, perhaps increasing awareness of an author's publication and subsequently increasing its citation value. But most of the benefits that ZooBank will be able to offer the community will likely involve integration with external data resources, which can help taxonomists locate and access information of direct relevance to their work. Steps are currently being taken to ensure ZooBank is tightly integrated with other informatics initiatives, but this is an area of ZooBank that needs further input from the community.

Another way to facilitate prospective registration is through cooperation with active journals that publish works containing information of nomenclatural importance. Two such journals (*Zootaxa* and *ZooKeys*) are actively engaged in the development of ZooBank, and have played an important role in keeping ZooBank moving forward. Other journals have expressed a strong interest in becoming similarly involved. Perhaps within the near future, a tight alliance with ZooBank will increase the value of publications, enabling such journals to draw more quality manuscript submissions and enhance their impact factor.

We believe that the most challenging difficulty in establishing a regimen of prospective registration is in establishing the right policies and protocols of ZooBank. In broad strokes, some of the core principles are relatively easy to define. But as alluded to above, the difficulty lies in the specific details. Three general scenarios have been described for the general function and workflow for prospective registration (Polaszek et al., 2008; Pyle and Michel, 2008). The only way to arrive at satisfactory answers to many of the outstanding questions is through maintaining active dialog among practicing taxonomists. Some of this dialog has already begun, but much more is still needed.

RETROSPECTIVE REGISTRATION

A separate, but equally important aspect of ZooBank is retrospective registration. This shares with prospective registration many of the detailed issues relating to minimum data requirements and validation. The task of populating ZooBank with complete legacy information on nomenclatural acts is not trivial. It should be kept in mind that there are an estimated 16,000–24,000 new additions for animal names yearly (N. Robinson, Zoological Record pers. comm., P. Bouchet, pers. comm.) to an estimated 1.7–1.8 million described animal species (Bouchet, 2006). As each species may have from one to ten (or even more) synonyms, the numbers of names to be checked for homonymy and primary synonymy is enormous. Estimates are often based on a few well-studied groups, but it should be underscored that the pattern of taxonomic work differs greatly among taxonomic groups.

There is a wide range of possible sources of such content, ranging from scanned literature with text derived from OCR software, such as the increasing body of content originating from the Biodiversity Heritage Library to citizen scientist initiatives, as have worked for GalaxyZoo in astronomy (www.galaxyzoo.org) or Herbaria@home in collections management (http://herbariaunited.org/atHome/), to highly robust nomenclatural databases for certain groups of animals, which could serve as seed content for retrospective registration content. Bringing together information from disparate sources is a great challenge for a project like ZooBank that aims ultimately to have only information of the highest reliability. However, in the long term, successful fusion of information will allow ZooBank to exceed the qualities of its predecessors, the published nomenclators of Sherborn, Neave, and von Schulze. Sherborn pointed to taxonomic catalogues as sources for the most informed data on scientific names. He commented, with a heartrending plea, that:

Although much time has been expended in trying to secure the endless combinations and permutations of specific names, it is felt to be impossible for one human being to attain completeness in this direction by reason of the colossal amount of literature to be dealt with. Those who wish to gain this desirable result are referred to such works as the British Museum Catalogues of Birds, Marsupials, and Fossil Fishes, Brady's Report on the Foraminifera of the Challenger, Dalla Torre's Hymenoptera, Bronn's Index Palaeontologicus, Stiles and Hassall's Indexes to Worms, etc., where such attempts have been carried to successful conclusion. Still, a great mass of references has been here included which it is hoped will have secured all generic and trivial names, and put the searcher on the track of a more complete synonymy. But even in this direction the methods adopted by many authors are such as to baffle the ingenuity of the recorder unless he happens to be a specialist in each group. Objection may be raised to those cases where the trivial name is referred back to a previous genus without a reference being given. I plead for compassion. Many hundreds of these cases have been pursued only to find that the author has quoted a previous author wrongly either by name or for the genus, and the time has been wasted. I am no longer young and, regrettable though it may be to me to leave such references unverified, I know my life is limited and I must press on. (Sherborn 1921, p. ix)

ZooBank will be able to harness the expertise of many sources without driving any single zoologist to a sorrowful end.

Once again, while the broad concept seems straightforward, the detailed implementation raises many issues. For example, what motivation would the owners and managers of such existing database content have in allowing portions of their databases—often having been built over decades of difficult effort—to be wholesale included within ZooBank? Further to this, how can ZooBank ensure substantial cross-linking back to original source content such as type collections, both in terms of providing appropriate credit for the data source, as well as enabling users of ZooBank to quickly access more detailed information that may be available on the original source? In the case of willing content providers, what standards of data quality should be applied to the imported content, and at what stage of validation could content be directly imported? How could such standards be applied across disparate data sets, which may have come about as a result of different priorities? For example, many nomenclatural databases include their own internal measures of quality and completeness; there would need to be protocols for translating these myriad quality metrics to the established ZooBank standards of quality (whatever those turn out to be).

The issues and associated questions are not trivial, and will likely be resolved and answered only on a case-by-case basis, following guidelines established collaboratively by the ICZN, the existing content holders, and the broader taxonomic community.

QUANTITY VS. QUALITY

At the heart of many of these outstanding issues lies a fundamental interplay between data quantity and data quality. Though oftentimes represented as a trade-off, these two measures of ZooBank source content are not necessarily mutually exclusive. As mentioned previously, there may be several levels of validation along the entire process of ZooBank registration. That larger quantities of content may be imported at a low level of validation does not preclude the establishment of rigorous policies and protocols for designating high levels of validation. As long as the current stage or level of validation is made clear on the ZooBank Website, both quality and quantity can coexist within the same system.

THE NEXT 250 YEARS

Though there is a persistent sense of urgency for ZooBank to quickly become adopted as the official online registry for zoological nomenclature as mandated through provisions in a future version of the ICZN Code, there is an equal (if not greater) need to "get it right." The Linnaean Enterprise has persisted for two and a half centuries, and the ICZN Code is itself more than a century old. While information technology and shifting paradigms of scientific discourse advance at an amazing pace, ZooBank must strike a careful balance between satisfying immediate needs while both honoring a robust historical legacy and anticipating needs for the next 250 years.

Several initiatives are currently in the early stages of development, which will certainly have direct bearing on the role and function of ZooBank. The Global Names Architecture (GNA)—an effort to establish a common set of indexes and Web services intended to crosslink taxonomic data from a broad array of sources—will rely heavily on nomenclatural authorities such as ZooBank and its associated content providers to build such crosslinks. Large initiatives such as GBIF and the Encyclopedia of Life have revitalized the perceived need to find harmony among the different nomenclatural codes.

There is no doubt that, as the ICZN moves forward with such initiatives as the acceptance of electronic forms of published works (ICZN, 2008) and a revised 5th edition of the Code, and the

Internet fundamentally transforms the way scientific information is communicated among research-ers, the Linnaean Enterprise is on the cusp of a fundamental paradigm shift. Whether it main-tains the same level of relevance it has enjoyed these past two and a half centuries during the 250 years to come, may well depend on how carefully the taxonomic community of today navigates the uncharted waters in which it now finds itself.

ACKNOWLEDGMENTS

We wish to thank Andrew Polaszek for organizing a very stimulating symposium, Jon Todd, Ken Johnson, and the many ICZN Commissioners and other interested taxonomists, especially on the ICZN list server, who have contributed many of the ideas included herein. Keri Thompson (Smithsonian Libraries) and David Remsen (GBIF) provided current usage statistics for *Index Animalium* and *Nomenclator Zoologicus*.

REFERENCES

Agassiz, L. (1848) *Nomenclatoris zoologici index universalis : continens nomina systematica classium, ordi-num, familiarum et generum animalium omnium tam viventium quam fossilium... homonymiis plan-tarum).* Soloduri : Sumptibus Jent et Gassman. 1135 pp.

Bouchet, P. (2006) The magnitude of marine biodiversity. In: Duarte, C. (Ed.) (2006). *The exploration of marine biodiversity: Scientific and technological challenges.* 31–62.

ICZN (2008) Proposed amendment of the *International Code of Zoological Nomenclature* to expand and refine methods of publication. *Zootaxa*, 1908: 57–67.

Linnaeus, C. (1758) *Systema naturae per regna tria naturae, secundum classes, ordines, genera, species, cum characteribus, differentiis, synonymis, locis.* Tomus I. Editio decima, reformata. 10th ed., Vol. 1, pt. 1. Holmiae, 1, ii+824 pp.

Marschall, A.F. (1873) *Nomenclator zoologicus continens nomina systematica generum animalium tam viven-tium quam fossilium, secundum ordinem alphabeticum disposita, aspiciis et sumptibus* C.R. Societatis zoologico-botanicae / conscriptus a comite. Ueberreuter (M. Salzer), 482 pp.

Melville, R.V. (1995) *Towards stability in the names of animals: A history of the International Commission on Zoological Nomenclature 1895–1995.* International Trust for Zoological Nomenclature, London. 104 pp.

Neave, S.A. (1939–1996) *Nomenclator Zoologicus: A list of the names of genera and subgenera in zoology from the tenth edition of Linnaeus, 1758, to the end of 1935 (with supplements).* Zoological Society of London, London. (also available online http://uio.mbl.edu/NomenclatorZoologicus/browse.html).

Polaszek, A., Agosti, D., Alonso-Zarazaga, M., Beccaloni, G., Bjørn, P. de Place, Bouchet, P., et al. (2005a) Commentary: A universal register for animal names. *Nature*, 437, 477. (http://www.iczn.org/Nature_Commentary.pdf).

Polaszek, A., Alonso-Zarazaga, M., Bouchet, P., Brothers, D.J., Evenhuis, N., Krell, F.-T., et al. (2005b) ZooBank: the open-access register for zoological taxonomy: Technical Discussion Paper. *Bulletin of Zoological Nomenclature*, 62, 210–220. (http://www.iczn.org/ZooBank_Paper.htm).

Polaszek, A., Pyle, R., and Yanega, D. (2008) Animal names for all: ICZN, ZooBank, and the New Taxonomy. pp. 129–142. In: Wheeler, Q.D. (Ed.). *The new taxonomy.* CRC Press, Boca Raton. 237 pp.

Pyle, R. (2008) Summary of session on ZooBank (session 5). *Bulletin of Zoological Nomenclature* 65: 257–260.

Pyle, R.L., Earle, J.L., and Greene, B.D. (2008) Five new species of the damselfish genus *Chromis* (Perciformes: Labroidei: Pomacentridae) from deep coral reefs in the tropical western Pacific. *Zootaxa*. 1671, 3–31. (http://www.mapress.com/zootaxa/2008/f/zt01671p031.pdf).

Pyle, R.L. and Michel, E. (2008) ZooBank: Developing a nomenclatural tool for unifying 250 years of biologi-cal information. In: Minelli, A., Bonato, L. and Fusco, G. (Eds.) *Updating the Linnean heritage: Names as tools for thinking about animals and plants. Zootaxa, 1950*: 39–50.

Remsen, D.P., Norton, C., and Patterson, D.J. (2006) Taxonomic informatics tools for the electronic *Nomenclator Zoologicus. Biological Bulletin* 210: 18–24.

Scudder, S.A. (1882) *Nomenclator zoologicus.* An alphabetical list of all generic names that have been employed by naturalists for recent and fossil animals from the earliest times to the close of the year 1879. *Bulletin of the United States National Museum* 19. 340 pp.

Sherborn, C.D. (1902–1933) *Index animalium; sive, Index nominum quae ab A.D. MDCCLVIII generibus et speciebus animalium imposita sunt.* Trustees of the British Museum. 7056 pp. (also available online http://www.sil.si.edu/digitalcollections/indexanimalium/).

17 Celebrating 250 Dynamic Years of Nomenclatural Debates

Benoît Dayrat

CONTENTS

By promulgating today these *Règles internationales*, the International Congresses of Zoology do not have the pretension to have accomplished a definitive work. Exactly as the rules that sufficed in Linné's time could not meet our needs, the code that seems to us to appropriately answer our current concerns will be regarded as inadequate by our successors. Science moves forward: it asks new questions, for which new solutions must be found. (Preface to the first *Règles internationales de la nomenclature zoologique*, Blanchard in ICZN 1905:11, tr.)

INTRODUCTION

Biological nomenclature is fundamentally important to all biological sciences and beyond, because it ensures that one taxon is designated by only one name and that one name refers to only one taxon. This is what Darwin understood when he accepted to be one of the eight members of the British Association committee that in 1842 devised the first code of zoological nomenclature (Strickland 1843). Biological nomenclature, which governs the application of taxon names across the Tree of Life, is one of the pillars of the language of taxonomy and, therefore, more broadly, all biological sciences—other important pillars of taxonomy are found in the philosophy of classification, especially set theory and logic, and, naturally, evolutionary theory. Whenever we refer to the common housefly as *Musca domestica* and not as 10-7-26-081-052-0325761, as was once proposed, we follow nomenclatural rules.

Most biologists would probably acknowledge that nomenclature helps guarantee clarity of communication about taxa, at least in principle. However, nomenclature is poorly understood. For instance, name changes, which taxonomists regard as a sign of progress, are a common source of frustration for many users of names (e.g., Blackwelder 1948; Barnett 1986; Erzinclioglu and Unwin 1986; Feldmann 1986; Tubbs 1986; Dubois 1998). Also, nomenclature is often perceived as complicated, perhaps too complicated, and this perception is not unfounded. In fact, it can be so complicated that even taxonomists deal with it reluctantly. It requires a great deal of time and patience to be acquainted with all the minute details of the entire literature published on a certain taxon as well as other necessary information (e.g., type material), and fewer and fewer of us can afford the luxury of being scholars. Also, solving nomenclatural issues can be quite tricky, and not everyone is familiar with nomenclatural rules: for instance, how many taxonomists could define allotype, cotype, genotype, hapantotype, holotype, lectotype, neotype, paralectotype, paratype, syntype, and topotype?

Also, most biologists probably perceive our codes of nomenclature as completely static, as if they were written in stone. This perception may result from the idea that since nomenclatural rules aim at promoting stability of taxon names, they ought to be as stable as possible. However, this perception is unfounded. As surprising as it might seem, nomenclature has always been and still currently is dynamic. In fact, one of the fathers of the *International Code of Zoological Nomenclature* (ICZN), Raphaël Blanchard (first president of the International Commission on Zoological Nomenclature from 1895 to 1919), made it clear in the Preface to the first edition of the *Règles Internationales de Nomenclature Zoologique*, that there is no reason that nomenclatural rules should remain unchanged over time (and many changes have actually been introduced in the codes throughout the years):

> By promulgating today these *Règles internationales*, the International Congresses of Zoology do not have the pretension to have accomplished a definitive work. Exactly as the rules that sufficed in Linné's time could not meet our needs, the code that seems to us to appropriately answer our current concerns will be regarded as inadequate by our successors. Science moves forward: it asks new questions, for which new solutions must be found. (Blanchard in ICZN 1905:11, tr.)

Beyond changes introduced in the codes, the main reason that nomenclature is dynamic is that it is characterized by a long history of intense debates, some of which led to major changes in nomenclatural practices. For instance, the starting point of zoological nomenclature, 1758, celebrated in this volume, has been actively discussed. As everyone knows, the starting point in today's zoological nomenclature is the 10th edition of Linné's (1758) *Systema Naturae*, "deemed to have been published on 1 January 1758" (Article 3; ICZN 1999). Animal taxon names published in the 10th edition of *Systema Naturae* have precedence over names published in or after 1758, with the exception of the names published in Clerck's (1757) *Aranei Suecici* [Spiders of Sweden] also "deemed to have been published on 1 January 1758." What fewer people know, however, is that it took zoologists 50 years to decide that the 10th edition of the *Systema Naturae* should be the starting point for zoological nomenclature. Many alternative starting dates (e.g., 1700, 1722, 1735, 1751, 1760, 1766, 1767, 1871) were proposed between 1842, when the British Association committee on zoological nomenclature devised the Strickland Code (in which the starting point is not 1758), and 1892, when the Second International Congress of Zoology (Moscow) finally adopted the 10th edition of the *Systema Naturae* as a starting point for zoological nomenclature. Naturally, some authors also thought that no starting date was needed.

One of the goals of the present contribution is to show that celebrating 250 years of Linnaean nomenclature means celebrating 250 dynamic years of debate. Nomenclatural concepts and rules with which we are familiar today have developed through a slow historical process, and their adoption was in most cases preceded by warm discussions in which opposite views were actively defended. The history of any major nomenclatural issue (principle of priority, typification, grammar and spelling, gender agreement, ranks, suffixes, etc.) illustrates that nomenclature has not always been what it currently is.

However, not all nomenclatural issues can be reviewed here. Therefore, to illustrate that nomenclature has always been dynamic, the present study focuses on: (1) the development of the Principle of Priority throughout the 19th century, from its first definition by de Candolle (1813) up to its application in the first *International Rules of Zoological Nomenclature* (ICZN 1905) and the implementation of a new article altering a strict application of priority under specific conditions in the first edition of the Code (ICZN 1961); and (2) the debates that took place in the 1960s, as a consequence of the emergence of numerical taxonomy and the early use of computers. The debates about priority are analyzed here in detail because priority is certainly one of the most fundamental concepts in nomenclature. Also, the history of the Principle of Priority is particularly interesting because disagreements over its application led to the coexistence of several codes of zoological nomenclature in the second half of the 19th century. The 1960s were a period of great nomenclatural creativity in which some of the most radical ideas were proposed, such as a system of numericlature in which numbers would replace names.

The two historical accounts presented in detail here show that debates have always been one of the main factors of progress in nomenclature, both from a practical and theoretical point of view. Indeed, it is because established nomenclatural practices have continually been challenged that new rules and recommendations have been introduced. Naturally, not all new proposals ended up as part of an actual rule or recommendation. However, regardless of whether a challenging idea had any impact on actual nomenclatural rules, all new ideas and debates have positively impacted our understanding of nomenclatural concepts, by uncovering new issues to address or simply stimulating discussion. For instance, theoretical work on the philosophy and logic of the concept of type specimen (Heise and Starr 1968) was published after Oldroyd (1966) suggested that type specimens be abandoned.

Another goal of this chapter is to discuss today's nomenclatural debates within a long-term historical context, by using as a general approach the lessons we learn from the history of nomenclature. History of nomenclature provides us with two main lessons. First, nomenclature has always remained dynamic: as a set of rules or recommendations, nomenclature has changed; also, established practices have always been challenged and debated. However, this dynamism takes place within a slow and long-term time frame: again, no fewer than 50 years were needed for zoologists to make a final decision on the starting point for zoological nomenclature. So, in nomenclature, things require time to happen (or not to happen, for that matter). Second, all new ideas and proposals, even those that seemed "pure folly," have participated in increasing our understanding of nomenclatural concepts. History thus provides us with a new, interesting approach to our current debates. Indeed, nomenclature is undergoing serious discussion, partly thanks to the emergence of the PhyloCode (Cantino and de Queiroz 2007 and references therein), which has generated some of the most heated debates in the field of systematics since the 1990s. Discussions have also been generated by supporters of the current rank-based zoological code (e.g., Dubois 2006, 2008), especially with the possible unification of the three rank-based codes into a BioCode (e.g., Greuter et al. 1996, 1998; Hawksworth 1997).

In the early history of the International Commission on Zoological Nomenclature, Blanchard thought that new nomenclatural questions might be discovered and need to be addressed in the future. New questions may come from new theories or methodologies: numerical taxonomy generated serious nomenclatural debates in the 1960s; phylogenetic nomenclature is regarded by its founders (de Queiroz and Gauthier 1990, 1992, 1994; for a complete bibliography, see Cantino and de Queiroz 2007) as an extension of Hennig's (1966) Phylogenetic Systematics. In any case, after the publication and official implementation of the PhyloCode, two codes will then coexist (ICZN Code and PhyloCode) for zoologists. Possible scenarios for the long-term future of this coexistence are discussed here and placed in a broader historical context.

A HISTORY OF THE DEVELOPMENT OF THE PRINCIPLE OF PRIORITY

EARLY NOMENCLATURE: FROM LINNAEUS'S APHORISMS TO DE CANDOLLE'S *THÉORIE ÉLÉMENTAIRE*

Linnaeus's Aphorisms

The development of Linnaeus's nomenclature is not presented in detail here. However, a few important points that deeply impacted the development of nomenclature during the 19th century are briefly mentioned, especially because several Linnaean dates (e.g., 1735, 1751, 1758, 1766, and 1767) were discussed as possible starting points for zoological nomenclature.

Linnaeus's (1751) aphorisms from the *Philosophia botanica* served as the basis for nomenclatural practices up to the 1840s in zoology, when the Strickland Code was devised (Strickland 1843), and up to the 1860s in botany, when de Candolle (1867) published the first botanical rules (Nicolson 1991). Agassiz (1842–1846) still cited them as an authority in his *Nomenclator Zoologicus*.

Linnaeus developed his nomenclatural aphorisms in several books, starting with the *Critica botanica* (1737), which constituted a development of the Aphorisms #210–324 from the *Fundamenta botanica* (1736). The *Critica botanica* is divided into four parts. The first, entitled *Nomina generica* (Aphorisms #210–255), deals with the formation of generic names. The second part, *Nomina specifica* (#256–305), does not deal with our current specific names (or epithets), but with the diagnostic phrases then used to express the *differentiæ* among species; in fact, this section is entitled "*Differentiæ*" in the *Fundamenta botanica*. The third part, *Nomina variantia* (#306–317), deals with varietal names. Finally, the fourth part, *Nomina synonyma* (#318–324) deals with synonyms, i.e., different generic, specific (as *differentiæ*), or varietal names given to the same plant.

A close examination of the fourth part is particularly interesting because it tells us how Linnaeus dealt with synonyms. In that regard, it clearly demonstrates that, as pointed out by several authors (e.g., Lewis 1875; Dall 1878; Nicolson 1991), the Principle of Priority as we understand it today was absent from Linnaeus's Aphorisms. Linnaeus's main interest was to find the *best* name for a plant: "If it is decided that none of the synonyms is really suitable for the plant, then necessity compels us to make up a new one." (Linnaeus 1737:259; tr. Hort 1938:209) In the first three sections, Linnaeus provided recommendations on how to form good generic, specific (again, as diagnoses), and varietal names. In any case, the notion that the oldest synonym should have precedence over more recent ones did not exist in Linnaeus's *Critica botanica* (1737); nor did it exist in his cornerstone *Philosophia botanica* (1751), in which binominal nomenclature was described.

In 1751, Linnaeus described binominal nomenclature as the association of a generic name and a uninominal "trivial" name. The "specific" names then still referred to the polynominal *differentiæ*, or specific differences; it is only later that Linnaeus referred to the trivial names as "specific" names, and Lamarck (1778:lxxxiv), for instance, insisted on the fact that trivial names should be more properly called specific names. However, specific names (or epithets) were commonly referred to as "trivial" names throughout the 19th century (e.g., Wallace 1874:259). Although Linnaeus associated generic and trivial names in binominals before 1751 (Stearns 1959), such as in his *Pan Suecicus* (Linnaeus 1749), he consistently adopted binominal nomenclature for the first time in *Species plantarum* (1753) for plants, and then in the 10th edition of the *Systema Naturae* (1758) for animals and minerals. By combining generic and trivial names, Linnaeus's revolutionary idea was to separate two roles of names, the designation and the diagnosis, which had remained mixed in the polynominal names in phrases until then (e.g., Stearns 1959).

In Aphorism #257 of the *Philosophia Botanica*, Linnaeus (1751) clearly defines a trivial name as a single word (*vocabulo uno*). However, more importantly, he also indicates that trivial names do not follow any laws and can be selected freely (*vocabulo libere undequaque defunto*). This largely explains why Linnaeus changed some specific (trivial) names between the 10th and the 12th edition of the *Systema Naturae*, which generated so many debates in the 19th century as to which edition should be used as a starting point in zoological nomenclature. Also, as pointed out by Dall (1878:14), binominal nomenclature remained an accessory matter for Linnaeus, who mainly focused on *differentiæ* (the polynominal specific names), which he regarded as more informative, being more complete than the trivial names. This is particularly obvious in Aphorism #257, where Linnaeus distinctly claims that trivial names are left aside from his aphorisms because specific differences are his main concern.

This leads to what is probably the most important aspect of Linnaeus's aphorisms, i.e., whether they were meant to be strict rules or just recommendations. Linnaeus only occasionally referred to his aphorisms as rules, such as at the end of the preface (no pagination) of the *Critica botanica,* where he claims that those rules (*regulas*) apply in the animal and mineral kingdoms as well as in the vegetable kingdom. However, as clearly illustrated by his treatment of synonyms, Linnaeus did not accept any objective rule for selecting a name among several synonyms; authors have to choose the *best* name, and even create a new name if no good name is available. Thus, most of Linnaeus's aphorisms explain how to form or choose a good name and in practice most of them are recommendations instead of firm rules. When Verrill (1869:92) complained about the fact that authors have ignored the "most essential laws" proposed by Linnaeus and "made sacred by the usage of the best naturalists of the past century," he missed the fact that, actually, Linnaeus's aphorisms were for the most part subjective and, in fact, directly allowed flexibility.

Since Linnaeus allowed name changes, there was no strict application of "precedence" in his aphorisms. This is an important point that deeply impacted nomenclatural debates throughout the 19th century. Indeed, following Linnaeus's model, up until the 1830s, naturalists freely replaced existing names by new—and supposedly better—names. Some authors, however, complained about those endless and unjustified name changes. In fact, one of the main purposes of the Strickland Code (1843) was to help end, or at least restrict, name changes by adopting the Principle of Priority. Then, starting in the 1840s, the question became whether priority should apply without or with exceptions.

Early Reactions to Linnaeus's Nomenclature

Although binominal nomenclature was widely used by most naturalists soon after Linnaeus proposed it, a few individuals refused to adopt it. Albrecht von Haller continued to use names in phrases even in his late works (e.g., Haller 1768). De Candolle (1813:224–225, tr.) mentioned that Haller's students were using the numbers that Haller had given to his plants instead of his phrases but that "everyone agreed that a specific name was better than an arbitrary and insignificant number." The same kind of argument was used in a different context by authors opposed to the numerical system of nomenclature proposed in the 1960s. Stearns (1959:12) mentions that Linnaeus and his students were using a similar system before 1753; during an excursion, in 1748, species were designated as "*Galium* 119," the latter number referring to an entry in Linnaeus's (1745) *Flora Suecica.*

According to De Candolle (1813:225, tr.), several naturalists, in particular Buffon, thought that species names should be "independent from classification" because species are too often placed in different genera by different authors. Generally speaking, however, Buffon was not interested in classification, nor was he interested in names. Buffon thought that all methods and systems of classification are arbitrary because intermediary forms prevent classifying organisms in distinct groups (Roger 1993). Also, in an article on the description of animals, Buffon (1753:114–115) argued that one should know organisms completely before one gives them a name, and that describing organisms is more fundamental and necessary than naming species and groups.

Michel Adanson, in his *Histoire du Sénégal* (1757:xvi-xix), adopted binominal nomenclature, but was strongly opposed to the use of meaningful specific names because, according to him, such names usually become erroneous once the species is better known. In fact, Adanson (1757:xvi) insisted that we should make sure that names are not built from any recognizable etymological roots. Thus, new names created by Adanson had no classic etymology, such as *Yetus* Adanson (1757:cxiii, 44), which refers to a seashell. This approach would have been a radical way to avoid the endless debates that took place in the 19th century on whether Latin grammar and orthography should be strictly enforced. Interestingly, Adanson seemed to regret unjustified name changes, an issue that caused much ink to flow in the 19th century: "We must conserve the old [names], especially those that seem good, and that have been adopted by the masters of the art" (1757:xv, tr.).

Finally, Bernard de Jussieu was another famous Parisian naturalist who refused to adopt binominal nomenclature, although he let his nephew Antoine-Laurent de Jussieu, and André Thouin, use it in *Jardin du Roi* (Dayrat 2003).

Nomenclature in Lamarck's *Flore Française*

Lamarck (1778:lxxxii–lxxxvii) gave a few general remarks on nomenclature in the *Discours préliminaire* of his *Flore française*. First, Lamarck (1778:lxxxii) complained about name changes, which make nomenclature the most difficult part of botany. Lamarck also provided a few recommendations on how to form appropriate names. Generic names should not be precise; for instance, *Potentilla* should be preferred to *Quinquefolium* because some species do not have compound leaves with five leaflets. Specific names (epithets) should be as precise as possible and help give a distinctive idea of the species referred to, such as *Menianthus trifolia*. Finally, misleading names should be avoided, such as in *Euphorbia spinosa* (which bears no spines). However, Lamarck did not say a word on what to do if a name is not appropriate. Should it be replaced, or kept? As we shall see, it took several decades before zoologists could agree on the action to be taken.

Fabricius's *Philosophia botanica*

Fabricius (1778:101–121) provided 46 aphorisms of nomenclature in his section on *Nomina* of his *Philosophia entomologica*, which were largely inspired by Linnaeus's (1751) aphorisms. Thirteen canons deal with *Nomina trivialia*, i.e., our specific names. In canon #45, Fabricius (1778:121) asserts that "Trivial names ought never to be changed without the most urgent necessity (*Nomina trivialia nunquam absque summa urgente necessitate mutanda sunt*)." This canon, which seems close to the concept of priority, was quoted by Strickland (1835:36) in the beginning of one of his first articles on nomenclature.

Bergeret's *Phytonomatotechnie universelle*

Jean-Pierre Bergeret's system of nomenclature, exposed in his *Phytonomatotechnie universelle* (Bergeret 1783–1785), now a rare book (Dayrat 2003), is radically different from anything proposed before and even after him. Bergeret's main idea was to build species names based on characters found in each species (each character state being coded by a particular letter, and the characters following each other in a precise way). Although Bergeret's system was quite logical, it yielded unusual "names" with nearly unpronounceable sounds, such as "*LUPXYGVEAHQEZ*," for our *Eranthis hyemalis* (Linné) Salisbury. Bergeret's method was never adopted. As Rozier (1783:224, tr.) pointed out in his review of Bergeret's *Phytonomatotechnie*: "The most well-made methods are sometimes very awkward in their application." Nonetheless, Bergeret's method remains an original attempt at making nomenclature more logical.

Du Petit-Thouars' Innovative (But Never Adopted) Nomenclature

Although Aubert du Petit-Thouars started presenting his studies before the members of the Institut de France as early as 1806, and even published some of them, the publication of his major work on orchids was delayed. A report evaluating his manuscript in 1806 indicated that it still was a draft and that du Petit-Thouars needed a great deal of time to complete it (du Petit-Thouars 1817:153–156). The referees who wrote this report also noted his unusual nomenclature (Desfontaines, Richard, Bosc in du Petit-Thouars 1817:156, tr.): "We find more disadvantages than advantages in adopting such a nomenclature, which, by the way, would turn science upside down."

Du Petit-Thouars' system of nomenclature was described in detail in his *Histoire particulière des plantes Orchidées* (1822). His goal was to build a system of nomenclature that would be more logical, based on a model of chemical nomenclature defended by French chemists (du Petit-Thouars 1817:168). His fundamental idea was to use the same suffix for each generic name that would belong to the same family. For the orchids, the suffix *–orchis* (which he also spells *–orkis*) was used to build names such as *Dryorchis, Amphorchis, Satorchis, Habenorchis*, and *Stellorchis*. Some of du Petit-Thouars' new names replaced existing names (e.g., *Habenorchis* replaced *Habenaria* Willdenow). Du Petit-Thouars also built species uninomials based on part of the specific name and part of the generic name. For instance, on Plate 4 of his *Histoire particulière des plantes Orchidées*, one reads three names: *Amphorchis* (the generic name), *Amphorchis calcarata* (the Linnaean binomial), and *Calcaramphis* (his newly invented uninominal species name), which supposedly helped retrieve or name more efficiently the species illustrated. Naturalists were not interested in the logic of du Petit-Thouars' new system. They saw only its disadvantages: "This ingenious method presents almost all the disadvantages of that of Bergeret." (De Candolle 1813:226, tr.)

De Candolle's (1813) First Definition of the Principle of Priority

Augustin-Pyramus de Candolle is a central author in nomenclatural history because he provided one of the first definitions of the principle of "priority" in his *Théorie élémentaire de la botanique* (1813), a cornerstone in the history of systematics (it also is in this book that de Candolle coined and defined "*taxonomie*," as the *théorie des classifications*, i.e., classification theory). Augustin-Pyramus de Candolle was the father of Alphonse-Pyramus de Candolle, one of the fathers of the code of botanical nomenclature.

De Candolle (1813:227) mentioned "*onomatologie*," a term then more or less synonymous with "dictionary" (it was often used in titles of medical dictionaries), as a synonym of nomenclature. De Candolle also gave Bergeret's "*phytonomatotechnie*," which literally refers to a "technique to form plant names," as a synonym of nomenclature. This suggests that, according to de Candolle, nomenclature had two meanings: (1) a list of names and (2) the formation of names, or the act of forming names. Today, nomenclature often also refers to the rules that govern the use of names. However, for de Candolle, "nomenclature" does not seem to cover the rules of nomenclature, to which he refers as the rules of nomenclature (*règles de la nomenclature*). So, when de Candolle (1813:227, tr.) claimed that "the goal of nomenclature of natural history is to be universal, common to scientists of all nations," he primarily meant that names should be universal and stable; by extension, the rules should also be common.

De Candolle (1813:227–228) mentioned three primary principles: (1) names must be in Latin; (2) names must follow basic grammar (e.g., names mixing Greek and Latin roots can be replaced); (3) the first name given to a species must remain unchanged, unless it is already in use for another species or contradicts some essential rules of nomenclature. De Candolle (1813:228–241) mainly focused on the formation of generic names (*noms des genres*); if based on a character, the latter cannot be in contradiction with any of the species contained in the genus; mythological names can be of some help if one is really in despair; the name of a scientist can be used, but only for those who contributed to science; anagrams or names made of letters randomly selected (as in Adanson's work) should be avoided. According to de Candolle (1813:241–246), finding a specific name (*nom d'espèce*) is much easier. A specific name could be changed only in cases where it was in complete contradiction with the plant itself: the name *Lunaria annua* should be changed into *Lunaria biennis*. Finally, de Candolle (1813:247–250) gave a few recommendations on family and variety names. All those rules were abundantly discussed by naturalists throughout the 19th century—in particular, should the "indelicate" names (mixing Greek and Latin) be changed?

More importantly, de Candolle (1813:250–251, tr.) defined for the first time the term priority (*priorité*): "One must always adopt the oldest name, except in the following cases..." Naturalists did not stop asking throughout the entire 19th century whether priority should apply with or without exceptions. De Candolle considered three main exceptions: a name that is completely inappropriate (such as in *Lunaria annua*); a name that is not binominal or in Latin; a name already in use (in animals or plants). Also, de Candolle insisted on the fact that when a species is transferred from one genus to another the specific name should not be changed, and that a plant name should be published and accompanied with a description (catalogues of names are excluded). As we shall see, Strickland's (1835, 1837a) priority rules were similar to those of de Candolle.

Although de Candolle (1813) provided a detailed analysis of the principle and application of priority, the latter was already in use in that period, both as a term and as a concept. For instance, Spinola (1811:145, tr.) asserted: "As for nomenclature, priority is my law. It also is that of M. Latreille and French entomologists." Lewis (1875:ii), strongly opposed to a strict application of priority argued:

Although the principle had been started some twenty years before, I believe it is the fact that until the British Association Rules of 1842, "priority" to intents and purposes remained a theory.

It certainly is true that replacing existing names that were not "good enough" was a common practice among naturalists in the first part of the 19th century. However, it is not true that priority remained just a theory during that period, in which several clear references to priority exist (e.g., Spinola 1811:145; Swainson 1821:pl. 42; De Candolle 1822:143, 1829:19–20; Bennett 1830:243–244). Brongniart (1827:80) also regretted that the law of priority did not apply more strictly in mineralogical nomenclature.

Dejean (1825:x) pointed out a conflict between two opposite trends in nomenclature: (1) the maintenance of names widely used, although more recent than older but forgotten names (which Dejean supported); (2) a strict application of priority and the use of the oldest name regardless of the situation (which Latreille supported, according to Dejean). Debates between proponents of "priority" and proponents of "convenience" remained stormy throughout the 19th century. It is in this context that Strickland originally dedicated himself to devising rules for zoological nomenclature.

TOWARD THE STRICKLAND CODE

Strickland's (1837) Rules

At the initiative of Hugh Strickland, it was resolved in 1842 by the British Association for the Advancement of Science that a committee be appointed to consider rules for zoological nomenclature. The history of this committee (its members, its meetings, etc.) is well documented (Sclater 1878; Verrill 1869; Melville 1995). Rules were discussed and adopted at a Manchester meeting in June 1842, and published the following year (Strickland 1843; Sclater 1878). Strickland, who deserves most of the credit for those rules, started working on nomenclatural rules in the 1830s.

In 1835, Strickland published a brief article in which he argued that the practice of replacing or altering generic and specific names without any compelling reason was detrimental to science:

> Surely, then, the evil of changing a name, which has once become current among naturalists, is much greater than any advantage supposed to result from substituting a term which is 'more appropriate.'" (Strickland 1835:38)

Indeed, naturalists then commonly replaced an existing name by another, knowing that both names referred to exactly the same entity. So, those replacement names were different from our synonyms, created in most cases by authors who ignore the existence of a former name (or disagree on species limits). Strickland (1835:39) referred to those replacement names as "aliases." Strickland thought that, in order to prevent the arbitrary replacement of names from worsening

> It would (…) be highly desirable if an authorized body could be constituted, to frame a code of laws for naturalists, instead of the present anarchical state of things in which every one does that which is right in his own eyes.

As Spinola (1811), de Candolle (1813), and others, Strickland (1835) adopted the Principle of Priority as his primary law of nomenclature, but accepted exceptions similar to those proposed by de Candolle.

Another issue that Strickland wanted to address was the fact that, in addition and in relation to the existence of "aliases," different names could refer to the same entity in different countries, which Westwood (1837:169), a member of the 1842 British Association committee, illustrated by several examples, such as: "The giant beetles, *Hercules*, *Actaeon*, etc., with which Linnaeus commences the insect tribes, are named *Scarabaeus* in France, *Geotrupes* in Germany, *Dynastes* in England." Westwood (1837) also linked the application of the Principle of Priority of generic names to type species, something that became heavily debated in the second part of the century, especially in the 1870s.

In 1837, Strickland published some Rules for Zoological Nomenclature, which have "little pretension to originality, but … are selected from the writings of several naturalists, especially from the birds of Mr. Swainson, many of whose aphorisms are adopted here" (Strickland 1837a: 173). Indeed, the previous year, Swainson published a 19-page book chapter, "On the nomenclature and description of birds," in which he proposed rules distributed in 24 paragraphs: "The necessity of nomenclature being regulated by fixed laws has been advocated by Linnaeus, Fabricius, and all the best systematic writers, both in zoology and botany" (Swainson 1836:228). Strickland's (1837a) first proposed rules were similar to the rules officially adopted by the British Association, especially regarding the law of priority, which would always apply except in the cases mentioned by de Candolle (1813). However, Strickland (1837a) did not mention any starting point for zoological nomenclature.

Ogilby and Strickland Debate (1837–1838)

Naturally, reactions to Strickland's (1837a) rules were soon published. In particular, a debate took place between Strickland and Ogilby (a future member of the British Association committee). Strickland (1837b:605) regretted that Ogilby did not adopt a "very convenient rule, now generally adopted by naturalists, that the name of a *family* … should be compounded of the name of the most typical or best known genus contained in it, with the termination *idae* or *adae*." Ogilby (1838a:150) reprobated the use of the suffix *–idae* for family names, which was not included in Strickland's (1837a) original set of rules, as well as the rest of Strickland's (1837a) "arbitrary, dogmatical, unfounded and unnecessary rules," such as the fact that names should be taken either from the Latin or Greek languages because it is "opposed to the genius and spirit of all languages, ancient and modern, and directly contradicted by the practice of the Greeks and Romans themselves" (Ogilby 1838a:153). This issue remained debated for many years. Naturally, Strickland (1838a:198) replied and tried to "do [his] utmost to preserve a temperate tone in the discussion." Strickland (1838a:200) explained better what he intended to achieve when submitting rules to naturalists "not as an *act* for their guidance, but as a *bill* for their approval." Strickland also proposed that naturalists

> form a congress, or in humbler phrase, a committee of naturalists from all parts of the scientific world, to draw up a code of zoological laws, not indeed to be like those of the Medes and Persians, yet to be adhered to for the sake of order and convenience. (1838a:199)

This constituted one of the first mentions of the necessity of an international body of experts for nomenclatural rules.

In the context of this debate, Ogilby (1838b:281) and Strickland (1838b:330) exchanged ideas on the definition of type species or type genus. They both thought that "type" was synonymous with "example." Ogilby (1838b:281) mentioned that "if the word type [is] merely synonymous with example, [he sees] no objection to it." Strickland reassured him that "by the type of a genus [he means] that species which is usually selected as an *example* of the genus." However, a few questions came next that were all abundantly debated for several decades. How do we determine which species (for the type of a genus) or genus (for the type of a family) "afford the best sample of the characters" (Strickland 1838b:330) on which the genus or family is based? Also, how can one determine which species should be the type when no type was designated originally?

Strickland (1838c) and Ogilby (1838c) exchanged a few more thoughts. Their conclusions are memorable. Ogilby (1838c:493) called for less dogmatism in nomenclature:

> If it tend to open the eyes of zoological legislators to their own fallibility, and to make them a little more moderate in their pretensions and a little less dogmatical and inaccurate in their statements, the controversy will have accomplished a very great *desideratum*, and fulfilled my intentions.

Strickland (1838c:555) wished to add "a few "more last words" on this exhausted subject," which, we know now, may never be exhausted. That first debate ended constructively: both gentlemen participated in the British Association committee on rules for zoological nomenclature.

The Strickland Code (1843)

The Strickland Code was first published in 1843 in the *Report of the British Association*, as well as in the *Philosophical Magazine*, and the *Annals of Natural History*. It was published again with a few revisions in 1865, under Jardine's supervision. Both the original set of rules and Jardine's Report were re-published by Sclater (1878), along with a list of all naturalists to whom Strickland had originally sent a draft of the rules for comments. It is not the object of this chapter to analyze all the rules of the Strickland Code. However, the treatment of priority, as well as rules related to it, are commented on.

The law of priority, the fundamental basis of the entire code, was defined as follows:

> The name originally given by the founder of a group or the describer of a species should be permanently retained, to the exclusion of all subsequent synonyms (with the exceptions about to be noticed). (Strickland 1843 in Sclater 1878:6)

Although priority had been adopted as a guiding principle by some naturalists for at least 30 years, no starting point had been proposed so far. In his personal draft, Strickland had already indicated that the starting point should be the *Systema Naturae*, but had not indicated which edition (Jardine 1866 in Sclater 1878:23). The Committee decided that the starting point was the 12th edition: "We ought not to attempt to carry back the principle of priority beyond the date of the 12th edition of the *Systema Naturae*." (Strickland 1843 in Sclater 1878:23) The main reason was that the 12th edition was the last edition published under Linnaeus's supervision. A precise date was not provided in the Strickland Code, at least in 1843; in the 12th edition, the volume on the Animal Kingdom is split between a Part one (1766) and a Part two (1767).

In 1865, the Committee in charge of revising the Strickland Code confirmed the choice of the 12th edition after "much deliberation" (Jardine 1866 in Sclater 1878:23), and gave the date 1766 for a starting point, which, strictly speaking, applies only to the first half of the animal kingdom. An alternative was the 10th edition, in which Linnaeus (1758) applied consistently the binominal nomenclature to all animals for the first time. However, the 12th edition was favored because Linnaeus changed quite a few names between the 10th and 12th editions of the *Systema Naturae*; selecting the 12th edition guaranteed that Linnaeus's last and supposedly preferred names would have precedence over former names that he discarded. The starting point of zoological nomenclature, which clearly is central in the application of the principle of priority, remained discussed extensively until the end of the 19th century, when a decision was finally made to select the 10th edition of the *Systema Naturae*.

Even though the British Association committee selected the 12th edition as a starting point, we owe to Strickland and the rest of the Committee to have pointed out the critical importance of a starting point for the application of priority. The need for a starting point probably became more pressing in the early 1840s. As de Selys-Longchamps (1842:iv, tr.) acknowledged:

> Recognizing the right of priority seems the only way to understand each other and prevent that zoology soon become chaos, a true Babel as long as it is fixed at 1760, the period [*époque*] of the establishment of the binary nomenclature by Linné and of the publication of Brisson's book for all the genera not adopted by Linné.

In 1842, two starting points had already been proposed: 1760 and the 12th edition of the *Systema Naturae*. ... It was just a start.

The Strickland Code is divided in two parts. The first includes "rules for rectifying the present nomenclature" and deals with names that have already been created. It contains only 14 articles (rules), most of which deal with the application of the Principle of Priority. The second part includes "recommendations [not rules] for improving the nomenclature in future" and deals with the formation of new generic and specific names.

Several articles of the first part of the Strickland Code set the stage for typification within genera. In particular, the Strickland Code provides ways of determining which is the type species in case no typical species was designated originally, especially when a genus is split into several genera. Articles related to the application of the law of priority mainly concern generic names. The notion of type specimen (for species) is not in the Code. A few acceptable exceptions to the law of priority are provided that are similar to the exceptions proposed by de Candolle (1813). For instance, under the Strickland Code, "a name whose meaning is glaringly false may be changed" (Strickland in Sclater 1878:10).

Article #10, which has remained misunderstood and debated, merits commenting on:

> A name should be changed which has before been proposed for some other genus in zoology or botany, or for some other species in the same genus, when still retained for such genus or species. (Strickland in Sclater 1878:10)

The last part of the article ("when still retained for such genus or species") indicates that Article #10 applies only when names are valid, but not for synonyms. In other words, a name does not have to be replaced (by a new name or a junior synonym) if the homonym is a synonym. In the current Code (ICZN 1999), names have to be replaced even if senior homonyms are synonymous. Also, note that in Article #10 both zoological and botanical names have to be considered, which was rejected by most future codes.

Article #12, which deals with another exception to priority, has also been widely debated: "A name which has never been clearly defined in some published work should be changed for the earliest name by which the object shall have been so defined" (Strickland in Sclater 1878:11). Although the Strickland Code provides a few guidelines about what constitutes a definition and a publication, many authors have followed their own interpretation, generating confusion in the application of Article #12.

Recommendations of the second part of the Strickland Code intend to render zoological nomenclature "palatable to the scholar and the man of taste." They are not rules because there is nothing that can be done "if authors insist on infringing the rules of good taste by introducing into the science words of … inelegant or unclassical character in future" (Strickland in Sclater 1878:13). Also, those recommendations are not retroactive because they would "undermine the invaluable principle of priority." However, some authors thought that "rules of good taste" should be enforced, and used that as a reason to replace many existing names, creating confusion and debates.

Within the long list of names that, according to the British Association committee should be avoided or at least used judiciously, one finds: geographical names (because *canadensis* may refer to a species distributed in Mexico); barbarous names (not from Greek or Latin); mythological or historical names because they most often can be regarded as "unmeaning and in bad taste"; names derived from persons "with no scientific reputation"; names of "harsh and inelegant pronunciation" (such as *Enaliolimnosaurus crocodilocephaloides*); hybrid names with parts taken from two different languages, which are "great deformities in nomenclature"; and, "nonsense names," such as names made of letters randomly composed.

The fact that the Strickland Code devoted many paragraphs to the "rules of good taste" in name formation was clearly a heritage from concerns of naturalists from the late 18th and early 19th centuries, such as Linnaeus (1751) and de Candolle (1813). After several decades of intense debates, it was finally resolved that most kinds of names banished from nomenclature in the Strickland Code (especially the so-called "barbarous" names) could be admitted (ICZN 1905).

Note finally that, perhaps because of the presence of Ogilby in the British Association committee, the use of the suffices –idæ for family names and –inæ for subfamily names is only a recommendation, not a rule. It also is recommended that all specific names be written with a small initial (including those based on persons), but that recommendation is removed from the second edition of the Strickland Code (1865).

The British Association committee was not international. All members were British, such as C. Darwin, R. Owen, W. Ogilby, and J.O. Westwood. However, Strickland thoughtfully sent a copy of the draft of the rules to many British and foreign naturalists and societies during the preparation of the Code (Sclater 1878). At least 180 individual naturalists, as well as 35 societies, journals, or libraries, were contacted. Unfortunately, it is not known how many comments were sent back. Strickland contacted all of the most famous naturalists of his time, such as Agassiz (Switzerland); Audubon, Gould, and Gray (United States); Blainville, Deshayes, Duméril, d'Orbigny, Milne-Edwards, and Valenciennes (France); Bronn, von Buch, Ehrenberg, Pfeiffer, and Rüppell (Germany); Fries and Lovén (Sweden); and Parlatore (Italy). Strickland's efforts for reaching the international community were remarkable and visionary.

Impact of the Strickland Code on Nomenclature

The Strickland Code was translated into French in the *Revue Zoologique* (Guérin-Méneville, 1843) and the *Bibliothèque universelle de Genève* (Anonymous 1844). In both cases, however, the starting date (12th edition of the *Systema Naturae*) was not mentioned. Given that most readers probably did not go back to the original English version, such incomplete translations participated in generating confusion among naturalists. This seems to suggest that it took a long time for zoologists to realize how central a fixed starting point is in order to build a stable universal nomenclature. This is also supported by the fact that the Kiesenwetter Code (1858) simply starts at Linnaeus, with no specific date. In 1845, the Strickland Code was adopted by the Association of American Geologists and Naturalists with only one minor suggestion (no need to recommend writing all specific names with a small initial). It was translated into Italian by Charles Lucien Bonaparte for the fourth meeting of Italian scientists, in Padova (Anonymous 1843; Minelli 2008). As expected, one of the main questions that was discussed in Padova was the application of priority. Interestingly, however, it was also proposed that both zoological and botanical rules be joined (Minelli 2008). The Strickland Code does not appear to have been translated into German.

The Strickland Code did not seem to solve nomenclatural issues, as acknowledged in the report introducing the revised version (Jardine 1866 in Sclater 1878:23):

> Whether it is from the Rules and recommendations not being sufficiently well known, or from an idea that no one has any right to interfere with or make rules for others, many gentlemen appear to cast them away, and do not recognize them at all, while others accept or reject just what pleases themselves; in consequence many very objectionable names have been given, and a very base coinage and spurious combinations have been going on.

This also was the opinion of Crotch (1870:59):

> Notwithstanding the "Rules for Zoological Nomenclature" sanctioned by the authority of the British Association, it would not seem that any perceptible improvement has taken place. (...) The laws of priority are of course assented to tacitly by all, but not applied.

Finally, Lewis (1875:v) even accused the Strickland Code to have generated confusion:

> The legislators of 1842 had made the discovery that the names employed here [in the United Kingdom] were different from those employed elsewhere [especially continental Europe], and they enacted a rule

to cure the evil. … We discovered between three and four years ago that the bare rule of priority (as construed now) has let in practices which promote and do not dissipate confusion.

In the mid 19th century, each naturalist held his own independent opinion on how the principle of priority should apply. By promoting priority so strongly, the Strickland Code forced naturalists to express their opinion about nomenclature. Disagreements arose slowly but, inevitably, a second code of zoological nomenclature was adopted in Germany, fifteen years after the Strickland Code.

THE KIESENWETTER CODE (1858)

Kiesenwetter proposed some nomenclatural rules at the German Congress of Entomology (Dresden, May 1858), attended by 45 entomologists (41 from Germany, 2 from Spain, 1 from Norway, and 1 from Prague, Bohemia). Adopted after some discussion, those rules were published in German in the *Berliner entomologische Zeitschrift* (Kiesenwetter 1858) and then translated into French (e.g., Mulsant 1860; Fauvel 1869). According to Fauvel (1869:91), Kiesenwetter's rules were "unanimously adopted by the most commendable authorities," at least among entomologists. Kiesenwetter meant to design rules of entomological nomenclature, not zoological nomenclature, and those rules constitute one of several specialized codes produced in the second half of the 19th century. In fact, several authors then thought that codes could apply to specialized groups instead of all animals (Edwards 1873). As surprising as it might seem, the Strickland Code was not mentioned by Kiesenwetter, suggesting that Kiesenwetter and other (German) entomologists refused to concede any authority to the British Association and did not think that the existence of several codes would be a source of any problem in nomenclature.

The Dresden rules are similar to the Manchester rules in many regards, such as: Latin is the language of nomenclature; the rules for existing names are distinct from the (non-retrospective) rules for newly created names; names should be considered only when accompanied with a diagnosis, a description, or a figure that can enable species re-identification. There are also notable differences, such as: the Dresden rules include only rules, no recommendations; descriptions and diagnoses must be published in a Romance or Germanic language, and a diagnosis in Latin is mandatory (Rule #11); the use of the suffix –idae for family names is a rule, not a recommendation (Rule #3); all spelling that contradicts Latin orthography or grammar must be corrected (Rule #6), whereas in the Strickland Code (Rule #14) the original spelling of a name is retained as long as it has been amended; the starting point for nomenclature simply is "Linnaeus," with no specific date, suggesting that entomologists who met at Dresden did not understand the critical importance of a starting date; finally, inaccurate specific names (e.g., *albus* for a black animal) do not have to be changed, whereas, under the Strickland Code, the law of priority may not apply in such cases.

As in the Strickland Code, priority applies strictly, except in cases of homonymy: "When several names equally acceptable have been introduced to refer to the same object, the oldest one is the highest authority and must be preferred" (Kiesenwetter 1858:xviii). Several authors claimed that, in the Kiesenwetter Code, the choice of preserving the oldest name among several was not absolute and should also be guided based on "greater or less convenience" (Edwards 1873:26; Lewis 1875:vi; Melville 1995:10). However, Kiesenwetter allowed flexibility in the priority rule only in one particular case (Rule #15): "If it is not possible to establish which of several names is the oldest, one is free to choose that which is the most appropriate" (Kiesenwetter 1858:xix, tr.).

Finally, Kiesenwetter agreed with Strickland that, when a genus was being split, then its application should be governed by the type species. Ways to identify the type species are even provided: (1) the species designated as typical by the original author; (2) the species that presents the characters indicated by the original author; (3) the species that is the most remarkable or that is the most commonly found; (4) the first species described or listed by the author. As expected, the determination of type species has remained actively debated, which was partly due to the fact that the solutions proposed were either arbitrary (first species mentioned) or subjective (species that presents most of the generic characters described by an author).

Interestingly, Kiesenwetter did not attach any particular importance to specimens from old authors, because it was thought that they were commonly mixed with non-original material (Rule #12):

> Contrary to the opinion of many of today's entomologists, one cannot admit any right of priority for a description that can only be interpreted thanks to supposedly typical specimens, whether this is correct or not. ... For the old entomologists, one must leave aside species names for which no certain data are available. (Kiesenwetter 1858:xvii, tr.)

Naturally, authors disagreed on whether supposedly typical specimens should be used when discussing the application of names, as illustrated in the debate between Waterhouse (1862) and Schaum (1862).

Descriptions versus Type Specimens: The Debate between Waterhouse and Schaum (1862)

Although both the Strickland Code and the Kiesenwetter Code required that names be accompanied by a description in order to be "eligible" for priority, authors disagreed on criteria to determine whether a description was acceptable, as illustrated in the review of Waterhouse's (1858) *Catalogue of British Coleoptera* by Schaum (1862). Schaum (1862:323) complained about the fact that Waterhouse rejected well-known names in favor of older synonyms that were not accompanied by any proper description:

> It is universally acknowledged that a species must have been described in order that priority can be claimed for its name; and no one thinks of assigning priority to a mere catalogue name, because we are unable thereby to recognize the species to which it has been given. This reason, however, compels us to put the query, *when* can it be said that a species has been described? Even the most fanatical advocate of the law of priority will not pretend that a species has been described, concerning which utterly false notices, or erroneous or unimportant indications, are given, which so completely fail in characterizing the species that no one is able to recognize it.

Also, Schaum (1862:324) thought that Waterhouse made the mistake of giving more importance to specimens than to written descriptions: "Mr Waterhouse did not ascertain these older names by the study of the books, but merely by the investigation of real or supposed typical specimens."

Waterhouse (1862) regretted that Schaum gave too much importance to descriptions (even incomplete) compared with type specimens. Although Waterhouse admitted that some type specimens might be mixed, he argued that it could still be most often possible to link a description (which Schaum regarded as incomplete) to a type specimen (which Waterhouse regarded as a key source to determine the meaning of a species name). Thus, Waterhouse (1862:338) concluded by arguing that:

> When a described species can, *by any means*, be determined, so as to leave no reasonable doubt of the identification, it appears to me desirable that the name given to the species by the describer be adopted, provided that name be the earliest the species has received.

The expression "*by any means*," emphasized by Waterhouse, indicates that, according to him, it was possible to determine the meaning of a name using a description *and* type specimens, whereas Schaum considered that only written descriptions should be used.

This debate is highly interesting because it shows that people started thinking of type specimens while they were debating about priority, but did not necessarily understand that both issues were tightly linked, and that the application of the Principle of Priority could not be solved unless types were taken into consideration. It took several codes and several decades of debates before the link between priority and type specimens was fully understood (see below). Most taxonomists would agree today that using type specimens to determine the application of a species name is a rigorous approach.

A Revised Edition of the Strickland Code (1865)

Gray's Comments (1864) on the Strickland Code

In 1863, the Strickland Code was re-published in the *Edinburgh New Philosophical Journal*, and comments were requested by Jardine from whoever was interested. The comments published by Asa Gray, the famous American botanist, are particularly interesting.

Gray's (1864:278–279) comments on Rule #2 ("The binomial nomenclature, having originated with Linnaeus, the law of priority in respect of that nomenclature, is not to extend to the writings of antecedent authors") are as follows:

> The essential thing done by Linnaeus in the establishment of the binomial nomenclature was that he added the specific name to the generic. He also reformed genera and generic names; but he did not pretend to be the inventor or establisher of either, at least in Botany. This merit he assigns to Tournefort … and he respected accordingly the genera of Tournefort, Plumier, etc. … While, therefore, it is quite out of question to supersede established Linnaean names by Tournefortian, we think it only right that Tournefortian genera, adopted as such by Linnaeus, should continue to be cited as Tournefort. So, as did Linnaeus, we prefer to write *Jasminum*, Tourn., *Circaea*, Tourn., *Tamarindus* Tourn., etc.

Gray raised the question of the authorship in cases where Linnaeus himself accepted Tournefort's authorship: Gray did not mean that Tournefort's names should have priority over Linnaeus's names. However, several authors, especially Raphaël Blanchard, first president (1895–1919) of the International Commission on Zoological Nomenclature, argued that zoological nomenclature should go back to writers who coined genera in the early 18th century that were later used by Linnaeus.

On Rule #10 ("A name should be changed which has before been proposed for some other genus in zoology or botany, or for some other species in the same genus, when still retained for such genus or species."), Gray (1864:279) writes:

> We submit that this rule, however proper in its day, is now inapplicable. Endlicher, who in a few cases endeavored to apply it, will probably be the last general writer to change generic names in botany because they are established in zoology. It is quite enough if botanists, and perhaps more than can practically be effected if zoologists, will see that the same generic name is used but once in each respective kingdom of nature.

Note that this does not mean that Gray supported the existence of two distinct sets of rules.

Gray's (1864:279) comment on Rule #12 ("A name which has never been clearly defined in some published work should be changed for the earliest name by which the object shall have been so defined.") seems to indicate an early difference between botanical and zoological practices, and perhaps partly explains why it was resolved in the revised version of the Strickland Code that Botany would not be introduced in the Code:

> And as to names without characters, may not the affixing of a name to a sufficient specimen in distrib-
> uted collections (a common way in botany) more surely identify the genus or species than might a brief
> published description!

This comment emphasizes the critical importance of specimens used by an author to create a new genus or species name (i.e., the type series or type species). As we already saw, some zoologists of that period gave more credit to the description than to specimens when deciding on the application of a name, as stated clearly in Kiesenwetter Rule #12.

The Minor Changes to the Revised Strickland Code (1865)

In 1860, it was resolved by the members of the British Association for the Advancement of Science that the surviving members of the committee appointed in 1842 meet again to discuss alteration and improvement potentially needed in the Stricklandian rules and recommendations. However, "from the difficulty of bringing such a committee together," nothing happened until 1863, when it was decided that:

> Sir W. Jardine [reporter], A.R. Wallace, J.E. Gray, Professor Babington, Dr Francis, Dr Sclater, C. Spence
> Bate, P.P. Carpenter, Dr J.D. Hooker [botanist], Professor Balfour, H.T. Stainton, J. Gwyn Jeffreys, A.
> Newton, Professor T.H. Huxley, Professor Allman, and G. Bentham [botanist], be a Committee, with power
> to add to their number, to report on the changes, if any, which they may consider it desirable to make in the
> Rules of Nomenclature drawn up at the instance of the Association. (Jardine 1866 in Sclater 1878:22)

The Code was then reprinted and zoologists were requested to send their comments to Jardine. At Bath (1864 meeting of the BAAS), the Committee did "little to complete further the code of Zoological Nomenclature" (Jardine 1866 in Sclater 1878:22). However, a meeting took place in London (at Jeffreys' house) with four members of the committee (Jeffreys, Wallace, Sclater, and Jardine) to review comments sent to Jardine, as well as opinions of the committee members (Babington sent his personal copy of the code with observations written on the margin). Finally, amendments to the code proposed by the committee were accepted, after discussion, at the Birmingham meeting of the BAAS (1865).

Overall, changes made were minor (Jardine 1866 in Sclater 1878:23–24). (1) It was claimed that Botany should not be introduced into the code, which certainly did not come as a surprise to naturalists since a split between zoology and botany already existed de facto, as illustrated in Gray's (1864) comments on the 1842 version of the code. (2) It was recommended by the committee that the code should not be altered:

> … the permanency of names and convenience of practical application being the two chief requisites in
> any code of rules for scientific nomenclature, it is not advisable to disturb by any material alterations
> the rules of zoological nomenclature.

Although committee members understandably wished to protect "their" code, several new codes were published in the second half of the 19th century. (3) The committee also was of the opinion

> … after much deliberation, that the XIIth edition of the "*Systema Naturae*" is that to which the limit of
> time should apply, viz. 1766. But as the works of Artedi and Scopoli have already been extensively used
> by ichthyologists and entomologists, it is recommended that the names contained in or used from these
> authors should not be affected by this provision.

That the date 1766 was kept "after much deliberation" shows that the question of the starting date for zoological nomenclature was far from being settled. In fact, it remained actively debated for the next 30 years. (4) Rule #13 ("specific names, when adopted as generic, must be changed") was altered so that it was actually the generic names that, in such cases, should be "thrown aside, not the old specific name." (5) The committee emphasized again the recommendation under which the use of a person's name for a specific name should not be taken lightly. (6) Finally, the recommendation that specific names should always start with a small initial was removed (as suggested by the American Association of Geologists and Naturalists in 1845).

Surprisingly, Kiesenwetter's (1858) entomological rules were not mentioned, although there is little doubt that the committee was aware of their existence. The reasons for this are unclear. In any case, this omission is not just a detail. It illustrates the state in which nomenclature was in that period. There was a lack of communication among national communities and, as a consequence, no common international efforts for discussing nomenclatural rules.

Ideas about critical issues, especially types and the starting date were still unsettled. Also, the application of priority was still poisoned by subjective opinions on so-called "indelicate" names. For instance, why should a specific name such as *tomcodus*, Latinized from the vernacular fish name Tom Cod, be rejected, as claimed by Jardine (in Sclater 1878:23)? Rejecting such a name would open a can of worms ... as the adoption (or creation) of a more recent name would violate the Principle of Priority for a subjective reason. Pascoe (1865) provided a good illustration of the negative and confusing impact of subjective opinions on the circumstances that should lead to a name change. Pascoe (1865:9543) argued that it is extremely difficult to decide objectively whether a name change is legitimate, such as in the case of Westwood replacing *Hyphaereon* into *Campylocnemis* because it was regarded as too close to *Hyperion*. As a result, Pascoe thought that name changes should be allowed only when spelling is strictly identical.

Generally speaking, it seems that naturalists did not have the time (or did not take the time) to debate enough about all those issues in the 1850s and 1860s. However, many debates took place in the 1870s that clarified many ideas and prepared the ground for the future codes proposed in the last quarter of the 19th century.

Verrill's (1869) Comments on the Revised Edition of the Strickland Code

In 1869, Verrill commented on the 1842 Strickland Code as well as Jardine's 1865 report on amendments to the code, in the *American Journal of Science and Arts*. Several of Verrill's comments directly relate to the application of the priority rule.

First, the law of priority should be limited to the Xth edition of *Systema Naturae*, because it is "more logical" (Verrill 1869:94). Verrill also emphasizes the importance of choosing a limit, whether it should be 1758 or 1767. Without being revolutionary (the Strickland Code indicated a date already), Verrill's comments show that mentalities were slowly changing. Indeed, in the 1850s and early 1860s, authors would pay little attention to the starting date for zoological nomenclature:

> If the Xth edition be taken as the limit, which seems to be the tendency among recent writers, especially in this country and in northern Europe, the date would be 1758. The second volume of the XIIth edition bears the date 1767. Disregard of this important and essential law has brought into conchology and some other branches of zoology an almost incredible amount of confusion within a few years, the indefinite names of Link, Klein, Brown, Columa and other ante-binomial and polynomial writers, having been revived and substituted for the well known names of Linnaeus and later authors. (Verrill 1869:97)

Second, application of Rule #4 ("The generic name should always be retained for that portion of the original genus which was considered typical by the author") seems complicated:

> This course has been systematically followed by some writers, and when carelessly done has often led to unfortunate and absurd results, especially when applied to the earlier writers, since it often happens that the actual position of the *first* species, in the restricted modern genera, cannot be determined with certainty. (Verrill 1869:98)

Verrill was right: several issues related to types had to be resolved before the law of priority could apply without generating confusion.

Third, Verrill (1869:101) thought that the application of Rule #10 ("A name should be changed which has before been proposed for some other genus in zoölogy or botany, or for some other species in the same genus, when still retained for such genus or species.") is objectionable, at least for generic names, and agrees with Gray (1864) that "all appear to admit the necessity of allowing the same [generic] name to be used once in either kingdom [botany and zoology]." This opinion was adopted in most of the codes proposed in the last quarter of the 19th century.

Finally, Verrill agrees with Rule #11 ("A name may be changed when it implies a false proposition which is likely to propagate important errors.") and cites abnormal structures caused by diseases or parasites as causes of errors:

> Thus, *Echinus gibbosus* Val. proves to be identical with our *Euryechinus imbecillis*, but the gibbosity is caused only by a parasitic crab lodged in the anal region, and is not present in normal specimens. (Verrill 1869:102)

This rule, which goes back to the late 18th century, when authors thought that specific names should mean something and appropriately describe the species, remained endlessly challenged; determining names that could "likely propagate important errors" was clearly subjective and thus jeopardized the universality of the priority rule.

THORELL'S *ON EUROPEAN SPIDERS*

In his imposing monograph *On European Spiders*, Thorell (1869) defended several ideas related to the application of the law of priority, which differed in several aspects from the Strickland Code. First, based on the fact that Linnaeus first defined binominal nomenclature in his *Philosophia botanica*, Thorell argued that 1751 should be the starting point for zoological nomenclature (specific and generic names). One advantage was that binomials published in Clerck's (1757) *Aranei suecici*, in which binominal nomenclature was used consistently, could be considered. In fact, under our current Code (ICZN 1999), Clerck's *Aranei suecici* is deemed to have been published on January 1, 1758, and all names published in it have precedence over the spider species names published in the 10th edition of the *Systema Naturae*, which is the only exception accepted by the Code.

There were no rules, neither in the Strickland Code nor in the Kiesenwetter Code, according to which "indelicate" names should be rejected. Strickland thought that no rule could prevent people from "infringing the rules of good taste by introducing into the science words of the same inelegant or unclassical character in future" (Strickland 1843 in Sclater 1878:13). In the Strickland Code, the "good taste" in forming new names was guided only by recommendations. However, in the 1860s, some zoologists clearly were of the opinion that "indelicate" names should be rejected (Jardine mentioned it in the Report on the revision of the Strickland Code, although no change was introduced in that regard in the Code). Thorell was opposed to keeping "indelicate" names:

> It is rightly observed by the British Committee, that a name once published is the property of the science, and cannot therefore be revoked or altered, not even by the person who has composed it. Exceptions however exist …: the Committee also admits, that there are names which ought unquestionably to be *discarded*, those namely, which in their signification are *absurd* or *false*. It would have been desirable that this sentence of reprobation had been extended also to certain classes of those names which the Committee only considers that naturalists ought in future to abstain from forming ("objectionable names"). Such are for instance mongrel names (compounded of two or more different languages) (…) and names manufactured by mutilating and mangling other names, e.g., *Cypsnagra* from *Cypselus* and *Tanagra*. To this class belong also the equally barbarous denominations that have arisen from the ridiculous practice of composing unmeaning generic names of arbitrarily combined letters, usually in the form of anagram: e.g., *Rocinela*, *Conilera*, *Cirolina*, *Anilocra*, formed from the letters in *Carolina*. We hope the time will come

when also such names as those just mentioned will be rejected, though this is not yet the case. (Thorell 1869:12)

However, as pointed out by authors opposed to changing names regarded as "indelicate" or "barbarous' (e.g., Pascoe 1865), it is difficult to draw a line between "delicate" and "indelicate" names. Also, if such a rule were to apply retrospectively, it would greatly jeopardize the application of priority, as perfectly understood by Strickland in 1842.

Thorell (1869:11) discussed the potential need for replacement of nearly identical names, such as *Ariadne* and *Ariadna*. However, he did not regard as nearly identical names of different gender, such as *Euryopis* and *Euryopa*, which shows that his opinion was fairly arbitrary, or at least not easy to follow. Thorell (1869:14) also was in favor of a strict application of Latin grammar and orthography, and suggested that all names erroneously formed be corrected, whilst retaining authorship with the original author. Thorell (1869:11–12) disagreed that names should be replaced (or the next synonym used when available) if they had been created in botany or zoology. As Thorell rightly pointed out, Linnaeus, who wanted no name to be used more than once in any of the three kingdoms, was dealing with far fewer species. Finally, Thorell (1869:17) used our modern way of citing authorship (at least in zoology): no parentheses if the species is still placed in the original genus, and parentheses if the species is placed in a different genus. Wallace (1871:lix) thought that Thorell applied Latin grammar too strictly:

> He is a strict purist, and alters the termination of every name he considers to be not classically constructed. He admits that there is often difference of opinion on these points, but does not seem to consider that the consequent confusion and instability of nomenclature is as great an evil as classical inaccuracy.

THE 1870S: A GREAT DECADE FOR NOMENCLATURAL DEBATES

The literature on nomenclatural rules was abundant in the 1870s. Many authors expressed deep, serious worries about the situation in zoological nomenclature. In particular, many authors were of the opinion that the Strickland Code failed in bringing stability to nomenclature. Agassiz (1871:355) was quite clear about it:

> The rules of nomenclature generally adopted are by no means satisfactory. The exceptions constantly taken to their application only increase the confusion, and the attempts made by the British Association to recommend a set of rules for the guidance of Naturalists have not been successful.

According to Edwards (1873:22), the Strickland Code "was not found to work altogether satisfactorily, and never did receive the general assent of Naturalists in their several departments." In his *Catalog der Lepidopteren*, Staudinger (in Staudinger and Wocke, 1871:xi, tr.) asserted: "The diversity of laws that govern nomenclature and the confusion that characterizes it will end up generating a true chaos if we cannot establish fixed rules." Other quotations could be cited.

The 1870s was a period of doubt. Important issues were still not solved, such as the starting date for zoological nomenclature, but one of the most serious discussions concerned the Principle of Priority. Although the vast majority of naturalists agreed that priority was a valuable and necessary principle, they also realized that its application could be subject to interpretation. The community began to split between those who remained loyal to priority and those who preferred to adopt a so-called "law of convenience." The latter started growing after several authors, such as Kirby (1871) and Scudder (1872), restored many old but completely unknown names to replace well-known, commonly used names:

> From 1861 to 1871 the tide went in the direction of restoring the earliest discoverable names. It is sufficient to mention the names of Gemminger and von Harold, the late Mr Crotch, Mr Scudder, and Mr Kirby to recall that the practice of "resurrection" resulted in the production of several volumes. (Lewis 1875:vii)

With the concept of priority's being directly attacked, nomenclature lived one of its most difficult periods. The codes written in the last quarter of the 19th century cannot be understood if the debates that took place in the 1870s are not considered, as those codes had to address all the issues raised in the 1870s. Many of the authors who participated actively in those debates were entomologists, such as W.H. Edwards, W.A. Lewis, O. Staudinger, and J.L. LeConte. Interestingly, it is during that period of doubt that some quite radical solutions were proposed (though not adopted).

Kirby's Controversial Synonymic Catalogue of Diurnal Lepidoptera

As soon as it was published, Kirby's (1871a) *Synomymic catalogue of diurnal Lepidoptera* was profusely criticized. Kirby held radical views on generic nomenclature, which he presented before the members of the Entomological Society of London in 1868 (Kirby 1868) and published in the *Journal of the Linnean Society, Zoology* (Kirby 1870). Generally speaking, Kirby was in favor of simple nomenclatural rules that applied strictly, with no exceptions. For that reason, he rejected any exceptions to the rigid limit of 1767 for a starting date of zoological nomenclature (the revised Strickland Code accepted names published by Scopoli and Artedi before the XIIth edition of *Systema Naturae*). Later, however, Kirby (1871b) argued that 1758 should probably be selected. In the same spirit, he argued that when no type species was designated by original authors for generic names, then the first species listed should be considered typical.

Although Kirby (1870:494) genuinely intended to solve or at least improve the "indescribable confusion [that] has arisen in generic nomenclature," several authors rejected what was his "revolution in generic nomenclature." (Kirby 1868:xliii) Westwood indicated that he had proposed exactly the same rules several years earlier but had to change his mind because "such a rule and the application of it retrospectively would cause so much confusion that the remedy would be worse than the disease." (in Kirby 1868:xliii)

Wallace (1871:lxiii) criticized heavily Kirby's work in his presidential address to the Entomological Society of London: "The most novel, and, as many will think, the worst feature of the book is the entire revision of the generic nomenclature." Wallace thought that Kirby had gone too far when applying some of the Strickland rules, and should have taken in consideration the limitations of those rules. For instance, under Rule #10, "A name should be changed which has before been proposed for some other genus in zoology or botany, or some other species in the same genus, *when still retained for such genus or species*," but, according to Wallace (1871:xliv), Kirby failed to take into account the last point of Rule #10 and replaced names even when their homonyms were completely forgotten.

However, the most important disagreement between Wallace and Kirby was related to the use of Hübner's names. In the 1870s, Hübner was regarded as an active "splitter" by the community of entomologists (Grote 1876:57). According to Wallace (1871:xliv), the consideration of generic names published in Hübner's "obsolete and useless catalogue" yielded a "wholesale change" of names that was not warranted by the Strickland rules. In fact, the Strickland rules (as well as the Kiesenwetter rules) clearly required that names be "clearly defined in some published work" (Rule #12). According to Wallace, Hübner did not describe properly his generic names, which explains why his work was "set aside (...) by most European entomologists." However, Kirby argued that the application of Hübner's generic names could be determined through type species.

Kirby's work shows that even when authors accepted the authority of the same rules (in this case the Strickland Code), they might still disagree on how to interpret those rules and apply them in practice, which probably explains why: "Notwithstanding the "Rules for Zoological Nomenclature" sanctioned by the authority of the British Association, it would not seem that any perceptible improvement has taken place" (Crotch 1870:59). For that reason, Kirby (1871b:42) claimed that "an international congress of naturalists would be very desirable to reconsider and, if necessary, revise [the rules and recommendations of the Strickland code]."

Although Kirby's approach to designating type species was rejected by the community, it forced naturalists to discuss the link between types and priority. In that regard, Kirby's (1868, 1870, 1871a, 1871b) contributions impacted positively on nomenclature. For instance, Crotch (1870:59), regretting that "Mr. Kirby, unfortunately, merely [pointed] out certain inconsistencies without suggesting any remedies or consistent plan of action," asserted:

> A genus, as far as I understand it, for the purposes of nomenclature, consists of but one species—its type.... No genus can be considered defined until a type is indicated, for characters must vary with our knowledge in every case; but when the type of a genus is not indicated, I am not inclined to cut the knot by the simple process of taking the first species, but to trace the genus historically until it has a type given to it.

As the starting date, Crotch (1870) proposed 1735, for the 1st edition of the *Systema Naturae*.

Staudinger's Bilingual *Catalog der Lepidopteren*

The *Catalog der Lepidopteren des Europæischen Faunengebiets* by Staudinger and Wocke (1871) was a bilingual work (German and French) that can also be referred to as the *Catalogue ou énumeration méthodique des Lépidoptères qui habitent sur le territoire de la faune européenne*. Staudinger wrote the Macrolepidoptera section, as well as the Introduction (in which nomenclature is discussed), and Wocke wrote the Microlepidoptera section. Although Staudinger accepted the XIIth edition of the *Systema Naturae* as a starting date for zoological nomenclature, he thought that 1758 (Xth edition) would make more sense, an idea shared by Scudder (1873). Staudinger was opposed to changing names for grammatical or orthographic reasons, as indelicate names do not hurt communication among naturalists. As an illustration of the confusion created by Latin orthographic improvement, Staudinger (in Staudinger and Wocke 1871:xiii) gave the example of *Nyctimera*, "improved" successively in *Nychtemera, Nyctymera, Nyctimera, Nychthemera,* and *Nyctimera*...

The 1871–1872 Issue of the *Entomologist's Monthly Magazine*

In the 1870s, many nomenclatural debates were published in the *Entomologist's Monthly Magazine*. In particular, the eighth volume is interesting because authors replied to each other, revealing their disagreements. Those articles are analyzed here chronologically.

Lewis (1871:1) opened the volume with harsh comments about nomenclature:

> It is a dreadfully frivolous work, I venture to think, this routing out from libraries of doubtful and obscure descriptions, and the spending of precious years in nicely balancing considerations upon the priority of a name. To give the title of scientific discovery to such a process would be to apply rich gilding to a very cheap and common sort of ginger-bread.

Lewis (1871:2) was strongly in favor of abandoning a strict law of priority:

> My proposal is that no name (whenever and wherever it may be discovered) be received henceforth, to the displacement of a universally recognized name; and this I humbly consider to be founded on strict common sense.

Lewis (1871:4) was so opposed to the ideal of accepting the oldest names in all cases that he even provocatively proposed June 1st, 1871 as a starting date for zoological nomenclature, so that all issues related to the priority of old names would immediately disappear.

McLachlan (1871:40), less radical than Lewis, argued that "resurrection men" could not be condemned because they resurrect names, but because, in some cases, "in their reverence for old names, [they] raise ghosts, ... names that should sink into oblivion, or rest quietly in the list of *'species indeterminatæ.'"* Again, the problem here seems to have been the criteria used to determine whether a name was published properly, i.e., in a way that could help future authors to re-identify the species being referred to. McLachlan (1871:41) also commented on Lewis's remark on naturalists spending some of their time dealing with synonymy:

> It is expedient there should be no more crime, no more deceit, in the world; and, as a consequence, no more prisons, police, and lawyers [Lewis was a professional lawyer]. But the evils exist, and the other *necessary* evils are required to keep them in check. Synonymy exists, and its existence renders necessary the evil that entomologists must waste precious time in unraveling it. The suppression of both crime and synonymy by a *fiat* is utterly impossible. I couple the words, but the existence of synonymy is too often owing to what are actual *crimes* against science. I hold that, when an entomologist describes an insect as new, without using every endeavor that is humanly possible to discover whether it be not *already* described, he commits one of the greatest crimes against science.

Briggs (1871:94) brought up an interesting argument against the use of priority in cases where it would replace a common species name that all naturalists agreed upon:

> I look upon the *accord* of entomologists as a "law" of itself, governing and paramount to both these special "laws"; where there is want of accord we have resource to one of them to obtain that accord, *but, where accord exists already, I say that we have no right to call either law into operation*; they are not wanted, and I look upon the present resurrectional movements as a tortuous application of a useful law.

Naturally, whether "indelicate" names should be replaced was still debated in 1872. Against authors who complained against "nonsense" names such as anagrams (Dunning 1872a), Sharp (1872:254) brought a quite new way of looking at scientific nomenclature:

> Mr. Dunning thinks *Lycaena Minimus* abhorrent; but I think it can be only because of some curious classical prejudices that he so considers it. Scientific nomenclature should be of no particular language, its object being to supply a universal language, and it is to be of assistance for this purpose that we make use of Latin and Greek words (as being more generally known than others); but we must handle them according to the rules of universal grammar.

This differs from the views expressed in both the Strickland and Kiesenwetter codes. Dunning (1872b:294) replies, citing Thorell, that without respect of Latin orthography and grammar, nomenclature will "gradually assume an appearance absolutely disgusting to a person possessing even the slenderest classical attainments."

Lewis: A Professional Lawyer Commenting on Nomenclature

Lewis was one of the strongest opponents to a strict application of the law of priority. His writings on priority impacted deeply the community of naturalists of that period. Like Briggs (1871), he thought there was no need to call for priority of older names when naturalists already agreed on a name for a given species because "the law of priority is a means to an end, and the end in this case is accord or common agreement to a name.... The object of the law is the important thing; not the law, which is only machinery" (Lewis 1875:viii). According to Lewis, giving priority to an ancient name over a more recent name made no sense because, in most cases, older descriptions are unrecognizable:

> Linné described 780 Lepidopterous insects, the number now known cannot be less than 30,000. Dozens of allied species all equally fit numbers of the old descriptions; and such descriptions are now necessarily of no value. (Lewis 1875:xii)

Other authors, especially Edwards (1873:24), made similar arguments against the law of priority.

This issue could have been partly solved by going back to type specimens. However, as most authors in that period, Lewis paid much more attention to written descriptions, necessarily incomplete, than to types. In fact, he agreed with Sharp, a strong defender of the strict priority rule, that "it is clear that [types] must not be exclusively or even strongly relied on" (Lewis 1875:xxv) when one wishes to determine which species is referred to by a species name. This opinion was widely shared in that period:

> Studying the types will tell me the palaeontologists of science [an expression not actually referring to palaeontologists, but, as a metaphor, for people interested in "dead" things in science]. But who does not know that? Those famous types are more or less spread out, sometimes exchanged, often destroyed, and, which is worse, replaced randomly. (Puton 1880:37)

Lewis (1875:xxxvii) proposed an interesting solution to the conflict between "convenience" and "priority":

> The objects are clear—(1) to exclude doubtful names; (2) to preserve accepted names. They are not identical, but both objects can and must be pursued together. It has been several times suggested that the enjoyment of universal acceptance for a period of years should give a name an indefeasible title to adoption. For a purpose which I have in view, I will fill in the number "thirty" and make the proposal read thus: No name for thirty years in universal acceptance should be displaced.

In this passage, Lewis clearly pointed to what we call now the *nomina dubia* (names of doubtful application) and conserved names (over oldest names). This approach was then quite revolutionary:

> This would be the working of the limitation. No name could be produced now for the first time from any book bearing date 1842 or previously.... That stops the evil spreading henceforward. No author can then bring up a name from Old Style books, unless the name has been kept alive by quotation as the true name in some work since 1842. Here is a measure there is no difficulty in applying, and its operation is simple. (Lewis 1875:xxxvii)

However, Lewis's method was not adopted, mainly because taxonomists discovered soon enough that going back to the type specimens is critical when determining the entity a name refers to, which cannot be answered if only written descriptions are considered.

Edwards, LeConte, Mead, and Others

Edwards (1873, 1876) largely agreed with Lewis, and was opposed to a strict application of priority: "The laws of priority are not inexorable, and such laws anywhere lead to absurdity and injustice." (Edwards 1876:87) As far as the starting point is concerned, he adopted Thorell's (1869) date, 1751, as long as authors consistently used binominal nomenclature. Interestingly, Edwards (1873:35) was in favor of developing a code for entomologists if no agreement could be found with other branches of natural history (although it does not seem that entomologists could agree with one another):

> I have heard ... that Entomologists have no right to separate themselves from other naturalists, and make a special Code for their own sole guidance. To this I would reply, why not?... Why should not each branch adopt Rules to suit its own case? If Botany may be excluded from the operations of a Code, why not Entomology?

In fact, several specialized codes were published in the last 25 years of the 19th century.

LeConte (1873, 1874a,b,c) was opposed to strict priority, because it is often difficult to determine the meaning of old names, which he proposed to deal with using a special starting date for "doubtful cases," at least in insects, which would be Olivier's (1789–1811) *Insectes* of the *Encyclopédie Méthodique*.

It is in this period of doubt that Sharp (1873), in order to address the fact that species names often change when moved from one genus to another, suggested that binomials be never changed after their creation, regardless of whether species would be transferred to a new genus or not. Wallace (1874) vividly criticized this idea based on the fact that generic names provided naturalists with valuable information. Interestingly, fixed binomials were re-invented in a different context (Michener 1963).

Finally, Mead (1874:108) coined the expression "law of stability" for the idea of keeping all names already widely adopted by naturalists, even if they were not the oldest ones, an expression that he preferred to "law of convenience."

Dall's Survey of Naturalists' Opinions on Nomenclature

Dall's (1878) Report was not a new code:

> This report does not form a proposition to be acted on by the Section, or by the Association in a final manner at the present time. It is merely an attempt at a complete presentation of the subject, without which no well advised action by naturalists as a body can be expected or is to be desired. (Dall 1878:10)

Dall wished to survey and present the diversity of North American opinions regarding nomenclatural issues, and to provide naturalists with a review of nomenclatural rules and recommendations proposed in zoology as well as botany up to that time.

One of the most interesting elements of Dall's Report was the results of his survey (Dall 1878:16–19). In 1876, he sent a circular including 27 questions to all North American naturalists (mainly zoologists) who had published systematic work in the last five years. Only "yes" or "no" were possible answers, in order to avoid ambiguity. Forty-five naturalists (three from Canada and forty-two from the United States) replied, including many of the naturalists already mentioned in this chapter, such as Lewis, LeConte, Edwards, Scudder, and A. Gray (for a complete list, see Dall 1878:20–21). According to Cope (1879:518), names on the list of people who replied include "an unquestionable majority of the best working naturalists of the country." Most questions received overwhelming agreement. The community, however, was divided over eight questions. Those disagreements, which are of great interest, are reviewed here.

Question I. No agreement could be found on the starting date: Xth edition (18 votes), XIIth edition (17), 1736 (1), 1753 (2 botanists), no answer (7). This confirms what is sensed from the literature of the 1870s.

Question IV. "If an author has not indicated his adoption of the binomial system by discarding all polynomial names in a given work, are any of his names therein entitled to recognition otherwise than in bibliography?" no (18), yes (18), doubtful (4), no answer (5). This is an important disagreement because it means that both generic and specific names from such works could become eligible for priority.

Question IX. "Is a name, when used in a generic sense, and otherwise properly constituted, subject to have its orthography changed by a subsequent author, on the ground that a proper construction from its classical roots would result in a different spelling?" no (21), yes (19) doubtful (8), no answer (2). Again, this confirms what is found in the literature from the 1870s; naturalists disagreed on whether names should be changed for grammatical or orthographic reasons. The most extreme opinions regarding that question certainly were expressed by Saint-Lager (1882, 1886), author of several contributions in which he suggested that all existing names be corrected by philologists and that, before creating any new name, naturalists should get approval of its correctness from a philologist (e.g., Saint-Lager 1886:54).

Question XVIII. "When a generic name has lapsed from sufficient cause into synonymy, should it be thenceforth entirely rejected from nomenclature? Or should it still be applicable to any new and valid genus?" reject (19), accept (23), doubtful (1), no answer (2). Answers to this question suggest that many naturalists considered that synonyms could be permanently rejected, which is something that does not exist in the current Code (ICZN 1999).

Question XIX. "Should a name which has been once used in one subkingdom, and has lapsed into synonymy, be considered available for use in any other if not entirely rejected from nomenclature?" no (20), doubtful (1), yes (18), no answer (6). This question relates to the previous one.

Question XXI. "Is it advisable to fix a limit of time, beyond which a name which had been received without objection during that time shall be held to have become valid, and no longer liable to change from the resuscitation of obsolete or uncurrent but actually prior names?" no (28), doubtful (1), yes (13), no answer (3). This question comes directly from Lewis's proposal (1875), which intended to mix both the law of priority and "convenience" or "stability" (by preventing well-established names from being replaced).

Question XXIII. Yes (30) to the following: "Should it be permitted to alter, or replace by other and different appellations, class, ordinal, and family names, which owing to the advance of Science and consequent fluctuation of their supposed limits have become uncharacteristic?" Yes (11) to the following: "Or should these also be rigidly subject to such rules of priority as might be determined on for generic or specific names?" No answer (4). Answers to this question suggest that a majority of authors (~75%) thought that supra-generic names should not be governed by a code of nomenclature, although about 25% thought that supra-generic names should be governed by rigid rules. The codes proposed so far had dealt only with genera and species (except for the use of the suffix –idae for family names). Scudder (1872) was one of the authors who pushed to apply the priority principle to all names, including higher groups.

Question XXVII. "Should a series of rules be recommended for adoption by the Association, would you be guided by these recommendations in cases where they might not agree with your own preferences?" yes (29), yes with reservations (15), no (1). Answers to this questions show that, in 1878, one third of North American zoologists did not seem ready to adopt common nomenclatural rules, whether they agreed on them or not. Obviously, opinions were strong.

Although his Report is not a code, Dall suggested that a starting date be adopted. He (1878:15) rightly pointed out that any starting point would be arbitrary (as clearly indicated in the current Code, see Article 3): "The system being of slow and intermittent growth, even with its originator, an arbitrary starting point is necessary." In the Report, Dall (1878:41–44) proposed that specialists from different areas of expertise (entomologists, malacologists, etc.) get together and agree on an "epoch-making" work, one for each group, i.e., Clerck's (1757) *Aranei Suecici* could be the epoch-maker for spiders, as suggested by Thorell (1869); the first edition of the *Systema Naturae* (Linné, 1735) could be the epoch-maker for bird generic names, etc. This being said, Dall (1878:44) also added that, in case specialists could not agree on a date (which clearly was a foreseeable possibility …), the XIIth edition be selected as the starting point because: (1) "it has been twice recommended by the British Association" and (2) "the usage founded on the B.A. rules should be maintained if possible." Dall's proposal was not adopted; in fact, several other possible starting points were proposed in the last two decades of the 19th century.

1880s: More New Codes

Chaper's (1881) Rules

Like Dall's (1878) Report, Chaper's (1881) *De la nomenclature des êtres organisés* was not a code, in the sense that it was not adopted as an official code.

> The work that it [the Committee] is honored to submit to you is not a code with any new principle, nor does it have the pretension to impose it to anyone under whatever authority. It simply is a circular where are put together, coordinated, and for everyone's interest and benefit, the rules and usages sanctioned by common sense and an already long practice. (Chaper 1881:37, tr.)

Chaper's set of rules was adopted by the Sociéte Zoologique de France simply to be presented at the International Congress of Geology, Bologna, along with Douvillé's rules. Chaper (1881:7) wished to make sure that zoologists influenced the code that palaeontologists intended to develop. However, Blanchard (1889) mentioned that, after Chaper's rules were adopted on June 14, 1881 by the Société Zoologique de France, they were followed in the *Bulletin* of the Society. More importantly, 1300 copies of Chaper's rules were sent to members of the Société, as well as to foreign naturalists and societies for comments. Blanchard (1889) used Chaper's rules as a basis for the rules that he presented at the first International Congress of Zoology.

Chaper's (1881) article was divided into two parts: (1) three pages with 17 nomenclatural rules, and (2) 30 pages with comments on those rules. Chaper wrote everything, in the name of a committee of eight members, including R. Blanchard, future first president of the International Commission on Zoological Nomenclature. Chaper's set of rules is remarkably concise and brief. In fact, it is so brief that one of the most needed items is missing, i.e., a starting point for zoological nomenclature. However, Chaper's rules were not supposed to serve as a code. For that reason, he discussed in detail the history of the binominal system in the Report, but did not express any preference for any date. However, Chaper argued that the genus concept was already present in several authors who preceded Linné, such as Tournefort, Lang, and Klein, suggesting that Chaper and the committee might have favored early dates as starting points. In fact, Blanchard (1889) selected 1700 as a starting point for generic names.

One detail in Chaper's (1881) rules is intriguing. Indeed, the first Article reads: (Chaper 1881:3, tr.): "The nomenclature adopted for all organized beings is binary and binominal [*binaire et binominale*]. It is essentially Latin. Each being is distinguished by a genus name followed by a species name." Chaper's ideas on the meaning of binary and binominal have remained unclear, especially because both adjectives were used interchangeably in the 19th century to refer to the Linnaean nomenclature. As Melville (1995:17) pointed out: "It is not clear why he used both adjectives when one would have sufficed." Blanchard (1889), whose rules were deeply inspired from Chaper's rules, also used both adjectives without giving a clear distinction between them.

Douvillé's Rules

Douvillé's (1882) rules were prepared by a committee of eight members (six palaeontologists and two mineralogists) in preparation for the International Geological Congress, Bologna (1881). Those rules are brief and add up to only two pages. As with Chaper's rules, they were not officially adopted. They were a working document. In fact, because Douvillé proposed identical rules for Botany and Zoology, two letters (Van Tieghem et al. 1882; De Candolle 1882) were sent to the president of the International Geological Congress to let him know that botanists already have a code and will not accept new rules. In the *Comptes-Rendus* of the Congress, it was mentioned:

> Our last session was filled with a serious study of the rules to be followed in the species nomenclature; and we are entitled to hope that what has been started by palaeontology, will be able to continue with help from botanists and zoologists. (Capellini 1882:188, tr.)

There was no follow-up. However, two features of Douvillé's rules are worth noting: (1) No starting point is proposed because any name that followed the binominal nomenclature could be considered; (2) An illustration is required for a specific name to be considered.

The Code of the American Ornithologists' Union (1886)

In April 1885, a Code of nomenclature was officially adopted by the American Ornithologists' Union. It was prepared by a committee of five members appointed in 1883 by the Union in the context of the revision of the checklist of North American birds. This 69-page Code, known as the "AOU Code," was published in 1886 (Coues et al. 1886). Although it was prepared by ornithologists, it was written to serve all zoologists. It included one important innovation, i.e., the introduction of rules governing trinomials, names referring to varieties (Coues et al. 1886:30–32), which were already widely used by ornithologists. As far as priority is concerned, the AOU Code adopted the 10th edition of *Systema Naturae* as the starting point (Coues et al. 1886:32). Also, importantly, the AOU Code rejected the idea that the *lex prioritatis* would not apply for names that had been in use for a certain period of time (25 and 30 years proposed by Dall and Lewis, respectively): Committee members thought that it was difficult in most cases to determine whether a name had been universally used:

> Unless perfect agreement could be obtained,—and of this there is very little probability,—the proposed rule would tend to increase rather than lessen the confusion it would be the design to remove. (Coues et al. 1886:39)

Thus, it was adopted that: "The law of priority is to be rigidly enforced in respect to all generic, specific, and subspecific names." (Coues et al. 1886:40) In the AOU Code, the application of priority is more flexible with supra-generic names; it applies only if family or sub-family names are strictly synonymous (i.e., identical delineation).

The AOU Code was commented on, especially by Sharpe (1886), the member of the committee in charge of establishing an authoritative list of birds of the British Islands for the British Ornithologists' Union, the "sister" of the AOU. Sharpe (1886:169) rejected a strict application of the law of priority: "The A.O.U. list does not simplify existing nomenclature to begin with, and it is the great love of change, which has been so characteristic of recent ornithological work in America, which makes us skeptical as to whether even the authority of the A.O.U. "List" will be sufficient to prevent further modifications in this direction." However, Sharpe (1886:170) constructively suggested that both sister unions communicate to see "if a common ground of agreement cannot be arrived at." This agreement took place in the context of the International Commission on Zoological Nomenclature in the next decade.

A Different Voice: Preudhomme de Borre

In 1886, Preudhomme de Borre, president of the Société entomologique de Belgique, expressed interesting opinions on how to get rid of nomenclatural confusion. According to him, the biggest problem in entomology was not nomenclatural but found its roots in taxonomic practices:

> The big ill of our period is not the catalogues; it is the isolated description of novelties [new species] which are discovered every day; it too often means the description of novelties which are not new. (Preudhomme de Borre 1886:cxcviii, tr.)

According to Preudhomme de Borre, changes of generic or specific names should be allowed only in monographs:

> It thus is, in spite of all obstacles, in the return to monographic works that must reside the hope of those who desire that the systematic inventory of living beings that we are dealing with could regain true order. (1886:cci, tr.)

Although his idea of restricting the right of changing names in monographs seems a bit extreme, it is undeniable that many synonyms and a great deal of nomenclatural confusion was and still is due to the fact that authors sometimes create new species names too lightly. This proposal was never adopted. Preudhomme de Borre (in Milne-Edwards 1889:414) raised his voice at the First International Congress of Zoology held in Paris in 1889, where he argued—in vain—during the discussion of Blanchard's proposed rules that name changes should be allowed only in monographs.

TOWARD THE *RÈGLES INTERNATIONALES DE LA NOMENCLATURE ZOOLOGIQUE* (1905)

Discussion of Blanchard's Rules at the First International Congress of Zoology

As first president of the International Commission on Zoological Nomenclature (1895–1919), Raphaël Blanchard played a critical role in the development of the *Code of Zoological Nomenclature*. Since his participation in Chaper's 1881 committee, he remained interested and involved in nomenclature. In particular, he used the opportunity of the first Congrès International de Zoologie (Paris, 1889), to present before the participants a series of rules for approval by the Congress (Blanchard 1889:397–404). Blanchard's (1889) contribution to the *Compte-Rendu des séances du Congrès International de Zoologie* is an imposing document that contains: a long essay (Blanchard 1889:333–397) explaining each rule he proposes; the actual rules he proposes (Blanchard 1889:397–404); the discussion that takes place at the Congress (Milne-Edwards 1889:405–418); and the rules adopted by the Congress (Blanchard 1889:419–424). Several documents related to the discussion on zoological nomenclature are added in an appendix (Chaper 1889; De Candolle 1889; Horst et al. 1889; Hubrecht et al. 1889; Oberthür 1889; Saint-Lager 1889). Finally, a list of conventional abbreviations for authors' names adopted by the Congress closes the *Compte-Rendu*.

The rules adopted by the Congrès International de Zoologie (Blanchard 1889:419–424) were very similar to the rules presented by Blanchard at the Congress. The 13-page-long document describing the discussion of each of Blanchard's proposed rules is fascinating because it shows the detailed reactions of participants. Although more people were probably present in the room during the discussion, the following zoologists expressed their opinions and were thus recorded in the *Procès-verbaux*: Bedel, Blanchard, Chaper, Dautzenberg, Fischer, Girard, de Guerne, Kraatz, McLachlan, Milne-Edwards, Preudhomme de Borre, Riley, de Selys-Longchamps, Simon, Trimen, Trouessart, and Vaillant. At the beginning of the discussion (Milne-Edwards 1889:405), Riley and de Selys-Longchamps suggested that any rule adopted by the Congress should simply be regarded as some kind of advice, but not as a law that would have authority over zoologists; Milne-Edwards argued that the International Congress of Zoology was the ideal opportunity to agree on common rules. Most articles were adopted without discussion or with only minor modifications. As for the articles relating to priority (Blanchard's articles 42 to 53), only Article #42 (which defines priority) was adopted in 1889. Due to a lack of time, the discussion was interrupted by the president, which is a pity because Blanchard's Articles #43–44 used 1722 as starting point...

Indeed, Blanchard (1889:386, tr.) held strong opinions about priority, and most particularly about the starting point for nomenclature:

> We too, are of the opinion not to apply priority beyond the 10th edition of the *Systema Naturae*, the first one in which Linné used binary nomenclature. However, we must express some reservation in favor of Tournefort, Lang, Klein, Clerck, and Adanson. The works of those authors are in conformity with the binary method; at least, they had an exact notion of the genus and they precisely delineated its limits: we cannot leave them aside without an outrageous injustice.... For us, the year 1700 is thus the most extreme date beyond which the "search of paternity" is not allowed. This date, however, does work only for Botany; zoological nomenclature only really starts in 1722, with Lang. Any generic, and specific name established since 1700 for plants and since 1722 for animals, in conformity with rules that precede, will hold priority and will have to be substituted to any more recent name, even admitted by Linné.

The discussion on priority took place in the second International Congress of Zoology, in Moscow (1892).

Blanchard had three more years to think about the starting point.... In particular, he could think about the comments published in the Appendix of the *Compte-Rendu du Congrès*: Dutch entomologists and zoologists favored 1751 as a starting point (Horst et al. 1889; Hubrecht et al. 1889); Oberthür (1889) suggested that a figure be mandatory for a new species name to be considered, and he interestingly argued that there was a great need in fixing the use of the word *typus* or *type* in natural history; as usual, Saint-Lager (1889) provided endless linguistic remarks; de Candolle (1889) agreed with Blanchard that Tournefort was the founder of the genus concept, but did not comment on Blanchard's starting point; he also argued that abbreviating the authors' last names was a bad idea, in Botany as in Zoology.

The Second International Congress of Zoology (1892): 1758 It Will Be

Blanchard's work between the first and second Congress was remarkable. This 82-page *Deuxième Rapport sur la nomenclature des êtres organisés* (Blanchard, 1893) includes: (1) precisions on the rules adopted in the first Congress (Blanchard 1893:3–13), which mainly concern spelling issues; (2) the main body of the Report (Blanchard 1893:14–65), which complements Blanchard's (1889) first Report; (3) the rules newly proposed by Blanchard (1893:65–72); (4) the rules adopted by the Paris and Moscow congresses (Blanchard 1893:72–83). Note that all rules were discussed and voted upon again, including the rules already adopted in Paris in 1889, because Blanchard changed the wording of several rules and added new rules (e.g., rules were added for naming hybrids).

All the discussions about Blanchard's proposed rules are available in the *Procès-verbaux du Congrès* for the two sessions in which the discussions took place, with A. Maklakov and E. Chantre as presidents. One gets a good idea of what happened in Moscow by comparing the rules proposed by Blanchard, the rules adopted, and the discussions. In brief, Blanchard was exceptionally successful at convincing most participants in accepting his rules. Most rules were accepted with no or little discussion. Blanchard's rules #13 and #30 were rejected, but with no major implications. Some rules were reworded, but with no change of meaning. Although more people were probably present in the room during the discussions, the following zoologists expressed their opinions and were thus recorded in the *Procès-verbaux*: Blanchard, Brusina, Bunge, Chantre, de Guerne, Janet, Jentink, Kapnist, Maklakov, Milne-Edwards, Oschanine, Schlumberger, Studer, and Zograf.

In the body of the Report, Blanchard (1893:37–45) defended again the idea that 1722 should be used as a starting point for zoology because Tournefort, Lang, Klein, etc. adopted a binary nomenclature before Linné. Logically, his proposed rules suggested that 1722 be used as a starting point (Blanchard 1893:68–69, tr.):

[Art. 15] Binary nomenclature was founded by Tournefort, in 1700; Lang was the first to apply it in Zoology, in 1722; it only is in 1758 that Linné used it in the classification of animals.... [Art. 16] The year 1722 thus is the date up to which zoologists must go back to search for the most ancient generic or specific names. Any pre-Linnaean name must be adopted if it is in conformity with article 35 of the rules adopted by the Congress of 1889....

Here is what happened during the discussion (Chantre 1893:xxxvii, tr.):

The Congress approves the terms of the Article 16 and expressively admits the exactness of the historical considerations and criticisms which it summarizes. Nonetheless, after having heard the observations of Mr the Prof. A. Milne-Edwards, who insists on the disadvantages that would result from rejecting of a large number of names presently in use and replacing them by more ancient names, from pre-Linnaean authors, the Congress adopts the following article, instead of articles 15 and 16 of the Report.

The text thus adopted reads (Blanchard 1889:81, tr.):

[Art. 45] The 10th edition of the *Systema Naturae* (1758) is the starting point of zoological nomenclature. The year 1758 is thus the date up to which zoologists must go back to search for the most ancient generic and specific names, acknowledging that they are in conformity with the fundamental rules of nomenclature.

Obviously, 1758 had more chances than 1722 to win the crowd. In fact, Blanchard (1893:44) admitted that most zoologists were in favor of 1758. Also, ornithologists adopted the date 1758 as their starting point during a Congress at Budapest, and the German Committee on nomenclature as well as Sherborn (author of the imposing *Index Animalium*) also adopted 1758. The Second International Congress of Zoology, Moscow, was thus a turning point in the history of "priority"; for the first time, 1758 was officially adopted as an arbitrary, but nonetheless common, starting point by zoologists. Blanchard's role was immense, and, at the end of the session in which his rules were discussed, the Congress warmly applauded and thanked him for his "painful, long, and fruitful work." It may be important to emphasize that: (1) Blanchard was smart enough not to push for his agenda (1722 as a starting point); and (2) Blanchard's proposed date was rejected through a quick but fair process. It is almost embarrassing to realize that 1758 was selected so easily given that, for about 50 years, the 12th edition of the *Systema Naturae* had (supposedly) been used as starting point.

Again Another Code: The Code of the Deutsche Zoologische Gesellschaft

In 1894, on the initiative of J.V. Carus, the Deutsche Zoologische Gesellschaft published a newly adopted Code, the *Regeln für die wissenschaftliche Benennung der Thiere* (Carus et al. 1894). As far as priority is concerned, this Code selected 1758 as the starting point. It also insisted on a strict use of Latin grammar and orthography, and required that all names misspelled when first created be changed subsequently. As always, this Code was commented on. In particular, it was discussed in March 1896 at a meeting of the Zoological Society of London (Allen 1896; Sclater 1896; Stebbing 1896). Stebbing (1896:254) summarized that discussion as follows: "No body of regulations on the subject of scientific nomenclature can possibly give universal satisfaction." Indeed, Stebbing (1896:257) asked about the necessity to comply with Latin grammar and spelling:

From among numerous examples offered, *Oplophorus* may be cited, which is to be corrected into *Hoplophorus*. Can anything be more superfluous? The Latins themselves were uncertain whether H was a letter or only a breathing. They fluctuated between the spelling of Adria and Hadrian, of Hannibal and Annibal. Why should we, then, be more Roman than the Romans?

As expected, Sclater (1896) stood by the Strickland Code on all important differences from this new German Code, especially the starting point, which he still claimed should be the 12th edition of the *Systema Naturae* instead of the 10th edition.

The Third International Congress of Zoology: The Creation of the International Commission on Zoological Nomenclature

The rules adopted by the International Congress of Zoology at Moscow, in 1892, were not the only rules available in the 1890s. For instance, ornithologists and other zoologists could use the Code of the American Ornithologists' Union (which introduced rules for variety names); the Code of the Deutsche Zoologische Gesellschaft also was available to zoologists. This confusion ended at the Third International Congress of Zoology (Leiden, 1895) when F.E. Schulze (1896) proposed before the Congress participants that a commission of five members be nominated to produce one single code that could be recommended in different countries and be written in three languages (French, English, German). Schultze's proposal was immediately adopted and the following five members, all experts on nomenclature, were nominated: Blanchard (Paris), V. Carus (Leipzig), F.A. Jentink (Leiden), P.L. Sclater (London), and W. Stiles (Washington, DC). Naturally, divergent opinions were soon expressed. At a meeting of the Zoological Society of London held in 1896, some participants argued that international committees be appointed, "not to draw up a Code of rules, but to produce an authoritative list of names –once and for all– about which no lawyer-like haggling should hereafter be permitted" (Allen 1896:328).

The decision made in Leiden marked the birth of the International Commission on Zoological Nomenclature. Now, the destiny of the Code would remain in the hands of both the Commission and the Congress. It would still be ten long years before the first Code would be published, as the *Règles internationales de la nomenclature zoologique* (ICZN 1905). As for priority, it was resolved at the Berlin Meeting (1901) that it would apply strictly, with no exceptions, although heated discussions about that issue did not cease.

Toward a Limitation of Priority

In today's Code (ICZN 1999), exceptions to the principle of priority exist, such as in the case of *nomina protecta*, i.e., junior synonyms or homonyms being maintained under certain conditions (Article 23.9.1). This compromise, as always in nomenclature, has been heavily discussed before and after implementation: "The history of the Statute of Limitation [i.e., for the limitation of the application of priority] is long and tortuous" (Mayr 1972:453).

Early on, it was realized by the International Congress of Zoology and the International Commission on Zoological Nomenclature that, in some specific cases, the application of some of the 1905 *Règles* might need to be suspended to avoid nomenclatural confusion and promote stability. It is the reason that, in 1913, the International Congress of Zoology (Monaco) conferred upon the ICZN the plenary powers to limit the application of the 1905 rules. One of the purposes of those plenary powers was that zoologists could submit requests to the Commission to protect a junior synonym over a senior name (implying a reversal of precedence). That is the reason that, when Chapman (1923) argued that a strict application of priority generated confusion in nomenclature by resurrecting earlier and forgotten names, ICZN commissioner Bather (1923:182) promptly replied:

> Mr. Chapman's letter overflows with good sense; but it has all been said before. His laments, however, will not have been entirely wasted if you will permit this consolatory reply—namely, that in the year 1913, at the International Congress of Zoologists in Monaco, an agreement was reached in the largely attended section on nomenclature and confirmed in plenary session, by which the International Commission on Zoological Nomenclature was given power, on certain conditions, to suspend the rules in those cases where their operation was contrary to the general convenience. The Commission has, on the request of various zoologists, already taken action in several cases.

The possibility of submitting requests for limiting the application of priority was not enough for some authors, who thought, as did some taxonomists from the 1870s, that the Principle of Priority should be abandoned in favor of a principle of "continuity" (i.e., junior synonyms that were widely used should be conserved over earlier names). In the first half of the 20th century, debates about priority continued among zoologists as well as botanists (e.g., Shear 1924; Mansfield 1939; Heikertinger 1940, 1943a, b, 1953; Richter 1942a, b; Friebel and Richter 1943; Smith 1945; Blackwelder 1948). Beyond the Principle of Priority, it was the notion of stability that was discussed, i.e., whether it was realistic to think that nomenclature could reach absolute and permanent stability. Against authors who argued that favoring usage over priority was the best way to build a stable nomenclature (e.g., Shear 1924; Heikertinger 1940, 1943a, b, 1953), others replied that in stability could not and should not be avoided:

> Any proposal that rates stability ahead of the advancement of the science of systematics or the development of one of its myriad components is a backward step and one doomed to ultimate failure and discard. (Blackwelder 1948:309)

Toward the end of the 1940s, it appeared to some zoologists that the ICZN needed to find more efficient ways to deal with regrettable name changes related to a strict application of priority, leading to confusion:

> A great deal of name-changing continued during the inter-war years, a process which, as … is evident from the correspondence reaching the Commission [Francis Hemming was then its secretary] from many sources, has led to an over-growing demand, both from systematists and from workers in the applied biological fields, that more effective means should be found to secure stability in zoological nomenclature. The whole question was considered with great care both by the International Commission and also by the International Congresses of Zoology at their joint meeting held in Paris in July, 1948. (Hemming 1950:371)

Much work and discussion led to the addition of an article limiting the application of priority in the first edition of the *Code* (ICZN 1961).

A major step toward a limitation of priority was the recognition (not the adoption) of the "principle of conservation" (of junior synonyms over senior synonyms) at the 14th International Congress of Zoology, Copenhagen, in 1953. Discussions that took place at the Colloquium on Zoological Nomenclature, in Copenhagen, are mentioned in the detailed Report edited by Hemming (1953), then secretary to the International Commission: "After considerable discussion, the Colloquium agreed, by a majority, to recommend the inclusion in the forthcoming edition of the Règles, of a provision recognizing the Principle of Conservation to a limited extent" (Hemming 1953:25). Several drafts for an article governing the application of the principle of conservation were then presented by: (1) E. Mayr, (2) C.L.Hubbs and W.I. Follett, (3) N.D. Ridley, and (4) W.I. Follett, E. Mayr, R.V. Melville, and R.L. Usinger. All those drafts are reproduced in Hemming's report (1953:119–122). However, "it was unfortunately found impossible to secure agreement upon any of these drafts" (Hemming 1953:25) and the precise wording was left to the responsibility of the International Commission.

A Statute of Limitation of the law of priority was adopted at the 15th International Congress of Zoology, London, in 1958, and was first introduced as Article 23b in the first edition of the Code (ICZN 1961). Article 23b reads as follows:

> A name that has remained unused as a senior synonym in the primary zoological literature for more than fifty years is to be considered a forgotten name (*nomen oblitum*). (i) After 1960, a zoologist who discovers such a name is to refer it to the Commission, to be placed on either the appropriate Official Index of Rejected Names, or, if such action better serves the stability and universality of nomenclature, on the appropriate Official List. (ii) A *nomen oblitum* is not to be used unless the Commission so directs. (iii) This provision does not preclude application to the Commission for the preservation of names, important in applied zoology, of which the period of general usage has been less than fifty years.

Article 23b was controversial and, naturally, immediately discussed: "An unexpected and highly controversial innovation of the 1961 Code of Zoological Nomenclature … is article 23(b), the "*nomen oblitum*" rule." (Smith and Williams 1962:11) While some authors thought that Article 23b would promote stability, others only saw chaos in it (Smith 1964). A new vote, which took place at the International Congress of Zoology in Washington DC, in 1963, confirmed the limitation of priority by a small majority (Mayr et al. 1971). The problematic original wording of Article 23b partly explains the negative reactions. The idea of limiting the application of priority has remained in the Code, but Article 23b was modified throughout the successive editions of the Code in order to address several issues (e.g., Mayr et al. 1971; Collette et al. 1972; Corliss 1972; Mayr 1972; Cornelius 1987; Ng 1991; Melville 1995): Article 23b (ICZN 1961) became Article 79 (ICZN 1985) and is now back to Article 23.9–12 (ICZN 1999).

Smith and Williams (1962) mentioned several issues in the original wording: (1) every case of *nomen oblitum* needed to be submitted to the Commission (it now is an automatic process); (2) the limitation of priority dealt only with synonyms, not homonyms (reversal of precedence applies to both synonyms and homonyms in the current Code); (3) "primary zoological literature" was poorly defined (this concept no longer is in the Code, where clearer criteria of publication are now considered; see Article 23.9.1, ICZN 1999:28). Ng (1991), in favor of a strict application of priority, provided a detailed analysis about interesting cases: a senior synomym became a *nomen oblitum* because of application of Article 23b; later on, the junior synonym and its senior synonym were thought to refer to two distinct units; a new name had to be created because the senior synonym had become a *nomen oblitum* and could no longer be used. The current Article 23.12.2 addressed that issue:

> A name which was rejected under the former Article 23b may, in absence of any other cause of invalidity, be used as valid if it is no longer considered to be synonym of another name, or if its synonyms are themselves invalid under the provisions of the Code. (ICZN 1999:30)

The basic principle of conservation according to which priority may not apply strictly (in order to preserve stability and if certain conditions are met) has survived for about 50 years, i.e., since it was first introduced in the Code (ICZN 1961). However, debates about stability in nomenclature have certainly not decreased since 1961 in the context of both the zoological and botanical codes (e.g., Erzinclioglu and Unwin 1986; Tubbs 1986; Feldmann 1986; Brummit 1987; Hawksworth 1988, 1991; Hawksworth and Greuter 1989; Ng 1991; Silva 1996; Dubois 1998). The question being: how stable can or should nomenclature be? Although nomenclatural changes at the species level clearly are regrettable, especially when exclusively due to nomenclatural rules, it is important to keep in mind that, at least at the species level, nomenclature cannot be perfectly stable. For instance, Brummit (1987:80) argued that the type method (typification) "has an inbuilt element of instability" that can hardly be avoided. Also, Dubois (1998:1) argued that:

> Name changes due to the mere application of nomenclatural rules are much less numerous than those due to the progress of taxonomic research, and they would be even much less common if zoologists and editors paid more attention to the international rules of nomenclature.

This last comment by Dubois leads us to discuss another issue related to the application of nomenclatural rules and stability of names: the enormous amounts of time it takes to get acquainted with the entire literature on a given taxon. Taxonomic work is poorly rewarded, and nomenclatural work is even more poorly rewarded. What people often do not understand, however, is that sound taxonomic work requires detailed nomenclatural knowledge. Regardless of the kind of rules supported by the Code (absolute priority versus priority and possible conservation of junior names), people talking about "stability" of zoological nomenclature need to understand that: (1) dealing with all nomenclatural situations requires valuable expertise; (2) in order to reach that level of expertise, taxonomists need to spend a great deal of time working on basic nomenclatural data (lists of names, type availability, etc.); (3) more importantly, this kind of work by taxonomists should be highly respected and regarded as a direct part of their mission and duties. Indeed, one of the reasons that nomenclature is unstable is that the history of taxon names is highly complex in all taxa; we need people to be experts of that history, regardless of our rules.

Let There Be Types!

As surprising as it might seem, a historical review of the type concept and the terms related to typification, as well as their practical and philosophical implications, still needs to be written. In particular, it is unclear when type specimens started being used by zoologists (and botanists for that matter). Although detailed studies clearly are needed on this topic, it seems that it was only in the 1890s that zoologists realized the critical importance of type specimens in relation to the application of the Principle of Priority for species names: in fact, many papers suddenly addressed that particular issue starting in 1890 (Bather 1897; Cook 1898, 1900, 1902; Marsh 1898; Merriam 1897; Thomas 1893, 1897; Schubert 1897). Blanchard's (1889, 1893) early rules included articles on type species (because of the abundant literature on this topic published in the second half of the 19th century) but did not include anything on type specimens. Because of the literature published in the late 1890s on type specimens, the first *International Rules* (ICZN 1905:39) provided a brief recommendation on type specimens in the appendix: "This diagnosis should state in what museum the type specimen has been deposited and should give the museum number of said specimen." Terms such as holotype, paratype, syntype, and neotype, became more widely used only later (e.g., Frizzell 1933; Schenk and McMasters 1936).

One important challenge at the beginning of the 20th century was to go through all the terms that had been proposed so far for types and select those appropriate for the Code. This was not an easy process; Frizzell (1933) inventoried no fewer than 233 terms available in the literature, most of which have been completely forgotten, such as: aedoeotypus, agriotype, chorotype, graphotype, hypotype, ideotype, and tautogenotype. This long list of terms shows that taxonomists, especially those interested in nomenclature, have always been profusely creative. More importantly, it also shows that the terminology related to types was abundantly discussed before terms were chosen for the Code (as always in nomenclature).

THE 1960S: ANOTHER GREAT DECADE FOR NOMENCLATURAL DEBATE

The Emergence of Numerical Taxonomy and Its Impact on Nomenclature

The 1960s is a fascinating period for nomenclatural history. Indeed, it saw the publication of the first edition of the Code (ICZN 1961) as well as some of the most radical changes ever proposed in nomenclature (most of which were published in *Systematic Zoology*), mainly as a consequence of the emergence of numerical taxonomy (Sneath 1961; Sokal and Sneath, 1963; Sokal et al. 1965; references therein) and early use of computers (Jahn 1961; Sokal and Sneath 1963, 1966). The 1961 Code was preceded by the *Copenhagen Decisions on Zoological Nomenclature* (Hemming 1953), which were abundantly discussed (e.g., Follett 1954, 1956; McMichael 1956; Bradley 1957). Naturally, the Code was also discussed (e.g., Smith 1962).

Ralph and Wood (1954) introduced some brief thoughts on the use of "punch cards" in taxonomy and the need for a "standardized biological taxonomy code" whenever species are referred to with punch cards from different laboratories. However, some of the most radical changes in nomenclature were proposed immediately after the 200th birthday of the 10th edition of the *Systema Naturae* was celebrated (e.g., Lindsey and Usinger 1959; Stearns 1959).

In the 1960s, the use of computers in taxonomy was referred to as "electronic data processing," or EDP, brilliantly summarized by Sokal and Sneath (1966:2):

> Computers can help the taxonomist in many ways. They rapidly compute similarity coefficients among taxa. They are able to generate classifications based on representation of taxonomic structure by any of a number of definable principles and can present the output of such computations as dendrograms or other schemes by a variety of graphic output devices. Computers can store information about single species, such as characters, collections, and distribution records. This information can be searched at tremendous speeds and summaries or reports obtained on any group classified by any of a variety of criteria. Computers can store accession lists of museums and regularly update these in terms of loans and new accessions. They can prepare lists of materials available for study in a museum and search the lists of museum holdings for special material of interest to a given investigator.

"The Machines Will Win, Whether the Taxonomists Fight or Not"

Jahn's (1961) article on "Man versus Machine: a future problem in protozoan taxonomy" constituted one of the first upfront radical attempts at challenging traditional nomenclature. In this contribution,

Jahn (1961:179–180) argued that a computer-based, fully logical and numerical system of nomenclature should be developed. Some of Jahn's statements merit citation here, such as:

(1) We could save a considerable amount of time and a larger amount of confusion for investigators and for students by the use of numbers as well as names for all taxa, and even by the substitution of numbers for names in many technical discussions. (2) The most efficient use of computers would require a completely logical taxonomic system. (3) The development of a completely logical system will require some severe revisions of our present system. (4) Names, although still used, will become less important. (5) The laws of priority will decrease in importance along with the names, and might eventually be discarded. (7) All of these developments will irritate most taxonomists. (8) The taxonomists will rise to fight the machines. (10) The machines will win, whether the taxonomists fight or not.

In Jahn's system, *Plasmodium hexamerium* is given the number "1310101D" (i.e., 1 for the Phylum Protozoa, 3 for the Class Sporozoa, 101 for the family Plasmodiidae, etc.). Jahn strongly favored logic above all concepts in nomenclature, including priority:

It may be necessary to ignore and/or discard the rule of priority, certainly, at the least, emend it, before we can devise logical systems of taxonomy. The literature on the superorder Lobida is complex enough to be used as an outstanding example of the fact that a completely "legal" system is likely to be illogical and that a completely logical system is therefore likely to be "illegal."… (Bovee and Jahn 1965:238)

Numericlature: Should Numbers Replace Names?

Several authors commented on and expanded Jahn's proposal during the 1960s. Michener (1963) applied Jahn's nomenclatural proposal to numerical taxonomy. Michener's basic idea was that, since numerical taxonomy constituted a new, objective, experimental way of classifying organisms, it also represented a great opportunity to reform nomenclature. Michener tried to imagine what kind of nomenclature would be developed by scientists (whom he refers to as "exobiologists') if they were to find organisms to classify on another planet. Michener argued that, as radically new as it might seem, such a new system could be implemented by "earth-oriented taxonomists." Under Michener's numerical system of nomenclature, *Musca domestica* is referred to as "0325761," i.e., a "meaningless reference number" in order to "avoid Linnaeus's error of incorporating into the designation of the organism information about its classification which is subject to change with improved knowledge or changing ideas." (Michener 1963:166) However, classification can be represented using a numbering system similar to that proposed by Jahn (1961), with the addition of a dash to separate numbers though, so that *Musca domestica* can be referred to as "10–7–26–081–052–0325761."

Little (1964) slightly modified Michener's system. Little introduced the possibility of adding subgroups between parentheses, so that "-001(002)-" would mean the subfamily 002 within the family 001; he also introduced a standard of three numbers within each category so that classification is represented as "…-010–007-…" instead of "…10–7-…" because, according to him, it would ease the development of such a system using computers.

Hull (1966:14) proposed what he calls "Phylogenetic Numericlature," i.e., a "system of identification, positional, and phyletic numbers for taxa that makes possible a significant relationship between numerical classification and phylogeny." Indeed, the numbering system of nomenclature or numericlature, developed by Michener, intended to represent phenetic similarities. Hull (1966) slightly modified Michener's system so that phyletic relationships could be represented.

Other authors, such as Rivas (1965), were in favor of implementing a numerical system to ease storage and retrieval of taxonomic information using computers, but such a system would only complement the current nomenclature, especially the binominal nomenclature, which should not be abandoned.

However, several authors (Crovello 1967; Parkes 1967; Randall and Scott 1967) expressed some doubts as to whether a numerical system was absolutely necessary to develop computer-based storage and retrieval of taxonomic data. According to those authors, storage and retrieval of taxo-nomic data could also be developed efficiently using the current nomenclatural system: "Michener and others under-estimate the ability of machine systems to conform to user conventions" (Randall and Scott 1967:281).

A debate about the syntax of nomenclature also took place. Indeed, Hull (1968:474) argued that Randall and Scott (1967) had overlooked an important aspect of numericlature: "The main reason for introducing numericlatural systems into taxonomy is to provide an explicitly formulated syntax adequate for the purposes of taxonomy." Also, according to Hull (1966:473), "In order for words to form a language (whether natural or artificial) there must be rules of syntax. The syntactical rules of Linnaean nomenclature are minimal." Randall and Scott (1969) claimed that Hull had failed to provide convincing arguments to demonstrate that Linnaean nomenclature was less syntactical than numericlature, to which Hull (1969:469) replied that "the syntax of nomenclature may be adequate for the purposes of electronic data processing, but it is inadequate for several other purposes." One of the purposes better served by numericlature than nomenclature was, according to Hull (1969:469), the "relationship between classification and phylogeny."

Yochelson (1966) agreed that computers could help increase data processing and argued that, therefore, the Code should be simplified, especially with respect to grammar and orthographic issues because it would greatly simplify the use of computers. Thus, according to Yochelson (1966), taxon names (whether they have been Latinized or not) should not be regarded as Latin names, but rather as symbols used for communication:

> Adoption of the underlying philosophy that nomenclature is a language in its own right will clear away much debris. Does it really matter in the space age whether –opsis is masculine, feminine, or neuter? Is it a matter of biologic significance that specific names should be changed from –a to –us? Why not forget all "latinization" and its accompanying sterile scholarship, and get on with the study of biology (Yochelson 1966:89)?

The idea that nomenclature should free itself from Latin grammar was not new. Some authors had argued similar ideas in the 1870s. It is interesting to note, however, that this idea was defended by Yochelson in a new context, as a consequence of the debate on the use of computers in taxonomy. As expected, some people disagreed and immediately reacted. Tortonese (1967:278) argues that "if we retain Latin, we really ought to have it in grammatically proper form and not complain if some time is necessary to effect this."

Dupraw (1965) pushed the application of numerical treatment a step further and got rid of taxonomic categories and nomenclature altogether in a system that he referred to as "Non-Linnean taxonomy":

> In the non-Linnean system, definition of conventional, hierarchical categories (species, genus, etc.) is eliminated; specimens are dealt with as individual points on a multivariate scatter diagram, which rep-resents the closest two-dimensional approximation to the specimen distribution in a "full-dimensional" character hyperspace. (Dupraw 1965:1)

In a controversial article entitled Efficiency in Taxonomy, Sokal and Sneath (1966) reviewed the recent literature on taxonomy and addressed some misunderstandings about numerical taxonomy. They also discussed nomenclatural issues. In particular, they argued that "any new nomenclatural system must clearly be EDP oriented…. [and] should allow for automated information storage and retrieval" (Sokal and Sneath 1966:10). As we saw, this was a central theme developed during the 1960s that originated from the recent availability of computers and punch cards. Sokal and Sneath (1966:10) also argued that "we should free ourselves from the notion that a name can fulfill a mul-tiplicity of functions." Sokal and Sneath (1966) supported Michener's proposal under which unique registration numbers could replace species names. Also, they thought that supra-specific affinities would better be expressed by "classification numbers" because "the usefulness of the old genus names has largely vanished" (Sokal and Sneath 1966:11). Naturally, some authors replied to Sokal and Sneath. In particular, Kalkman (1966) argued that names are much more easily remembered than numbers and that, in the future of nomenclature, numbers should not replace names, in spite of the clearly beneficial use of computers for other purposes.

Oldroyd's Views on the Future of Taxonomy

Oldroyd (1966) wrote a controversial article entitled The Future of Taxonomic Entomology to which several people replied vigorously. In this contribution, Oldroyd focused on necessary changes in the Code and the role of the International Commission on Zoological Nomenclature in exploring and devising changes. In particular, Oldroyd (1966:253) argued that "the rigid framework of a mandatory Code, and in particular its type system are incompatible with quantitative methods based on a long series of specimens," and that, therefore, the Commission should "plan for a date, not too far ahead, after which all traditional taxonomy shall cease at the species-level and below." According to Oldroyd, new practices of quantitative taxonomy (based on large numbers of individuals) should replace traditional taxonomy (based too often, according to Oldroyd, on too few specimens) because of the new development of powerful computer-based tools that help analyze large amounts of data when delineating species. Oldroyd (1966:254) claimed that the type system indirectly encourages taxonomists to describe new species based on too few specimens: "The greatest evil of the type-concept, however, is that it actively encourages the practice of describing monotypic units: species based upon single specimens, or genera based upon single species."

Several authors reacted against the idea of abandoning types (e.g., Gurney 1967; Parkes 1967; Steyskal 1967) and reminded Oldroyd that holotypes, lectotypes, neotypes, and syntypes are just name-bearers. But Oldroyd (1967) insisted: "We have made a fetish of types, and going somewhere to 'see the types' has become a taxonomic pilgrimage. In fact, the type of *Felis leo*, if it still exists, is a sacred cow." I have spent (and intend to continue to spend) quite some time tracking "sacred slugs and snails." Oldroyd actually confused two distinct issues. The first issue is taxonomic, i.e., the number of individuals used by taxonomists to delineate species, and most taxonomists would agree that so-called "typological" approaches in which only one or few specimens are considered should be avoided as much as possible. The second issue is nomenclatural, i.e., the use of single (holotype, lectotype, neotype) or several (syntypes) specimens as reference points for a species name.

Heise and Starr's Philosophy of Nomenifers

A thorough analysis of the concept of "type specimen" was provided by Heise and Starr (1968). In this article, the authors explored philosophical questions related to the use of type specimens, which they referred to as "nomenifers," (literally "name-bearers"), a term first coined by Schopf (1960):

> The name associated with a type specimen (a nomenifer) is a class name, not a proper name, which names the taxon of which the nomenifer is a member. The nomenifer is not merely a member of the taxon than is any other member, which is to say that the nomenifer does not have a privileged logical status. By convention, the nomenifer's name (unlike that of any other member of the taxon) cannot change as long as there is a taxon with that name. The nomenifer serves epistemologically as an "official," certain, or unquestionable (but not necessarily typical) instance of the taxon whose name it bears. (Heise and Starr 1968:467)

This illustrates well the idea that debates in nomenclature help generate better understanding of the theory and philosophy of nomenclature, as Heise and Starr's analysis helped clarify the philosophy surrounding type specimens after Oldroyd argued against the use of type specimens.

Uninominal Species Names

An Early Challenge to Binominal Nomenclature: Amyot

Amyot seems to be the first author who described precisely a uninominal species nomenclature, to which he referred as *"méthode mononymique."* He also courageously applied it throughout his entire monograph on the *Rhynchota* (Amyot wrote all Latin taxon names in italics), a group that includes hemipteran insects. According to Amyot, species names were too unstable and species nomenclature should become independent from classification:

> This species is indicated in your collection under the most modern name; it really is the specific name that is inherited from Linné, Fabricius, etc., but the name of the genus has changed several times.... You have today's generic name, but tomorrow will come a new author who will change it again, then after him other ones, and you will thus have to change the name of the genus endlessly. (Amyot 1848:4, tr.)

Under Amyot's nomenclature, species were named using a single Latin word beginning with a capital letter, such as *Solenostethium*, *Irochrotus*, *Odontotarsus*, etc. However, those uninomials were not the specific names of the Linnaean binomials designating those species. To guarantee uniqueness and avoid homonymy (too many species would be called simply *nigra*, *rubra*, etc.), Amyot created new uninomials for species.

Amyot's higher classification was also quite unusual. His classification was based on a hierarchy of groups subordinated to each other, but included very few of the well-known ranks. Amyot used the terms *"Ordre"* (Order) and *"Tribu"* (Tribe) for two ranks in his hierarchy, but most other ranks were indicated by new terms: the Section, Subsection, and Trisection levels were above the Order; just below the Order was the Tribe; below the Tribe were the Division, Subdivision, and Tridivision levels. Also, Amyot rejected the term "genus," which he thought had nothing to do with its etymological root (*generare* in Latin, which means "to reproduce"). Thus, Amyot did not use generic names in his monograph.

Amyot's system of classification was highly innovative: species could be placed within divisions, subdivisions, or tridivisions, depending on the level of subordination needed to classify a certain taxon. For instance, the second Tribe of the Order *Hemiptera* was called *Breviscuti* and included three Divisions: the 31 species of the Division I (*Supericornes*) were classified in tridivisions; the 77 species of the Division II (*Infericornes*) were placed in subdivisions because tridivisions were not needed; and finally, the two species of the last Division III (*Caecigenae*) were directly placed within the Division because no subdivisions or tridivisions were needed. This system is interesting; Amyot classified species within taxa of any categorical rank. It clearly simplified the classification and avoided having to deal with long lists of taxon names when a suborder or a superfamily includes only one species. Although Amyot used ranks, he used them very freely, in a way similar to a rank-free system. The drawback of Amyot's innovative nomenclatural system was that, because he rejected genera as well as other taxonomic ranks, he created new names for many higher taxa in addition to all the new names he created for the species. Amyot's system was never adopted, probably because of confusion due to the abundance of new names.

One of Michener's Proposals for "Earthly Nomenclature"

One of the "possible reforms in earthly nomenclature" proposed by Michener (1963) was a uninominal system of nomenclature for species names. In doing so, Michener applied ideas defended by others before him. For instance, Michener cited Cain (1959):

> It seems clear that with the great alteration in the status of the genus since the time of Linnaeus, an effectively uninominal system will come into use. This will then be free from that part of the present instability caused by generic changes, which are mostly matters of opinion, and from the inclusion of a particular way of classifying things in the names of things to be classified.... The necessity for avoiding

homonyms and synonyms will, of course, remain as before, since it is a character of all reference systems. (Cain 1959 in Michener 1963:164)

Michener also cited Berio (1953), according to whom a "complete freeing of nomenclature from systematics would provide the ideal means for stabilizing nomenclature, for in the main, names are changed for reasons of a systematic nature." (Berio 1953 in Michener 1963:164) However, as mentioned above, the idea of using uninominals had been defended by earlier writers, especially by Amyot (1848), and, according to De Candolle (1813), by Buffon.

Michener's uninominal method was simple: the combination between the generic and specific names should remain fixed, even if the species is transferred to a new genus, and the original date of specific name is added to the fixed combination: *Apis mellifera* thus becomes *Apis-mellifera*, 1758 or can be abbreviated simply as *A.-mellifera*. Michener (1964) applied his uninominal method to the tribe Paracolletini (Hymenoptera, Apoidea) and showed that a new classification of the 332 species listed in this tribe would require 288 new combinations and 16 homonyms, all of which could be avoided using a uninominal system. The idea of fixing binomials to avoid generic name changes was first proposed by Sharp (1873).

Naturally, authors expressed their disagreement with Michener's innovative ideas on species nomenclature. Steyskal (1965:348) defended the current system because, after all, it was "already serving quite well," which is an argument that new proposals commonly have to face. And Johnson (1970:234) also argued that the current system of nomenclature should be retained chiefly "*faute de mieux.*"

An Answer to Michener: Amadon's "Suneg" Concept

Amadon's (1966) paper certainly shows how creative taxonomists were in the 1960s. Amadon thought that Michener's fixed binomials were too confusing because *Binghamiella-antipodes* 1853 would be a species in the genus *Callomelitta* 1853 (example from Michener 1964:187). However, Amadon agreed that species binomials were not stable enough. Thus Amadon (1966) proposed a new system, based on a new concept, the "suneg," ("genus" spelled backward) and defined as follows:

A term identical with the type genus of a family, subfamily, or tribe, but to be printed in Roman, not italics, and to be used, except in strictly taxonomic publications, in lieu of generic names for all the species in any given family, subfamily, or tribe. (Amadon 1965: 55)

Amadon's "suneg" names referred to taxa that are more or less similar to family names in the sense that they are much larger (more inclusive) than genera. Thus, using them as the first part of species names reduced the risk of name changes due to the transfers of species from one genus to another. Amadon (1966:56) illustrates his system by proposing the names Alca *impennis*, Alca *torda*, Alca *brevirostre*, and Alca *grylle atlantis*, for, respectively, *Pinguinus impennis*, *Alca torda*, and *Brachyramphus brevirostre*, and *Cepphus grylle atlantis*.… There seems to be a simpler way to solve that issue, which is to use family names as the first part of species binomials whenever generic names cannot be determined.

Lanham's (1965) Uninominal Form of Species Names

Lanham (1965) agreed with Michener that species names are not stable enough because species are too often transferred from one genus to another, but he proposed another uninominal method for species names. This method, which was presented and commented upon on several occasions (Dayrat et al. 2004; Dayrat 2005a; Dayrat and Gosliner 2005), is a form of species names allowed under the *PhyloCode* (Cantino and de Queiroz 2007), at least once species names are established under the authority of the appropriate rank-based code (Dayrat et al. 2008). It is based on the association of the specific name, the author's name, the original date, and the page number, such as in *montana* Cockerell, 1928:365.

Lanham (1965), as well as Berio (1953), Cain (1959), and Michener (1963, 1964), and even Amyot (1848) and Sharp (1873) in earlier times, all favored a species nomenclature that would exclusively serve a designatory purpose, i.e., unique identification of species, but would not be used for indicating supra-specific relationships. Lanham (1965:144) also argued that "consideration of supra-specific categories should be deleted from the Code." According to Lanham:

> An arbitrary list of families, used with the understanding that it is only a rough approximation to evolutionary relationships, and used only as a filing system, would fill many requirements. Perhaps the currently used array of supraspecific categories would serve quite well to express phylogeny, with informal considerations of usage and usefulness, instead of the present pseudo-legal technicalities, serving to give a degree of stability. (1965:144)

Also, as Lanham rightly pointed out, uninomials are "well suited to computer method of storing information" because they do not change over time (except for changes related to the taxonomy).

Whitehead's Comments on Uninominal versus Binominal Systems

According to Whitehead (1972), the binominal system is the interface between low-level (species) and high-level (supra-specific) taxonomy. Whitehead (1972:222) was of the opinion that it is a great thing that binominal nomenclature "provides the means for expressing each new advance in taxonomic understanding," and that adopting a uninominal system would simply "remove the evolutionary content from the very point in the nomenclatural system where it has greatest significance, i.e., at species level." Another interesting idea introduced by Whitehead was that, regardless of whether the binominal system was abolished, the interface between low- and high-level taxonomy would still exist, and that it probably was better to face it (binominal system) rather than avoid it (uninominal system).

The adoption of a uninominal system has not been considered in the context of the current rank-based codes. However, this does not solve the issues mentioned by Michener, Lanham, and others, i.e., the fact that species names are poorly stable because species are regularly moved from one genus to another. On the one hand, a uninominal system would radically and logically solve that issue. On the other hand, many taxonomists undeniably wish to maintain an indication of supra-specific relationships within the species name. The question thus becomes: could the current system be modified so that both goals could be served? A possible answer is discussed next.

TODAY'S NOMENCLATURAL DEBATES

THE EMERGENCE OF THE *PHYLOCODE*

The *PhyloCode* (Cantino and de Queiroz 2007) or *International Code of Phylogenetic Nomenclature* (*ICPN*), provides rules for defining clade names through explicit reference to phylogeny. Phylogenetic nomenclature constitutes a new approach to nomenclature mainly because, in contrast to the rank-based codes (*ICZN, Code; International Code of Botanical Nomenclature*, or *ICBN; Bacteriological Code*, or *BC*), it is independent of mandatory ranks and types. The *PhyloCode* has been through seven drafts since 2000. The current version (4b) differs from previous drafts because it includes a new article on species names (Dayrat et al. 2008). The *PhyloCode* will be published in the near future, along with the companion volume, which will include a first series of definitions of clade names across the Tree of Life. The date of publication of the *PhyloCode* will be its official starting date too: i.e., systematists will be able to refer to clades and publish new clade names using the *PhyloCode* instead of a rank-based code.

After the publication of the *PhyloCode*, two codes will then coexist (the Code and *PhyloCode* for zoologists, *ICBN* and *PhyloCode* for botanists, *BC* and *PhyloCode* for bacteriologists). It is difficult to predict how the coexistence of two codes will actually affect the work of systematists, biologists, as well as other users of taxon names. In any case, there is no particular reason to think that taxonomy will suddenly become "chaotic" when the *PhyloCode* is officially implemented. In that regard, it is interesting to note that, in the second half of the 19th century, zoologists had several competing codes to choose from, each code promoting different rules (including major differences such as distinct starting points and distinct applications of the Principle of Priority), which ultimately yielded a constructive outcome, i.e., the creation of the International Commission on Zoological Nomenclature (1895) and the first *International Rules for Zoological Nomenclature* (ICZN 1905). The existence of more than one nomenclatural code can thus be regarded as a positive sign of the current dynamism in nomenclature and taxonomy, instead of as a source of chaos, although other authors might argue that taxonomists should have other concerns and priorities in this era of biodiversity crisis.

What Future for the Coexisting Codes?

It is difficult to predict how this coexistence will evolve and end in the long-term future mainly because we cannot know in advance what practices will be favored by taxonomists. Also, there are other factors that are even harder to predict and that will certainly play an important role, such as the sociological behavior of taxonomists (e.g., some influential systematists might start a "trend" of using—or rejecting—the *PhyloCode*). At this stage, we can draw at least five possible scenarios. (1) The rank-based codes will prevail, and the *PhyloCode* will disappear without influencing them. (2) The rank-based codes will prevail, and the *PhyloCode* will disappear but only after influencing them and causing nomenclatural changes. (3) The rank-based codes will be replaced by the *PhyloCode*. (4) The rank-based codes and the *PhyloCode* will continue to coexist, overlap, and remain independent. (5) The rank-based codes and the *PhyloCode* will merge and be reconciled into a single code of nomenclature (specificities of each rank-based code could be maintained, such as the treatment of type specimens).

Clearly, different people would vote or hope for different scenarios. The last scenario would have my preference as it would seem to be a good solution. A possible way of merging the *PhyloCode* and the Code (for zoologists) would be that the *PhyloCode* rules govern all clade names above the species level (with, e.g., no mandatory ranks and types), and the ICZN rules govern the application of species names (with, e.g., type specimens). This would bring together the best of each code. In fact, the latest version of the *PhyloCode* requires that new species names be created under the authority of the appropriate rank-based code, although it also provides rules and recommendations on how to cite species names once those are made available.

Given the usual pace at which nomenclature evolves, it is probably going to take many years before we can determine how the interactions between the *PhyloCode* and the rank-based codes are going to evolve and, perhaps, end. Again, in nomenclature, things take a great deal of time: It took zoologists not less than 50 years to choose a common starting point for zoological nomenclature; 16 years went by between the time the first International Congress of Zoology (Paris, 1889) started to discuss the necessity of common and international rules of zoological nomenclature and the time those rules were published (ICZN 1905).

Before possible interactions between the *PhyloCode* and the Code are discussed in more detail, it is necessary to summarize their similarities as well as their differences. An abundant literature can be found on that subject (Cantino and de Queiroz 2007 and references therein); only the major points are summarized here. The *PhyloCode* shares several similarities with the rank-based codes, including the *BioCode*. First, both systems share the same fundamental goals, i.e., providing unambiguous methods for applying names to taxa, selecting a single accepted name for a taxon among competing synonyms, and promoting nomenclatural stability. Also, both systems deal with taxon names, *not* taxa, that is, they govern the application of names, not how taxa are delineated. Both systems use the principle of priority, or precedence, to determine the correct name of a taxon when synonyms exist. However, both systems have conservation mechanisms that allow a later established name to have precedence over an earlier name.

The *PhyloCode* also differs significantly from the rank-based codes. The phylogenetic system is independent of mandatory taxonomic ranks. In the phylogenetic system, "species" and "clade" are not ranks but different kinds of biological entities. Phylogenetic nomenclature uses phylogenetic definitions (node-based, branch-based, apomorphy-based) to determine the application of clade names, whereas the rank-based codes use mandatory ranks and types. Under the *PhyloCode*, only clades (and species) can be named, whereas the rank-based codes allow the creation of names for all taxa regardless of their phylogenetic status. The establishment of a name under the *PhyloCode* requires both publication and registration, which is not mandatory in the rank-based codes. Finally, the *PhyloCode* applies to the entire Tree of Life (as does the *BioCode*).

The methods of application of taxon names constitute a major incompatibility between the phylogenetic system and the rank-based system. At this stage, it is difficult to envision a code in which the application of names would be governed through both mandatory ranks and types (type species, genera, etc., not type specimens) and phylogenetic definitions. Ranks could potentially be used in the context of the *PhyloCode*, as indicated in *PhyloCode* Article #3: "This code does not prohibit, discourage, encourage, or require the use of taxonomic ranks." However, the use of *mandatory* ranks is incompatible with the *PhyloCode*. The possible introduction of optional ranks into the *PhyloCode* has been poorly explored so far; new interesting proposals may be provided in the future.

THE SPECIAL CASE OF SPECIES NAMES

Beyond and despite incompatibilities, it is worth asking whether there are areas in which the *PhyloCode* and the rank-based codes could be reconciled. Few articles address that question (e.g., Kuntner and Agnarsson 2006). However, as surprising as it might seem, the form of species names is an area in which the rank-based codes could potentially benefit from the recent findings in phylogenetic nomenclature. An Article (#21) dealing with species was recently added to the *PhyloCode*, after more than ten years of debates. At the first meeting of the International Society for Phylogenetic Nomenclature (ISPN) in Paris, July 2004, it was decided that a species code separate but compatible with the code for clade names be drafted by Philip Cantino, Julia Clarke, Kevin de Queiroz, and me. This species code was supposed to incorporate an epithet-based form of species name similar to the form of species names first proposed by Lanham (Lanham 1965; Dayrat et al. 2004, 2008; Dayrat 2005a; Dayrat and Gosliner 2005). However, after two years of work on this species code, it was realized that species names could be better dealt with in a different way. Problems encountered in the development of a species code included: (1) the coexistence of two species nomenclatures; (2) the fact that new species names established under the *PhyloCode* would have to be regarded as available (Code) or validly published (*ICBN, BC*) under the rank-based codes; (3) some uncertainty on how to address infra-specific names; (4) typification inconsistencies among the rank-based codes; and (5) an absence of major benefits justifying the development of a new species code. The alternative solution, presented at the second meeting of the ISPN (Yale University, June 2006), led to the adoption of Article #21 in May 2007 and its introduction into the *PhyloCode* in September 2007.

The historical development that led to Article #21 as well as its content have already been presented and discussed (Dayrat et al. 2008). Only the main properties of Article #21 are summarized here. (1) It requires compliance with the corresponding rank-based codes for creating new species names and determining precedence among synonyms. (2) The two parts of Linnaean binomials are interpreted in a way that is consistent with phylogenetic nomenclature. In particular, the first word of a binomial is interpreted as a "prenomen" instead of a generic (ranked) name, once it is established under the appropriate rank-based code. (3) The use of a species prenomen (i.e., ranked generic name under the *International Code of Zoological Nomenclature*) is not mandatory once a species name is established. A uninominal form of species name can then be used (based on the citation of the specific name or epithet as well as, at least, the author's name and the original date), and supra-generic names can be associated with the specific epithet (see below). In any case, if a prenomen is used, it is not tied to any rank under the *PhyloCode*. (4) Symbols are recommended to convey phylogenetic information about generic names (prenomina under the *PhyloCode*).

Article #21 recently added to the *PhyloCode* presents several advantages. (1) One description satisfies the requirements of both phylogenetic nomenclature and the rank-based codes: more importantly, the authority of rank-based codes is recognized for the creation of new species names. (2) The *PhyloCode* now is complete in addressing both clade and species names. (3) Species names (once established) can be used in a way that is consistent with the principles of phylogenetic nomenclature. (4) It does not disrupt established and critical taxonomic practices, especially the use of name-bearing type specimens (holotypes, syntypes, lectotypes, neotypes) for determining precedence among competing synonyms. (5) Finally, better communication of taxonomic (phylogenetic) knowledge is promoted because, for instance, species can be referred to clades that correspond to supra-generic taxa if no monophyletic taxon of generic rank can be found.

The latter point is a significant improvement in species nomenclature. I have been arguing for several years (since the 69th Annual Meeting of the American Malacological Society, Ann Arbor, Michigan, 2003) that the mandatory use of a generic name as the first part of Linnaean species binomials is problematic in cases where species cannot be assigned to any monophyletic taxon of generic rank (Dayrat et al. 2004, 2008; Dayrat 2005a; Dayrat and Gosliner 2005). This situation is particularly common in groups for which phylogenetic relationships are poorly known. That is the reason that, in 2005, a new species was described and named using a uninominal form of species names, "*aliciae* Dayrat, 2005." The specific name was combined with a family name instead of a generic name as the first part of a binomial, *Discodorididae aliciae* Dayrat 2005, because the only clade in which *aliciae* could be placed was the "family" Discodorididae. It was intentional that the binominal nomenclature promoted by the *International Code of Zoological Nomenclature* was not followed, although the genus name "*Discodoris*," which then referred to a metaphyletic taxon, was proposed as a possible generic name for those who would want to follow the strictly Linnaean species nomenclature. In fact, Berhens and Hermosillo (2005) cite that species as *Discodoris aliciae* Dayrat 2005.

The combination of supra-generic taxon names and specific names (or epithets) is allowed under the *PhyloCode*, once species are established. However, because the *PhyloCode* relies on the rank-based codes for the creation of new species names, it cannot allow the use of a supra-generic name instead of generic names in a newly created species binomial. So, in other words, the creation of a new name such as *Discodorididae aliciae* would not be allowed under the *PhyloCode*, because it is not allowed under the *International Code of Zoological Nomenclature*. But could the Code allow, when necessary, the use of a supra-generic name in lieu of a generic name as the first part of a species binomial?

Toward a More Flexible Linnaean Binominal Nomenclature?

In several publications from the 1960s (Michener 1963; Lanham 1965; Sokal and Sneath 1966) authors supported the use of a uninominal species nomenclature to solve a major issue in nomenclature, i.e., the instability of species names when species are moved from one genus to another. This instability is directly related to an issue already discussed by me, i.e., the fact that species cannot always be placed in a clade of generic rank because relationships are poorly known (Dayrat et al. 2004, 2008; Dayrat 2005a, submitted; Dayrat and Gosliner 2005). The major source of the problem here is that the use of a generic name as the first part of a Linnaean binomial is *mandatory*. This is too rigid and sometimes forces taxonomists to place species within genera based only on mere intuitions, although there is no phylogenetic evidence indicating that that given genus is monophyletic. Because intuitions differ, species are moved from one genus to another or, if they are not moved, are temporarily placed in genera that are not clades. In any case, the rigid mandatory use of a generic name generates confusion, as it tends to pretend that we know about relationships when we actually do not know.

The fact that a uninominal form of species names has only seldom been used or supported (Dayrat 2005a; Angielczyk 2007; Béthoux 2008) suggests that taxonomists may still not be ready to accept a uninominal method of species names, mainly because most taxonomists wish to conserve an indication of supra-specific relationships within the species name. So, could the binominal system be modified so that it could be both more stable and indicate more accurately evolutionary relationships? The conflict between those two goals (stability and indication of supra-specific relationships) is largely due to the fact that the current binominal system is too rigid; the mandatory use of a name of genus rank as the first part of the Linnaean binomials is the main source of conflict.

As pointed out by Amadon (1966) in his critique of Michener's (1963) uninominal form of species names, Linnaeus's genera correspond to much broader taxa than our current genera. There is no need to adopt a new concept, such as Amadon's "suneg." When no clade of genus rank can be determined, the specific name just needs to be combined with a supra-generic name, such as a family name, such as in the name *Discodorididae aliciae* (Dayrat et al. 2004, 2008; Dayrat 2005a, submitted; Dayrat and Gosliner 2005).

The name *Discodorididae aliciae* is binominal. It does not comply with the current rules of nomenclature however, because Discodorididae is a family name and not a generic name under the *International Code of Zoological Nomenclature*. The name *Discodorididae aliciae* is phylogenetically rigorous because, although we know that *aliciae* belongs to the clade Discodorididae, we do not know which "genus" it belongs to. The species name *Discodorididae aliciae* thus conveys accurate phylogenetic information.

In order to allow species names such as *Discodorididae aliciae*, the Code would have to be modified. Binominal nomenclature would need to become more flexible: supra-generic names would have to be allowed as first parts of species binomials. It does not mean that all species names would have to be changed. This solution could be used only in cases where we positively know that using a generic name is misleading and confusing. This flexible binominal nomenclature would probably be more stable in the long term because species would not have to be moved from one taxon to another, unless significant progress is made about systematic relationships.

Some readers probably think that this proposal is heretical and marks a complete rejection of the Linnaean species nomenclature. Although it is clear that it would mean an important change in the current species nomenclature of the Code (and other rank-based codes), it is not certain that it would mean a betrayal of Linnaeus's own taxonomic use of the binominal nomenclature. Indeed, if Linnaeus had discovered *aliciae*, he would have placed it in his genus *Doris*, which then included *all* dorids (Discodorididae, Dorididae, Chromodorididae, etc.) and corresponds to at least a superfamily or a suborder under the Code. Thus, it could be argued that, although a name such as *Discodorididae aliciae* is not strictly Linnaean under the Code, it is Linnaean in spirit because Linnaeus used names of much broader taxa as first parts of species binomials.

In a way, it appears that our problems come from the fact that the binominal species nomenclature has been maintained as a strict nomenclatural form of species names without taking into account the changes in taxonomic practices, especially the fact that Linnaeus's broadly inclusive genera are taxonomically very different from our more restricted genera. Our current rank-based nomenclatural system includes dozens of ranks, whereas Linnaeus used very few ranks. In most cases, Linnaeus's genera would now be taxonomically ranked at least as families or superfamilies. The only thing left in common between our current binominal nomenclature and Linnaeus's binominal nomenclature is the combination of a generic name with a specific name. However, taxonomically speaking, our current binominal nomenclature and Linnaeus's binominal nomenclature are quite distinct. Therefore, introducing some changes in the binominal nomenclature so that it could convey taxonomic information more accurately is not seen here as such a radical break from Linnaeus's legacy.

Allowing the use of supra-generic taxon names as first parts of Linnaean binomials would generate new homonymy issues, as specific names not found together in different genera may be found together in the same family. Possible solutions exist. For instance, in cases where homonyms would need to be avoided, former generic names could be cited as anchors. For newly created names, it could be recommended that, in the future, taxonomists choose specific names that would avoid potential homonymy with species placed in neighboring genera. Other ways to handle homonymy issues could certainly be proposed: e.g., the authorship could become part of the species name.

CONCLUSION

Studying the history of nomenclature reveals that established practices have always been challenged and new ideas have always been proposed. In that regard, the recent emergence of phylogenetic nomenclature and the debates it has generated (Cantino and de Queiroz 2007; references therein) are not a new phenomenon. Nomenclature went through periods of much more radical discussions. For instance, in the 1870s, the principle of priority, one of the most fundamental concepts in nomenclature (including in the *PhyloCode*), was directly attacked by many authors who suggested that a principle of "convenience" be adopted. In the 1960s, some authors suggested that a system of numericlature (numbers) replace nomenclature (names). For 250 years, nomenclature has remained dynamic: it has always generated intense debates and it has changed.

Changes in nomenclatural practices, however, are perceived only over long periods of time, which probably explains why many biologists view nomenclature as being static, permanent. That it took 50 years for zoologists to select 1758 as the starting point for zoological nomenclature is one of the best illustrations of this long-term development of nomenclature. Two main reasons might explain why changes are slow in nomenclature. First, it seems that taxonomists disagree more easily than they can agree, and some nomenclatural issues have always generated endless debates; as reviewed in this chapter, whether "barbarous" names (derived neither from Latin nor Greek) should be allowed was debated during the entire 19th century until 1905, when they were finally allowed in the first *Règles internationales de nomenclature zoologique* (ICZN 1905). Second, taxonomists tend to be quite protective of existing practices, and, in fact, radical proposals have rarely been adopted.

Quite a few examples of new nomenclatural proposals have never been adopted and did not cause any practical nomenclatural change because they did not become recommendations or rules in the codes (e.g., Bergeret 1783; du Petit Thouars 1822; Amyot 1848; Sharp 1873; Jahn 1861; Lewis 1875; Michener 1963; Oldryod 1966). Although all those innovative contributions sometimes seemed "pure folly" to some taxonomists, they have played a critical role. By challenging the established nomenclature, they kept taxonomists thinking about nomenclature and largely participated in increasing our understanding of nomenclatural theory and concepts.

So, for these reasons, I am of the opinion that not only is there nothing wrong with the fact that phylogenetic nomenclature challenges so deeply our established nomenclatural practices, but also that phylogenetic nomenclature will, in the long term, be beneficial to nomenclature, which will grow stronger regardless of how the coexistence of the Code and the *PhyloCode* (for zoologists) evolves in the future.

Although phylogenetic nomenclature shares some similarities with the historical debates reviewed here (development of priority in the 19th century and debates in the 1960s largely due to numerical nomenclature), it has some undeniable specificities. First, in contrast to most innovative proposals made in the past, it has been transformed into an actual set of rules and recommendations, the *PhyloCode*, which will soon be available to taxonomists. Also, it is important to mention the existence of the International Society of Phylogenetic Nomenclature, which represents an international body of experts and elects a Committee on Phylogenetic Nomenclature in charge of revising and addressing issues related to the *PhyloCode*.

The second specificity is that the coexistence of codes in the 19th century and the coexistence of codes that will soon take place (when the *PhyloCode* is published) are the results of two distinct processes. Taxonomists then had to deal with disagreements among *too many* codes in a period where no scientific society or group of zoologists could claim any authority over zoological nomenclature. Several codes coexisted in the 19th century because, before the creation of the International Commission on Zoological Nomenclature in 1895, no agreement had been reached. The upcoming coexistence of two codes (ICZN and *PhyloCode* for zoologists) results from a different historical process precisely because there is an International Commission on Zoological Nomenclature. So, several codes are soon going to coexist because a new nomenclatural code has branched out or split from an existing one (actually, from three rank-based codes).

Although they are the result of two distinct historical processes, the coexistence of several codes in the 19th century and today's coexistence of codes (e.g., *International Code of Zoological Nomenclature* and *PhyloCode* for zoologists) are similar situations: taxonomists are free to use different nomenclatural rules, and thus may possibly use different taxon names. We know how the story ended in the 19th century: The community of zoologists took the opportunity of the International Congress of Zoology to design a common set of rules and recommendations. However, it is unknown how today's coexistence of codes will end. Ideally, there should be only one code, and the *International Code of Zoological Nomenclature* and the *PhyloCode* could be united into a single code by taking the best from each code: all the species-level rules and recommendations from the Code (e.g., type specimens), and the higher-level rules and recommendations from the *PhyloCode* (e.g., clade name definition); at least this is what I regard as "the best" of each code at the present time; different people certainly hold different opinions. It is nearly impossible to predict whether this will happen, because too many factors are uncertain.

Regardless of whether the *PhyloCode* and the current rank-based codes will be united in the future, the main objective of this chapter is to argue that there is no reason for taxonomists to think that nomenclatural rules and practices should remain unchanged. They have continually changed in the past, and there would be nothing wrong if nomenclature were to change again: "In the world, nothing is permanent and laws, as living beings, perpetually evolve. We believe, however, that the present Report and the rules proposed therein meet the current nomenclatural needs" (Blanchard 1889:334, tr.). Does our current nomenclature meet our needs? Is it adequate? If yes, there is no reason to change it. If not, we may need to change it, and there would be nothing wrong.

One aspect of the future of nomenclature that is particularly unclear is how, in practice, communication can be guaranteed between the committees in charge of editing the rank-based codes and the Committee on Phylogenetic Nomenclature. It is because incompatibilities do exist between the *PhyloCode* and the rank-based codes that we need to make sure that regular exchanges take place among the people who are the most actively involved in the development of nomenclature in order to avoid misunderstanding and try to build a constructive environment for the future of nomenclature. It would be regrettable for the future of nomenclature and taxonomy that those communities of experts would not communicate with each other. Such exchanges could take place at nomenclature workshops, which sooner or later will be needed.

Another important aspect of nomenclature that we should also keep in mind when we discuss potential changes of nomenclatural practices is that nomenclature is a means of facilitating, not an end result, of taxonomy. In fact, one can read on the front page of the American Ornithologists' Union Code (Coues et al. 1886): "Zoölogical nomenclature is a means, not an end, of zoölogical science." This implies that nomenclature should remain open to changes that facilitate the work of taxonomists. This is exactly why numerical taxonomists argued in the 1960s that nomenclature should be changed so that it could serve better the modern context of numerical taxonomy. This also is why it has been more recently argued in the context of the *PhyloCode* that nomenclature should be changed to better serve the modern context of phylogenetic systematics.

New nomenclatural questions may arise from new techniques too. Early writers (e.g., Strickland 1835, 1843) rightly argued that names should be eligible for priority only if the species designated by that name had been properly described. As shown here, what constituted a "good" description was actively discussed throughout the 19th century, in relation to the Principle of Priority, but, basically, descriptions should include as much information as possible so that species can be re-identified. However, standards of species description change over time because new techniques are available. For instance, when Blanchard and the other members of the International Commission devised the first *International Rules of Zoological Nomenclature* in the late 19th century, the ability of comparing nucleotide sequences for taxonomic purposes did not exist. Should the Code recommend that DNA sequences of a few standard markers be included in species descriptions (Dayrat 2005b), as, in the 19th century, some authors thought that illustrations of specimens should be mandatory when creating a new name?

It seems that the future of nomenclature is going to be fascinating. It is difficult to predict how the coexistence of the rank-based codes and the *PhyloCode* will evolve in the long term. However, it will be interesting to follow how systematists will react to the official implementation of the *PhyloCode*. After the publication of the *PhyloCode*, it will be critical that members of the committees in charge of editing the rank-based codes and the Committee on Phylogenetic Nomenclature work constructively together to try to explore the possibility of unifying nomenclature. Taxonomists should not fear that changes may be needed in nomenclature, as long as those changes can help to better serve modern needs in systematics. Nomenclature has remained dynamic for the last 250 years, and will be dynamic for the next 250.

ACKNOWLEDGMENTS

I am very grateful to Andrew Polaszek for inviting me to participate in *Systema Naturae 250*, the symposium celebrating 250 years of zoological nomenclature, Paris, August 2008, and for giving me the opportunity to talk and write about debates in the history of nomenclature. I also thank Philippe Bouchet for a constructive review of the manuscript.

REFERENCES

Adanson, M. 1757. *Histoire du Sénégal*. Paris: Bauche.

Agassiz, A. 1871. Systematic zoology and nomenclature. *American Naturalist* 5:353–356.

Agassiz, L. 1842–1846. *Nomenclator Zoologicus*. Soloduri: Jent et Gassmans. [2 vols.].

Allen, J. A. 1896. Sclater on rules for naming animals. *The Auk* 13:325–328.

Amadon, D. 1966. Another suggestion for stabilizing nomenclature. *Systematic Zoology* 15:54–58.

Amyot, C. J. B. 1848. *Entomologie Française. Rhynchotes. Méthode mononymique*. Paris: Baillière.

Angielczyk, K. D. 2007. New specimens of the Tanzanian dicynodont "*Cryptocynodon*" *parringtoni* von Huene, 1942 (Therapsida, Anomodontia), with an expanded analysis of Permian dicynodont phylogeny. *Journal of Vertebrate Paleontology* 27:116–131.

Anonymous, 1844. Rapport d'une commission nommée par l'Association britannique pour l'avancement des sciences dans lebut d'aviser auxmoyens de render lanomenclature zoologique uniforme et permanente, rédigé par Mr. Strickland (Philos. Magaz., Août 1843). *Bibliothèque Universelle de Genève* 49:184–188.

Barnett, J. A. 1986. The stability of biological nomenclature: Yeasts. *Nature* 322:599.

Bather, F. A. 1897. A postscript on the terminology of types. *Science* 5:843–844.

Bather, F. A. 1923. The rule of priority in nomenclature. *Nature* 111:182–183.

Behrens, D. W. and A. Hermosillo. 2005. *Eastern Pacific nudibranch, a guide to the opisthobranchs from Alaska to Central America*. Monterey, California: Sea Challengers.

Bennett, E. T. 1830. *The gardens and menagerie of the Zoological Society delineated. Quadrupeds*. London: Thomas Tegg.

Bergeret, J. P. 1783–1785. *Phytonomatotechnie Universelle*. Paris: Chez l'auteur, Didot le jeune, Poisson. [3 vols.].

Berio, E. 1953. The rule of priority in zoological nomenclature. *Bulletin of Zoological Nomenclature* 8:30–40.

Béthoux, O. 2008. Revision and phylogenetic affinities of the lobeattid species *bronsoni* Dana, 1864 and *silvatica* Laurentiaux & Laurentiaux-Vieira, 1980 (Pennsylvanian; Archaeorthoptera). *Arthropod Systematics and Phylogeny* 66:145–163.

Blackwelder R. E. 1948. The principle of priority in biological nomenclature. *Journal of the Washington Academy of Sciences* 38:306–309.

Blanchard, R. 1889. De la nomenclature des êtres organisés. In *Compte-Rendu des Séances du Congrès International de Zoologie*, Ed. R. Blanchard, 333–424. Paris: Société Zoologique de France.

Blanchard, R. 1893. Deuxième rapport sur la nomenclature des êtres organisés. Deuxième Partie. In *Compte-Rendu des Séances du Congrès International de Zoologie*, Ed. J. Dumouchel, 1–80. Moscou: Laschkevitsch.

Bovee, E. C. and T. L. Jahn. 1966. Mechanisms of movement in taxonomy of Sarcodina. III. Orders, suborders, families, and subfamilies in the Superorder Lobida. *Systematic Zoology* 15:229–240.

Bradley, J. C. 1957. Cooperative or personal decisions on zoological nomenclature? *Systematic Zoology* 6:101–106.

Briggs, T. H. 1871. Law of priority versus accord. *The Entomologist's Monthly Magazine* 8:93–96.

Brongniart, A. 1827. Roches. In *Dictionnaire d'Histoire Naturelle*, Ed. F. Cuvier, 1–140. Paris: Levrault. [Volume 46].

Brummit, R. K. 1987. Will we ever achieve stability of nomenclature? *Taxon* 36:78–81.

Buffon, G. 1753. *Histoire Naturelle. Tome quatrième*. Paris: Imprimerie Royale.

Cain, A. J. 1959. The post-Linnean development of taxonomy. *Proceedings of the Linnean Society of London* 170:234–244.

Cantino, P. D. and K. de Queiroz. 2007. *International Code of Phylogenetic Nomenclature*, version 4b. http://www.phylocode.org.

Capellini, G. 1882. La deuxième session du Congrès. In *Congrès Géologique International, Compte-Rendu de la Deuxième Session, Bologna, 1881*, Ed. G. Capellini, 188. Bologne: Fava et Garagnani.

Carus, J. V., Döderlein, L., and K. Möbius. 1894. Berathung des zweiten Entwurfes von Regeln für die zoologische Nomenclatur im Aufrage der Deutschen Zoologischen Gesellschaft. *Verhandlungen der deutschen zoologischen Gesellschaft auf der dritten Jahresversammlung zu Göttingen, den 24 bis 26 Mai 1893* [1893]:84–88 [with] Bütschli, O., Carus J. V., Döderlein, L. Ehlers, E. Ludwig, H., Möbius, K. Schulze, F. E. and J. W. Spengel. Anhang: dritte Entwurf von Regeln für die wissenschaftlichen Benennung der Thiere. *Verhandlungen der deutschen zoologischen Gesellschaft auf der dritten Jahresversammlung zu Göttingen, den 24 bis 26 Mai 1893* [1893]:89–98.

Chantre, E. 1893. Séance du lundi 19/27 août. In *Compte-Rendu des Séances du Congrès International de Zoologie*, Ed. J. Dumouchel, xxxvii–xxxix. Moscou: Laschkevitsch.

Chaper, M. 1881. *De la nomenclature des êtres organisés*. Paris: Société Zoologique de France.

Chaper, M. 1889. Rapport fait au nom de la commission de nomenclature de la Société Zoologique de France. In *Compte-Rendu des Séances du Congrès International de Zoologie*, Ed. R. Blanchard, 437–466. Paris: Société Zoologique de France.

Chapman, F. 1923. The rule of priority in nomenclature. *Nature* 111:148–149.

Clerck, C. 1757. *Aranei Suecici*. Stockholm: L. Salvius.

Collette, B. B., Cohen, D. M., and J. A. Peters. 1972. Stability in zoological nomenclature. *Science* 177:452–453.

Cook, O. F. 1898. The method of types. *Science* 8:513–516.

Cook, O. F. 1900. The method of types in botanical nomenclature. *Science* 12:475–481.

Cook, O. F. 1902. Types and synonyms. *Science* 15:646–656.

Cope, E. D. 1878. The report of the committee of the American Association of 1876 on biological nomenclature. *American Naturalist* 12:517–525.

Corliss, J. O. 1972. Priority and stability in zoological nomenclature: Resolution of the problem of article 23b at the Monaco Congress. *Science* 178:1120.

Cornelius, P. F. S. 1987. Use versus priority in zoological nomenclature: A solution for an old problem. *Bulletin of Zoological Nomenclature* 44:79–85.

Coues, E., Allen, J. A., Ridgway, R., Brewster, W., and H. W. Henshaw. 1886. *The code of nomenclature and check-list of North American birds adopted by the American Ornithologists' Union*. New York: American Ornithologists' Union.

Crotch, M. A. 1870. On the generic nomenclature of Lepidoptera. *Cistula Entomologica* 2:59–70.

Crovello, T. J. 1967. Problems in the use of electronic data processing in biological collections. *Taxon* 16:481–494.

Dall, W. H. 1878. Report of the Committee on Zoological nomenclature to section B, of the American Association for the Advancement of Science, at the Nashville Meeting, August 31, 1877. In *Proceedings of the American Association for the Advancement of Science twenty-sixth meeting, Nashville, Tennessee, 1877*, Ed. F. T. Putnam, 7–56. Salem: Salem Press.

Dayrat, B. 2003. *Les botanistes et la flore de France, trois siècles de découvertes [Botanists and the French Flora: three centuries of discoveries]*. Paris: Muséum National d'Histoire Naturelle.

Dayrat, B. 2005a. Advantages of naming species under the *PhyloCode*: How a new species of Discodorididae (Gastropoda, Euthyneura, Nudibranchia, Doridina) may be named. *Marine Biology Research* 1:216–232.

Dayrat, B. 2005b. Towards integrative taxonomy. *Biological Journal of the Linnean Society* 87:407–415.

Dayrat, B. (submitted). *A monographic revision of discodorid sea slugs (Gastropoda, Opisthobranchia, Nudibranchia, Doridina)*. [Monograph accepted for publication in the *Proceedings of the California Academy of Sciences* by internal—institutional—reviews, and is now in external peer reviews].

Dayrat, B., Cantino, P., Clarke, J., and K. de Queiroz. 2008. Naming species in the *PhyloCode*: The solution adopted by the International Society of Phylogenetic Nomenclature. *Systematic Biology* 57:507–514.

Dayrat, B. and T. M. Gosliner. 2005. Metaphyly and species names: A case study in *Discodorididae* (*Mollusca, Gastropoda, Euthyneura, Nudibranchia*). *Zoologica Scripta* 34:199–224.

Dayrat, B., Schander, C., and K. Angielczyk. 2004. Suggestions for a new species nomenclature. *Taxon* 53:485–491.

De Candolle, Alp. P. 1867. *Lois de la nomenclature botanique adoptées par le Congrès international de botanique tenu à Paris en août 1867.* Paris: J.-B. Baillière.

De Candolle, Alp. P. 1882. Lettre de M. de Candolle à Monsieur Capellini. In *Congrès Géologique International, Compte-Rendu de la Deuxième Session, Bologna, 1881,* Ed. G. Capellini, 181–185. Bologne: Fava et Garagnani.

De Candolle, Alp. P. 1889. Lettre de Mr. Alph. De Candolle à M. R. Blanchard sur la nomenclature des êtres organisés. In *Compte-Rendu des Séances du Congrès International de Zoologie,* Ed. R. Blanchard, 482–485. Paris: Société Zoologique de France.

De Candolle, Aug. P. 1813. *Théorie élémentaire de la botanique.* Paris: Déterville.

De Candolle, Aug. P. 1822. Mémoire sur la tribu des Cuspariées. *Mémoires du Muséum d'Histoire Naturelle* 9:139–154, pls 8–10.

De Candolle, Aug. P. 1829. *Collections de mémoires pour servir à l'histoire du Règne Végétal. Mémoire sur la famille des Ombellifères.* Paris: Treuttel et Wurtz.

Dejean, A. 1825. *Species général des Coléoptères. Tome Premier.* Paris: Crevot.

De Queiroz, K. and J. Gauthier. 1990. Phylogeny as a central principle in taxonomy: Phylogenetic definitions of taxon names. *Systematic Zoology* 39:307–322.

De Queiroz, K. and J. Gauthier. 1992. Phylogenetic taxonomy. *Annual Reviews of Ecology and Systematics* 23:449–480.

De Queiroz, K. and J. Gauthier. 1994. Toward a phylogenetic system of biological nomenclature. *Trends in Ecology and Evolution* 9:27–31.

Douvillé, H. 1882. Règles à suivre pour établir la nomenclature des espèces. Rapport du secrétaire de la Commission. In *Congrès Géologique International, Compte-Rendu de la Deuxième Session, Bologna, 1881,* Ed. G. Capellini, 592–608. Bologne: Fava et Garagnani.

Dubois, A. 1998. List of European species of amphibians and reptiles: Will we soon be reaching "stability"? *Amphibia-Reptilia* 19:1–28.

Dubois, A. 2006. New proposals for naming lower-ranked taxa within the frame of the *International Code of Zoological Nomenclature. C. R. Biologies* 329:823–840.

Dubois, A. 2008. A partial but radical solution to the problem of nomenclatural inflation and synonymy load. *Biological Journal of the Linnean Society* 93:857–863.

Dunning, J. W. 1872a. Anagrams and nonsense-names in scientific nomenclature. *The Entomologist's Monthly Magazine* 8:253–254.

Dunning, J. W. 1872b. On the relation between generic and specific names. *The Entomologist's Monthly Magazine* 8:291–294.

Dupraw, E. J. 1965. Non-Linnean taxonomy and the systematics of honeybees. *Systematic Zoology* 14:1–24.

Edwards, W. H. 1873. Some remarks on entomological nomenclature. *Canadian Entomologist* 5:21–36.

Edwards, W. H. 1876. Notes on entomological nomenclature. *Canadian Entomologist* 8:41–52, 81–94, 113–119.

Erzinclioglu, Y. Z. and Unwin, D. M. 1986. The stability of zoological nomenclature. *Nature* 320:687.

Fabricius, J. C. 1778. *Philosophia Botanica.* Hamburg: C. R. Bohn.

Fauvel, A. 1869. Faune Gallo-Rhénane ou description des insectes qui habitent la France, la Belgique, la Hollande, les provinces rhénanes et le Valais. Chapitre V. Lois de la Nomenclature. *Bulletin de la Société Linnéenne de Normandie* 3:91–94. [French translation of Kiesenwetter's (1858) entomological rules].

Feldmann, R. M. 1986. The stability of zoological nomenclature. *Nature* 322:120.

Follett, W. I. 1954. Copenhagen decisions on zoological nomenclature. *Systematic Zoology* 3:172–174.

Follett, W. I. 1956. Modifications of the International Rules of Zoological Nomenclature proposed since the 1953 Copenhagen Congress. *Systematic Zoology* 6:33–36.

Friebel, E. and R. Richter. 1943. Can nomina becoming homonyms be preserved? *Senckenbergiana* 26:321–324.

Frizzell, D. L. 1933. Terminology of types. *The American Midland Naturalist* 14:636–668.

Gray, A. 1864. Nomenclature. *American Journal of Science and Arts* 37:278–281.

Grote, A. R. 1876. On genera and the law of priority. *Canadian Entomologist* 8:56–60.

Greuter, W., Hawksworth, D. L., McNeill, J., Mayo, M. A., Minelli, A., Sneath, P. H. A., et al. 1996. Draft BioCode: the prospective international rules for the scientific names of organisms. *Taxon* 45:349–372.

Greuter, W., Hawksworth, D. L., McNeill, J., Mayo, M. A., Minelli, A., Sneath, P. H. A., et al. 1998. Draft BioCode (1997): The prospective international rules for the scientific names of organisms. *Taxon* 47:127–150.

Guérin-Méneville, F. E. 1843. Rapport d'une commission nommée par l'Association britannique pour l'avancement de la science, chargée d'examiner les règles d'après lesquelles la nomenclature zoologique pourrait êtres établie sur une base uniforme et permanente (in-8. Londres, 27 juin 1842). *Revue Zoologique* [1843]:202–210.

Gurney, A. B. 1967. Descriptions of new species and the use of holotypes still valuable. *Systematic Zoology* 16:264–265.

Hawksworth, D. L. 1988. Improved stability for biological nomenclature. *Nature* 334:301.

Hawksworth, D. L (Ed.). 1991. *Improving the stability of names: Needs and options*. Königstein: Koeltz Scientific Books.

Hawksworth, D. L. (Ed.). 1997. *The new bionomenclature: The BioCode debate*. Paris: International Union of Biological Sciences (Biology International, Special Issue 34).

Hawksworth, D. L. and W. Greuter. 1989. Improvement of stability in biological nomenclature. *Biology International* 19:5–11.

Heikertinger, F. 1940. Was jeder Zoologe von den Nomenklaturfragen wissen soll Eine kurzgefasste Einfuhrung in eine vernachlassigte Disziplin. *Zoologischer Anzeiger* 130:139–155.

Heikertinger, F. 1943a. Kann Kontinuät der Tiernamen mit der Prioritätsregel erreicht werden? *Zoologischer Anzeiger* 141:35–52.

Heikertinger, F. 1943b. Das Nomenklaturproblem in Botanik und Zoologie. *Berichte der Deutschen Botanischen Gesellschaft* 61:42–48.

Heikertinger, F. 1953. The tragicomedy of zoological nomenclature. An introduction for every zoologist. *Verhandlungen der zoologisch-botanischen Gesellschaft in Wien* 9:35–45.

Heise, H. and M. P. Starr. 1968. Nomenifers: Are they christened or classified? *Systematic Zoology* 17:458–467.

Hemming, F. 1950. Stability in zoological nomenclature. *The Auk* 67:370–374.

Hemming, F. 1953. *Copenhagen decisions on zoological nomenclature. Additions to, and modifications of, the Règles internationales de la nomenclature zoologique approved and adopted by the fourteenth International Congress of Zoology, Copenhagen, August, 1953*. London: International Trust for Zoological Nomenclature.

Hennig, W. 1966. *Phylogenetic systematics*. Urbana: University of Illinois Press.

Horst, R., Leesberg, A. F. A., and C. Ritsema. 1889. De la nomenclature entomologique. Rapport présenté à la Société entomologique néerlandaise. In *Compte-Rendu des Séances du Congrès International de Zoologie*, Ed. R. Blanchard, 468–471. Paris: Société Zoologique de France.

Hubrecht, A. A. W., Hœk, P. P. C., van Hasselt A. W. M. and F. M. van der Wulf. 1889. Lettre des Sociétés néerlandaises de zoologie et d'entomologie au sujet de la nomenclature. In *Compte-Rendu des Séances du Congrès International de Zoologie*, ed. R. Blanchard, 466–467. Paris: Société Zoologique de France.

Hull, D. L. 1966. Phylogenetic numericlature. *Systematic Zoology* 15:14–17.

Hull, D. L. 1968. The syntax of numericlature. *Systematic Zoology* 17:472–474.

Hull, D. L. 1969. Reply to Randall and Scott. *Systematic Zoology* 18:468–469.

International Commission on Zoological Nomenclature. 1905. *Règles internationales de la nomenclature zoologique, International rules of zoological nomenclature, Internationale Regeln der Zoologischen Nomenklatur*. Paris: Rudeval.

International Commission on Zoological Nomenclature. 1961. *International code of zoological nomenclature*. London: International Trust for Zoological Nomenclature. [1st edition].

International Commission on Zoological Nomenclature. 1985. *International code of zoological nomenclature*. London: International Trust for Zoological Nomenclature. [3rd edition].

International Commission on Zoological Nomenclature. 1999. *International code of zoological nomenclature*. London: International Trust for Zoological Nomenclature. [4th edition].

Jahn, T. L. 1961. Man versus machine: A future problem in protozoan taxonomy. *Systematic Zoology* 10:179–192.

Jardine, W. 1866. Report of a Committee appointed to report on the changes which they may consider desirable to make, if any, in the rules of Zoological Nomenclature, drawn up by Mr. H. E. Strickland, at the instance of the British Association at their meeting in Manchester in 1842. *Report of the thirty-fifth meeting of the British Association for the Advancement of Science* [1865]:25–42. [Reprinted in Sclater 1878:21–24].

Johnson, L. A. S. 1970. Rainbow's end: The quest for an optimal taxonomy. *Systematic Zoology* 19:203–239.

Kalkman, C. 1966. Keeping up with the Joneses. *Taxon* 15:177–179.

Kiesenwetter, E. von 1858. Gesetze der entomologischen Nomenclatur. *Berliner Entomologische Zeitschrift* 2:xi–xxii.

Kirby, W. F. 1868. On the application of the law of priority to genera in entomology. *Transactions of the Entomological Society of London* [1868]:xlii–xlviii.

Kirby, W. F. 1870. On the necessity of a reform in the generic nomenclature of diurnal Lepidoptera, illustrated by a review of the genera proposed from the time of Linnaeus to the year 1816. *Journal of the Linnean Society, Zoology* 30:494–503.

Kirby, W. F. 1871a. *A synomymic catalogue of diurnal Lepidoptera.* London: John van Voorst.

Kirby, W. F. 1871b. On the rules and use of synonymy: In reply to Mr. W. A. Lewis. *The Entomologist's Monthly Magazine* 8:41–42.

Kuntner, M. and I. Agnarsson. 2007. Are the Linnean and phylogenetic nomenclatural systems combinable? Recommendations for biological nomenclature. *Systematic Biology* 55:774–784.

Lamarck, J.-B. 1778 [1779]. *Flore Françoise.* Paris: Agasse. [3 vols.]

Lanham, U. 1965. Uninominal nomenclature. *Systematic Zoology* 14:144.

LeConte, J. L. 1873. Annual address of the president of the Entomological Society of Ontario. *Canadian Entomologist* 5:182–183.

LeConte, J. L. 1874a. On some changes in the nomenclature of North American Coleoptera, which have been recently proposed. *Canadian Entomologist* 6:186–196.

LeConte, J. L. 1874b. On entomological nomenclature. Part I. On the law of priority. *Canadian Entomologist* 6:201–206.

LeConte, J. L. 1874c. On entomological nomenclature. Part II. On generic types. *Canadian Entomologist* 6:223–226.

Lewis, A. 1871. On the application of the maxim *"communis error facit jus"* to scientific nomenclature. *The Entomologist's Monthly Magazine* 8:1–5.

Lewis, A. 1875. On entomological nomenclature and the rule of priority. *Transactions of the Entomological Society of London* [1875]:i–xlii.

Lindsey, E. G. and R. L. Usinger. 1959. Linnaeus and the development of the International Code of Zoological Nomenclature. *Systematic Zoology* 8:39–47.

Linnaeus, C. 1735. *Systema naturae.* Lugduni Batavorum: Haak. [1st ed.].

Linnaeus, C. 1736. *Fundamenta botanica.* Amsterdam: Schouten. [1st ed.].

Linnaeus, C. 1737. *Critica botanica.* Lugduni Batavorum: Wishoff. [Translated by A. Hort, 1938, London: Ray Society].

Linnaeus, C. 1745. *Flora suecica.* Lugduni Batavorum: Wishoff.

Linnaeus, C. 1749. *Pan suecicus.* Uppsala.

Linnaeus, C. 1751. *Philosophia botanica.* Stockholm: Kiesewetter.

Linnaeus, C. 1753. *Species plantarum.* Stockholm: Salvius. [2 vols.]

Linnaeus, C. 1758–1759. *Systema naturae.* Stockholm: Salvius. [10th ed., 2 vols.]

Linnaeus, C. 1766–1768. *Systema naturae.* Stockholm: Salvius. [12th ed., 4 vols.]

Little, F. J. 1964. The need for a uniform system of biological numericlature. *Systematic Zoology* 13:191–194.

Mansfeld, R. 1939. Zur Nomenklatur der Farn- und Blutenpflanzen Deutschlands (VI). *Repertorium Specierum Novarum Regni Vegetabilis, Berlin* 47:137–163.

Marsh, O. C. 1898. The value of type specimens and importance of their preservation. *American Journal of Science* 6:401–405.

Mayr, E. 1972. The history of the Statute of Limitation. [Letter to the editor]. *Science* 177:453.

Mayr, E., Simpson, G. G., and E. Eisenmann. 1971. Stability in zoological nomenclature. *Science* 174:1041–1052.

McLachlan, R. 1871. Some considerations as to Mr. Lewis's views concerning entomological nomenclature. *The Entomologist's Monthly Magazine* 8:40–41.

McMichael, D. F. 1956. Problems of family nomenclature. *Systematic Zoology* 5:141–142.

Mead, T. L. 1874. On specific nomenclature. *Canadian Entomologist* 6:108–109.

Melville, R. V. 1995. *Towards stability in the names of animals. A history of the International Commission on Zoological Nomenclature 1895–1995.* London: International Trust for Zoological Nomenclature.

Merriam, C. H. 1897. Type specimens in natural history. *Science* 5:731–732.

Michener, C. D. 1963. Some future developments in taxonomy. *Systematic Zoology* 12:151–172.

Michener, C. D. 1964. The possible use of uninominal nomenclature to increase the stability of names in biology. *Systematic Zoology* 13:182–190.

Milne-Edwards, A. 1889. Discussion sur la nomenclature des êtres organisés. In *Compte-Rendu des Séances du Congrès International de Zoologie*, Ed. R. Blanchard, 405–418. Paris: Société Zoologique de France.

Morrison, H. K. 1873. Specific nomenclature. *Canadian Entomologist* 5:70–71.

Mulsant, E. 1860. Règles de la nomenclature entomologique. *Opuscules entomologiques* 11:7–20. [French translation of Kiesenwetter's (1858) entomological rules].

Ng, P. K. L. 1991. How conservative should nomenclature be? Comments on the principle of priority. *Bulletin of Zoological Nomenclature* 48:87–91.

Nicolson, D. H. 1991. A history of botanical nomenclature. *Annals of the Missouri Botanical Garden* 78:33–56.

Oberthür, C. 1889. Considérations sur la nomenclature zoologique. In *Compte-Rendu des Séances du Congrès International de Zoologie*, Ed. R. Blanchard, 471–476. Paris: Société Zoologique de France.

Ogilby, W. 1838a. Observations on "Rules for Nomenclature." *Magazine of Natural History* 2:150–157.

Ogilby, W. 1838b. Further observations on "Rules for Nomenclature." *Magazine of Natural History* 2:275–284.

Ogilby, W. 1838c. Letter from William Ogilby, Esq. in reference to Mr. Strickland's observations on the application of the term Simia. *Magazine of Natural History* 2:492–494.

Oldroyd, H. 1966. The future of taxonomic entomology. *Systematic Zoology* 15:253–260.

Oldroyd, H. 1967. On comments concerning the future of taxonomic entomology. *Systematic Zoology* 16:274–275.

Olivier, G. A. 1789–1811. *Insectes. Encyclopédie Méthodique*. Paris: Panckoucke. [5 vols.]

Parkes, K. C. 1967. A qualified defense of traditional nomenclature. *Systematic Zoology* 16:268–273.

Pascoe, F. P. 1965. A note on generic names having nearly the same sound. *The Zoologist* 23:9542–9543.

Petit-Thouars, A. du. 1817. *Le Verger Français*. Paris: Treuttel et Wurtz, Arthus-Bertrand.

Petit-Thouars, A. du. 1822. *Flore des îles Australes de l'Afrique. Famille des Orchidées*. Paris: Arthus-Bertrand, Treuttel et Wurtz.

Plate, L. von. 1893. Studien über opisthopneumone Lungeschnecken, II, Die Onciidien. *Zoologische Jahrbücher, Anatomie* 7:93–234.

Preudhomme de Borre, A. 1886. Le Président prend la parole et s'exprime comme suit. *Annales de la Société entomologique de Belgique* 30:cxcviii–ccv.

Puton, A. 1880. Quelques mots sur la nomenclature entomologique. *Annales de la Société Entomologique de France* 10:33–40.

Ralph, A. and G. C. Wood. 1954. Punch-card taxonomy. *Systematic Zoology* 3:143.

Randall, J. M. and G. H. Scott. 1967. Linnaean nomenclature: An aid to data processing. *Systematic Zoology* 16:278–281.

Randall, J. M. and G. H. Scott. 1969. Has nomenclature a syntax? *Systematic Zoology* 18:466–468.

Richter, R. 1942a. Kontinuät der zoologischen Nomenklatur gegen die Regeln oder mit ihnen? *Zoologischer Anzeiger* 139:115–127.

Richter, R. 1942b. Is there a possibility of unchangeable names of species? *Senckenbergiana* 25:340–356.

Rivas, L. R. 1965. A proposed code system for storage and retrieval of information in systematic zoology. *Systematic Zoology* 14:131–132.

Roger, J. 1993. *Les sciences de la vie dans la pensée française au XVIIIe siècle*. Paris: Albin Michel.

Rozier, F. 1783. Phytonomatotechnie universelle, c'est-à-dire, l'Art de donner aux plantes des noms tires de leurs caractères; par M. Bergeret, Chirurgien. *Observations sur la Physique, sur l'Histoire naturelle, et sur les Arts* 23:223–227. [Review of Bergeret's *Phytonomatotechnie universelle*].

Russell, N. H. The development of an operational approach in plant taxonomy. *Systematic Zoology* 10:159–167.

Saint-Lager, J.-B. 1882. *Des origines des sciences naturelles*. Lyon: Association typographique.

Saint-Lager, J.-B. 1886. *Le procès de la nomenclature botanique et zoologique*. Paris: J.-B. Baillière.

Saint-Lager, J.-B. 1889. Lettre de Mr. Le Dr. Saint-Lager à M. R. Blanchard sur la nomenclature des êtres organisés. In *Compte-Rendu des Séances du Congrès International de Zoologie*, Ed. R. Blanchard, 476–482. Paris: Société Zoologique de France.

Schaum, H. 1862. On the restoration of obsolete names in entomology. *Transactions of the Entomological Society of London* [1862]:323–327.

Schenk, E. T. and J. H. McMasters. 1936. *Procedures in taxonomy*. Stanford: Stanford University Press. [1st ed.].

Schopf, J. M. 1960. Emphasis on holotype (?). *Science* 131:1043.

Schubert, C. 1897. What is a type in natural history? *Science* 5:636–640.

Schulze, F. E. 1896. La parole est donnée à monsieur F. E. Schultze. In *Compte-Rendu des séances du troisième Congrès International de Zoologie, Leyde, 1895*, Ed. P. P. C. Hoek, 93–97. Leyde: E. J. Brill.

Sclater, P. L. 1878. *Rules for zoological nomenclature*. London: John Murray.

Sclater, P. L. 1896. Remarks on the divergencies between the "Rules for naming Animals" of the German Zoölogical Society and the Stricklandian Code of Nomenclature. *Proceedings of the Zoological Society of London* [1896]:306–319.

Scudder, S. H. 1872. A systematic revision of some of the American butterflies, with brief notes on those known to occur in Essex County, Mass. *Annual Report of the Trustees of the Peabody Academy of Science* [1871]:1–83.

Scudder, S. H. 1873. Canons of systematic nomenclature for the higher groups. *Canadian Entomologist* 5:55–59.

Selys-Longchamps, E. de 1842. *Faune Belge*. Liège: Dessain.

Sharp, D. 1872. On the relation between generic and specific names. *The Entomologist's Monthly Magazine* 8:254–255, 290–291.

Sharp, D. 1873. *The object and method of zoological nomenclature*. London: E. W. Janson, Williams, and Norgate.

Sharpe, R. B. 1886. The A.O.U. Code and checklist of American birds. *Nature* 34:168–169.

Shear, C. L. 1924. The failure of the principle of priority to secure uniformity and stability in botanical nomenclature. *Science* 60:254–258.

Silva, P. C. 1996. Stability versus rigidity in botanical nomenclature. *Nova Hedwigia* 112:1–8.

Smith, A. C. 1945. The principle of priority in biological nomenclature. *Chronica Botanica* 9:114–119.

Smith, H. M. 1962. Commentary on the 1961 Code of Zoological Nomenclature. *Systematic Zoology* 11:85–91.

Smith, H. M. and K. L. Williams. The nomen oblitum rule of the 1961 International Code of Zoological Nomenclature. *Herpetologica* 18:11–13.

Smith, J. L. B. 1964. The statute of limitation—stability or chaos? *Rhodes University Department of Ichthyology Occasional Papers* 1:1–16.

Sneath, P. H. A. 1961. Recent developments in theoretical and quantitative taxonomy. *Systematic Zoology* 10:118–139.

Sokal, R. R. and P. H. A. Sneath. 1963. *Principles of numerical taxonomy*. San Francisco: W. H. Freeman.

Sokal, R. R. and P. H. A. Sneath. 1966. Efficiency in taxonomy. *Taxon* 15:1–21.

Sokal, R. R., Camin, J. H., Rohlf, F. J., and P. H. A. Sneath. 1965. Numerical taxonomy: Some points of view. *Systematic Zoology* 14:237–243.

Spinola, M. 1811. Essai d'une nouvelle classification des Diplolépaires. *Annales du Muséum d'Histoire Naturelle* 17:138–152.

Staudinger, O. and M. Wocke. 1871. *Catalog der Lepidopteren des Europæischen Faunengebiets. Catalogue ou énumeration méthodique des Lépidoptères qui habitant sur le territoire de la faune européenne*. Dresden: O. Staudinger und Herman Burdach.

Stearns, W. T. 1959. The background of Linnaeus's contributions to the nomenclature and methods of systematic biology. *Systematic Zoology* 8:4–22.

Stebbing, T. R. R. 1896. Rules of nomenclature in zoology. *Natural Science* 8:254–256.

Steyskal, G. C. 1965. Notes on uninominal nomenclature. *Systematic Zoology* 14:346–348.

Steyskal, G. C. 1967. Another view of the future of taxonomy. *Systematic Zoology* 16:265–268.

Strickland, H. E. 1835. On the arbitrary alteration of established terms in natural history. *Magazine of Natural History* 8:36–40.

Strickland, H. E. 1837a. Rules for zoological nomenclature. *Magazine of Natural History* 1:173–176.

Strickland, H. E. 1837b. Objections to the nomenclature employed by Mr. Ogilby. *Magazine of Natural History* 1:604–605.

Strickland, H. E. 1838a. Reply to Mr. Ogilby's "Observations on rules of nomenclature." *Magazine of Natural History* 2:198–204.

Strickland, H. E. 1838b. Remarks on Mr. Ogilby's "Further observations on rules of nomenclature." *Magazine of Natural History* 2:326–331.

Strickland, H. E. 1838c. A few words of explanation in reference to Mr. Ogilby's Letter at p. 492. *Magazine of Natural History* 2:555–556.

Strickland, H. E. 1843. Report of a committee appointed to "consider of the rules by which the nomenclature of zoology may be established on a uniform and permanent basis." *Report of the twelfth meeting of the British Association for the Advancement of Science* [1842]:105–121. [Reprinted in Sclater 1878:5–19].

Swainson, W. 1820–1821. *Zoological illustrations. Volume 1*. London: Baldwin.

Swainson, W. 1836. On the natural history and classification of birds. In *The cabinet cyclopædia*, Ed, D. Lardner, 1–365 [Volume 1]. London: Longman.

Thomas, O. 1893. Suggestions for the more definite use of the word "Type" and its compounds, as denoting specimens of a greater or less degree of authenticity. *Proceedings of the Zoological Society of London* [1893]:241–242.

Thomas, O. 1897. Types in natural history and nomenclature in rodents. *Science* 6:485–487.

Thorell, T. 1869. Review of the European genera of spiders. *Nova Acta Regiae Societatis Scientiarum Upsaliensis* 7:1–242.

Tortonese, E. 1967. Again about specific names. *Systematic Zoology* 16:278.

Tubbs, P. K. 1986. The stability of zoological nomenclature. *Nature* 321:476.

Van Tieghem, P., Poissus, J., Fournier, E., and E. Bureau. 1882. Lettre de la Société botanique de France. In *Congrès Géologique International, Compte-Rendu de la Deuxième Session, Bologna, 1881*, Ed. G. Capellini, 178–181. Bologne: Fava et Garagnani.

Verrill, A. E. 1869. The Rules of Zoological Nomenclature. From the Report of a Committee "appointed to report on the changes which they may consider desirable to make, if any, in the rules of Zoological Nomenclature, drawn up by Mr. H. E. Strickland, at the instance of the British Association, at their meeting in Manchester in 1842. With notes by A. E. Verrill. *American Journal of Science and Arts* 48:92–110.

Wallace, A. R. 1871. The president's address. *Transactions of the Entomological Society of London* [1871]:li-lxxv

Wallace, A. R. 1874. Zoological nomenclature. *Nature* 9:258–260.

Waterhouse, G. R. 1858. *Catalogue of British Coleoptera*. London: Taylor and Francis.

Waterhouse, G. R. 1862. Observations upon the nomenclature adopted in the recently published "Catalogue of British Coleoptera," having reference more especially to remarks contained in Dr. Schaum's paper "On the restoration of obsolete names in entomology." *Transactions of the Entomological Society of London* [1862]:328–338.

Westwood, J. O. 1837. On generic nomenclature. *Magazine of Natural History* 1:169–173.

Whitehead, P. J. P. 1972. The contradiction between nomenclature and taxonomy. *Systematic Zoology* 21:215–224.

Yochelson, E. L. 1966. Nomenclature in the machine age. *Systematic Zoology* 15:88–91.

18 250 Years of Swedish Taxonomy

Fredrik Ronquist

CONTENTS

HISTORICAL PERSPECTIVE

In hindsight, it is fascinating how little Linné (Linnaeus) knew about the diversity of the planet. Even as an internationally established naturalist in his forties, he wrote (Linné, 1749):

> If we estimate the plants to approximately 10,000, the worms to 2,000, the insects to 10,000, the amphibians to 300, the fishes to 2,000 and the tetrapods to 200, then there are 26,500 species of living beings in the world. Our own country has almost 3,600 of these; of domestic plants there are approximately 1,300 and 2,300 species of animals inhabit the country as far as we know.

If these numbers had been anywhere close to the true diversity of the planet, Linné would have named an astonishingly large portion of the existing species himself. In reality, Linné had barely scratched the surface in most organism groups, even in his home country of Sweden. It is true that he described the bulk of Swedish vascular plants and vertebrates, but he clearly lacked an understanding of, or interest in, the diversity of other plants and animals. This is well illustrated by the insects, which I will be returning to throughout this chapter for examples illustrating the last 250 years of Swedish taxonomy. Insects account for about 80% of Swedish animal species and number in the tens of thousands but, in his *Fauna Suecica*, Linné (1761) listed only some 1,500 species that are still recognized today—a tiny fraction of the total.

Despite his taxonomic biases, or perhaps partly because of them, Linné inspired a large number of biologists in Sweden and around the world to study poorly known organism groups. This resulted in swift progress in the knowledge of the fauna and flora over the next century and half. For instance, Tullgren and Wahlgren (1922) estimated that some 15,000 species of insects were known from Sweden a century ago, an order of magnitude larger than the number Linné had listed in *Fauna Suecica*. Later in the 20th century, however, the pace of discovery decreased significantly. This is particularly notable given the simultaneous increase in the number of professional biologists, and it had a drastic effect on the exploration of European faunas and floras. For instance, in the peak period in the first half of the 19th century, entomologists described hundreds of new Swedish insect species every year, but the rate had dropped an order of magnitude by the late 20th century (Figure 18.1).

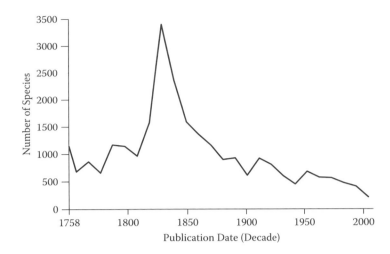

FIGURE 18.1   Date of description of the approximately 26,000 insect species known to occur in Sweden in 2009. Publication dates are grouped in decades after 1758 to even out yearly fluctuations. After the peak in the first half of the 18th century, the discovery rate dropped almost an order of magnitude toward the end of the 20th century. The STI will undoubtedly result in a significant boost in the coming decade, as more than 600 newly discovered species now await formal description.

Of course, there is often a significant time lag between the description of a new species and the first time it is collected in a particular region of interest. For instance, by the time Tullgren and Wahlgren published their overview of the Swedish insect fauna, more than 21,000 of our insect species had already been described but only 15,000 of those were known at the time to be Swedish. In other words, Tullgren and Wahlgren were unaware of a quarter of the Swedish species that had already been named. Much of the activity of Swedish entomologists in the 20th century consisted of adding these species to the Swedish list. By the turn of the century, Gärdenfors et al. (2003) estimated that 24,700 species of insects were known from the country. Fully two thirds of the species added since 1922 had already then been described based on material from other countries.

One might suspect that the decline in species discovery rates merely reflected the inventory nearing its completion, especially in Sweden with its depauperate and well-researched fauna. However, other factors were also at play. In the 20th century, many biologists started questioning the importance of inventorying species and instead turned their attention to the general principles of biology, drawing funds and students away from taxonomy. The resources that remained were not spent very efficiently with respect to the goal of completing the inventory. Conservatism combined with the prevalent emphasis on curiosity-driven research resulted in an inordinate concentration of attention on comparatively well-known organisms. This created the impression that floras and faunas were better known than they actually were, especially in Europe. Thus, toward the end of the 20th century, it was still unclear how close biologists actually were to completing the inventory of macroscopic life in Sweden or any other region containing a nontrivial number of species.

Tides Turn

In the 1990s, tides turned. Several prominent biologists had been arguing for some time that we were facing a biodiversity crisis ultimately threatening the survival of mankind, and politicians finally responded. By linking biodiversity with economic interests, the Convention on Biological Diversity (http://www.cbd.int) succeeded in 1992 in putting taxonomy on the political agenda. Several of the parties to the convention, among them Sweden, worked in the following years to promote inventorying efforts and taxonomic capacity-building in the tropics through the Global Taxonomy Initiative (GTI), which was officially launched in 1998.

The Swedish backing of the GTI inspired a critical look at the actual knowledge of the Swedish fauna and flora and the state of taxonomy in Sweden. At the same time, the work on the Swedish Red List, led by the Swedish Species Information Center (Artdatabanken), emphasized the large number of species-rich groups that could not be judged according to the official International Union for Conservation of Nature (IUCN) criteria because of the lack of taxonomic knowledge and easily accessible identification guides. Thanks to these factors and a favorable political situation, a group of biologists were able to convince the Swedish government in 2001 of the need for an initiative focused on the Swedish flora and fauna.

THE SWEDISH TAXONOMY INITIATIVE

Announced by the Swedish government in the spring of 2001, the goal of the Swedish Taxonomy Initiative (STI) is to complete the inventory of the multicellular species occurring in the country within twenty years. All species are to be documented scientifically, and the species that can be identified by amateur naturalists are to be presented in a well-illustrated biodiversity encyclopedia commissioned by the Swedish Parliament. The project is coordinated by the Swedish Species Information Center but involves a large number of participating academic institutions, natural history museums, and biologists in Sweden and abroad.

The focus on multicellular species was a strategic decision based on the goal of reaching out to amateur naturalists and the general public. Whereas Swedish amateur naturalists play an important role in monitoring the macroscopic flora and fauna, they are unlikely to be able to contribute productively when it comes to microscopic species, which are often identified using molecular genetic tools. Macroscopic species are also more susceptible to environmental changes and extinction because of their smaller population sizes, making it more critical to monitor their abundance. This is not to deny that macroscopic diversity is dwarfed by the incredible diversity of bacteria and unicellular eukaryotes, which is worthy of a charting initiative of its own. However, such an initiative would have to be designed completely differently.

The STI was officially launched in 2002 and, after a start-up phase of three years, the project has been fully funded since 2005. Of the current funding, about 30 million SEK (2.6 M €) per year is used for project administration and production of the encyclopedia, while about 15 million SEK (1.3 M €) is granted in support of taxonomic research and inventory projects on poorly known organism groups. In addition, about 20 million SEK (1.7 M €) is used to help Swedish natural history collections participate in the project. Over a twenty-year period, this sums up to 1.3 billion SEK or just over 1 billion €. It may sound like a massive investment, but it is still almost an order of magnitude smaller than the projected cost of the Large Hadron Collider (Achenbach, 2008), the world's largest particle accelerator, and the investment is spread over a longer time period.

The early years of the STI have been dominated by two major inventorying efforts. A marine inventory was started in 2006 and finished its last field season in 2008. On land, poorly known insect groups were targeted in the Swedish Malaise Trap Project, which operated a number of Malaise traps across the country from 2003 to 2006. Sorting and identifying this material will occupy biologists for years to come. In addition to the large-scale inventories, a number of taxonomic research projects on poorly known organism groups have also been funded and about ten training grants for PhD students have been awarded.

The first volume of the *Encyclopedia of the Swedish Flora and Fauna* (*Nationalnyckeln*), covering butterflies, was officially presented to the Swedish Parliament and Her Royal Highness Crown Princess Victoria, patron of the STI, in April of 2005 (Figure 18.2). Six additional volumes have been published since then, covering groups ranging from mosses to millipedes. The main text is in Swedish and summarizes the characteristics of each species, as well as its distribution, biology, and conservation status. The key facts are also given in English and the identification keys are bilingual.

FIGURE 18.2   An important product of the Swedish Taxonomy Initiative is the *Encyclopedia of the Swedish Flora and Fauna (Nationalnyckeln)*. The richly illustrated volumes include identification keys to all species and summarize essential information about diagnostic characters, life history, distribution, and conservation status of each species.

The volumes are richly illustrated by some of the best scientific illustrators and photographers in the country. More than 100 volumes are planned in total, making this the largest publishing project attempted in Sweden so far.

MORE DIVERSITY THAN EXPECTED

At the start of the STI, it was estimated that some 50,000 species of multicellular organisms were known from the country, almost half of which were insects (Gärdenfors et al., 2003). Even those of us involved in launching the STI did not expect the discovery of a vast number of additional species. I remember giving a talk in 2001, in which I ventured a guess that the project might add a couple of thousand new species records, and result in the discovery of a couple of hundred species new to science. We emphasized that the most important aspect of the project was that it enabled school kids, laymen, and amateur naturalists to discover and identify a larger fraction of the rich biological diversity of their home country, even in their own back yards. In this way, we could both improve the quality of environmental monitoring and increase public appreciation of biological diversity. In particular, many of us hoped to see the STI result in Swedish biodiversity monitoring, to a large extent relying on amateur naturalists, target a broader and more representative sample of species.

Relatively soon, however, it became clear that the Swedish flora and fauna were less well known than most of us had thought, as STI-funded activities resulted in the discovery of a surprisingly large number of new species. During a workshop in 2007 focused on meiofauna, for instance, twelve international experts were able to add some 100 species to the Swedish list, eighteen of which were new to science, in just two weeks. This represented a 2% increase in the number of marine species known from Sweden. Since 2002, a significant number of new species have also been added to the Swedish list of many other organism groups, but the results are most striking for the insects.

FIGURE 18.3   The Swedish Malaise Trap Project ran 61 Malaise traps from 2003 to 2006 at 44 sites representing a wide variety of habitats across the country, a sample of which is seen here. In the winter of 2004–2005, eleven more traps were deployed at eight additional sites. Traps were emptied year round and, at least in the southern third of the country, they produced significant numbers of insects even in the middle of the winter.

THE SWEDISH INSECT FAUNA

Already from the start of the STI, it was clear that the remaining challenges in completing the insect inventory mainly involved groups in the orders Hymenoptera and Diptera. Many of these groups are known to be easily collected using Malaise traps. The design of these traps was originally described by René Malaise, an avid collector and former curator at the Swedish Museum of Natural History in Stockholm. Legend has it that the idea came to Malaise when he observed insects gathering underneath the roof of his tent and eventually escaping through a hole. Many entomologists have refined the design since, but the trap is still essentially a tent-like construction of thin fabric, in which the two long sides have been removed and replaced with a central wall (Figure 18.3). Many insects that are trapped try to escape by moving upward and toward the light. The Malaise trap exploits this fact by having a roof shaped like an inverted funnel and ending in a large plastic jar half-filled with ethanol. Malaise traps are not only efficient in catching small flying insects, mostly Hymenoptera and Diptera, but many other insects also end up in the traps, including wingless species that crawl up the trap and into the collecting jar.

The scientific committee of the STI was persuaded early on that a large-scale Malaise trap inventory was warranted, and the Swedish Malaise Trap Project was launched in 2003. A total of sixty-one Malaise traps were deployed for three years at forty-four different localities across the country, representing a wide variety of habitats (Karlsson et al., 2005). Eleven additional Malaise traps at eight additional sites were run for about half that time period. The traps were emptied year round and, in the southern third of the country, they produced significant numbers of insects even in the middle of the winter.

When the traps were removed in 2006, it was estimated that the total catch contained some 40 million specimens. At the time of the last comprehensive tally a year ago, about 20% of this material had been sorted to major groups and 5% had been identified to species. The identified material contained some 600 species new to Sweden, more than 400 of which were new to science. In total, the Malaise trap program and other taxonomic research projects have expanded the Swedish list of insects with almost 1,900 species since 2003. About 600 of these are new to science, most of them still awaiting formal description.

THE NEW SPECIES

The new species are far from a random sample of insects. If we analyze previous estimates of the Swedish insect fauna, it is obvious that they have been significantly biased in terms of feeding niches (Figure 18.4). The insects known to Linnaeus were predominantly phytophagous; he obviously collected insects mainly in or around plants. The larger the fraction of the fauna known to us, the larger the proportion of parasitic species (the light areas in the pie charts). The part of the pie consisting of saprophagous species and their parasitoids, the community associated with the decomposition of organic compounds, has also grown significantly with increased knowledge of the fauna.

The new species added since 2003 essentially continue this trend: they tend to be saprophagous species or parasitoids. A significant portion of the new species identified thus far are scuttle flies (Diptera: Phoridae). They are extremely abundant and ubiquitous—one in every 30 specimens collected in the Swedish Malaise Trap Project is a scuttle fly. About 300 species of phorids were recorded from Sweden in 2003; we now have seen material of some 800 and expect that the total may be around 1,100. The biology is known only for a few species, some of which are parasitoids, but one might suspect that the latter represent the exception and that most scuttle flies are saprophagous or fungivorous.

Another group with many new species are the Mycetophilidae (fungus gnats). They are definitely saprophagous or fungivorous, like many new species in the family Cecidomyiidae (gall midges and relatives). The Figitidae represent yet another example of a group that has been previously overlooked. They are small parasitic wasps attacking various insect larvae, often Diptera larvae developing in decomposing organic matter. The genus *Trybliographa* (Figure 18.5) alone is now suspected to contain more than 50 species in Sweden, about half of which are undescribed.

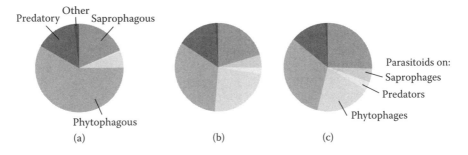

FIGURE 18.4 Larval or nymphal feeding niches of the known Swedish insect fauna at different points in time: (a) Linné (1761); (b) Tullgren and Wahlgren (1922); (c) Gärdenfors et al. (2003). As the knowledge of the fauna has increased, it has become obvious that previous estimates were significantly biased toward phytophagous species and against parasitoids and saprophagous species. Note, for instance, that Linné mainly knew phytophagous species and very few parasitoids. Results from the STI suggest that the trend toward larger fractions of parasitoids and saprophagous species will continue as the last portions of the fauna are charted.

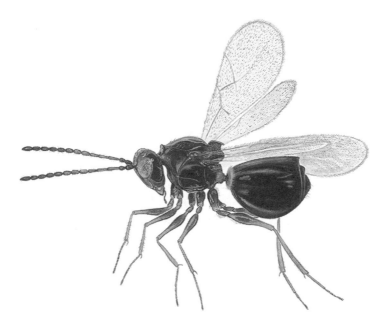

FIGURE 18.5 A representative of the genus *Trybliographa* (Hymenoptera: Figitidae: Eucoilinae). The species belonging to this genus are parasitoids of anthomyid larvae developing in fungi or decomposing parts of green plants. Thus, being small parasitoids of saprophagous species, they combine many of the characteristic properties of the new species discovered in the STI. About half of the fifty or so currently known Swedish species of *Trybliographa* are still undescribed. Illustration by Martin Holmer.

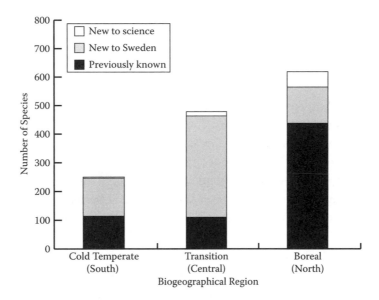

FIGURE 18.6 A large fraction of the new species discovered in the STI are from northern Sweden, which is not surprising given the focus on the southern third of the country in previous collecting. This figure illustrates this through a geographical breakdown of the species of fungus gnats and relatives (Diptera: Mycetophiloidea) currently known from Sweden (Kjaerandsen et al., 2007; Kjaerandsen, pers. comm.). New species to science and to Sweden are those that were undescribed or not previously recorded from the country in 2003, at the start of the STI. Thanks to material collected by Karl Müller, the mycetophilid fauna of northern Sweden was unusually well documented before the start of the STI.

TABLE 18.1
The Most Prolific Authors of Swedish Insect Species

Rank	Author	No. Species
1	Linné (Linnaeus)	1,470
2	Thomson	1,291
3	Zetterstedt	1,001
4	Fabricius	977
5	Meigen	929
6	Gravenhorst	734
7	Walker	596
8	Fallén	560
9	Holmgren	456
10	Haliday	324
11	Dennis	311
12	Schiffermüller	309
13	Kieffer	293
14	Gyllenhal	292
15	Erichson	276
16	Hübner	268
17	Loew	261
18	Graham	259
19	Förster	242
20	Zeller	233

Note: Our current estimates suggest that the total number of Swedish insects left to be described is in the thousands, still making it feasible for ambitious young taxonomists to make it to the top twenty on this list.

Another notable but perhaps obvious feature of the new species is that they tend to be small and inconspicuous. Otherwise, they would hardly have escaped the attention of amateur naturalists and professional entomologists for so long. A fairly large proportion of the new species are also from northern Sweden (Figure 18.6). This is hardly surprising given that previous collecting efforts have focused on the heavily populated southern third of the country, which shares a large fraction of its fauna with intensely inventoried neighbors such as Denmark and Germany.

So how many species of insects are there actually in Sweden? Using a combination of total diversity estimates based on the samples from the Swedish Malaise Trap Project and expert guesses, we are currently trying to answer this question. Our preliminary numbers point toward a total fauna of around 31,000 species. If this is correct, then there are still somewhere between 4,000 and 5,000 insect species left to discover. So far, about one third of the new species we have found have been undescribed, but it seems likely that this proportion will go up, if anything, as the inventory proceeds to the most obscure groups. Although it is hard to believe, this actually means there is still room for young ambitious taxonomists to make it to the top-twenty or even top-ten list of the most prolific authors of Swedish insect species (Table 18.1). Indeed, still in 2009, there may be more undescribed species of insects in Sweden than the total number known to Linné.

ORGANIZING BIODIVERSITY INVENTORIES

Perhaps the most important lesson to draw from the STI results is that our current knowledge of floras and faunas, even in such comparatively well-investigated regions as Europe, is still seriously biased. For instance, we still know very little about species performing critical ecosystem services, such as saprophagous and parasitic insects. There are likely to be other biases as well that are less obvious and therefore impossible to correct for. This compromises our ability to monitor environmental changes effectively, underpinning the need for completing the biodiversity inventory as quickly as possible.

From a scientific perspective, completing the inventory of life on Earth is probably best accomplished by having specialists focus on the global fauna or flora of one organism group at a time. This is the approach taken by the National Science Foundation in its recent Planetary Biodiversity Inventories program. The STI represents the alternative method of focusing on a geographic area instead, and completing the inventory across all organism groups. This is known as an ATBI (All Taxa Biodiversity Inventory). Although not ideal from a scientific standpoint, national ATBIs may be easier to organize and fund than taxon-based projects. In fact, the parties to the Convention of Biological Diversity are all committed to inventorying and monitoring their biological diversity. One possible interpretation, justified by the STI and similar efforts, is that this involves a commitment to fund national ATBIs.

Floras and faunas are almost always shared to a large extent across national borders, which raises difficult questions concerning the fair share of costs for national ATBIs. For instance, if the STI were successful, it would complete most of the ATBIs of a number of neighboring Northern European countries as well. This means that an isolated national ATBI places an unreasonable burden on a single country, involving a cost that will almost certainly be prohibitive in biodiversity-rich regions. At the same time, it also means that regional collaboration, if organized efficiently, could potentially reduce the cost of national ATBIs very significantly.

In fact, even though the Swedish flora and fauna are depauperate, it is doubtful whether the STI itself would succeed in completing the inventory without any participation of neighboring countries. Therefore, it was welcome news when the Norwegian government announced in 2008 that it would launch a taxonomic initiative of its own. The Finnish government has also actively supported inventorying efforts in recent years through the PUTTE project (Research programme of deficiently known and threatened species), thereby contributing to the Nordic collaboration needed for the STI to be successful.

IMPLICATIONS FOR THE FUTURE

In addition to savings through regional collaboration, the cost of national ATBIs could also be reduced significantly in the future thanks to the development of new informatics tools. As mentioned earlier, the production of the biodiversity encyclopedia accounts for almost half the cost of the STI. The reason is that the production is so labor-intensive: information needs to be manually extracted from the literature and summarized, distribution maps need to be assembled, illustrations need to be completed, etc.

In the future, taxonomists are likely to have access to convenient tools for producing ontology-tagged, machine-readable phylogenetic analyses and species descriptions (for recent manually produced examples, see Buffington and van Noort, 2007; Liljeblad et al., 2008; Pyle et al., 2008). They may also be mandated by the nomenclatural codes to make specimen records, scientific names, and illustrations available through open Web databases or Web portals, such as GBIF (http://www.gbif.org), ZooBank (http://www.zoobank.org) and Morphbank (http://www.morphbank.net), when describing new taxa.

Machine harvesting of this information would in principle allow automatic production of Web pages matching the content now presented by STI in the *Encyclopedia of the Swedish Flora and*

Fauna, but dynamic so that the information would always be up to date. The technology to do this is now being developed by the Encyclopedia of Life project (http://www.eol.org), which may thus slash the cost of national ATBIs in half. Of course, there is still an enormous amount of legacy data to deal with before an automated system can compete successfully with traditionally produced books, but a growing number of digitization initiatives are now busy addressing this challenge.

Molecular genetics techniques may also contribute to making national ATBIs more feasible and powerful, if not cheaper. A number of STI-funded research projects now routinely use DNA barcoding and mining of existing sequence databases to help identify species more rapidly. With the material collected in the marine and terrestrial inventories funded by the STI, there is also a unique opportunity to barcode the bulk of Swedish multicellular species in one fell swoop in the coming years. This would enable a number of interesting future developments, among them the possibility of continuous monitoring of biological diversity with automated traps using DNA-based identification of collected specimens.

DNA barcoding cannot completely replace traditional taxonomy, however. One reason is that amateur naturalists derive so much of their inspiration from the reward of being able to identify species with no tools other than simple optical aids. The general public also tends to be less excited by DNA sequences than by the morphological adaptations and life histories of the organisms themselves. Ultimately, the support for biodiversity conservation depends on this kind of popular involvement, especially as long as the biological environment is still rich enough to make the immediate consequences of biodiversity loss less obvious.

The development over the next few decades will be critical to the well-being of generations to come, maybe even to the survival of mankind. Biodiversity inventorying is likely to be one of many crucial efforts helping us address these challenges. One would imagine that Linné would be happy to know that taxonomy is again considered a societal priority and proud that Sweden is playing an important role in completing the task that he originally set before himself.

ACKNOWLEDGMENTS

A large number of people contributed to the data presented here. In particular, I would like to thank Carl-Cedric Coulianos Mattias Forshage, Kajsa Glemhorn, Bert Gustafsson, Ulf Gärdenfors, Rasmus Hovmöller, Mathias Jaschhof, Dave Karlsson, Jostein Kjaerandsen, Sibylle Noack, Ingemar Struwe, Rikard Sundin, and Sven-Olof Ulefors. In addition, I am grateful to the more than 100 people who assisted with various aspects of the Swedish Malaise Trap Project, most of them volunteers. Financial support was provided by the Swedish Research Council (Vetenskapsrådet), the Swedish Research Council Formas, and the Swedish Taxonomy Initiative (Svenska Artprojektet).

REFERENCES

Achenbach, J. 2008. The God Particle. *National Geographic* March 2008. http://ngm.nationalgeographic. com/2008/03/god-particle/achenbach-text (accessed Feb. 28, 2009).

Buffington, M. L. and S. van Noort. 2007. A world revision of the Pycnostigminae (Cynipoidea: Figitidae) with descriptions of seven new species. *Zootaxa* 1392: 1–30.

Gärdenfors, U., Hall, R., Hallingbäck, T., Hansson, H. G., and L. Hedström. 2003. Djur, svampar och växter i Sverige 2003. *Förteckning över antal arter per familj*. Uppsala: Artdatabanken.

Karlsson, D., Pape, T., Johanson, K. A., Liljeblad, J., and F. Ronquist. 2005. Svenska Malaisefälleprojektet, eller hur många arter steklar, flugor och myggor finns i Sverige? *Entomologisk Tidskrift* 126:43–53.

Kjærandsen, J., Hedmark, K., Kurina, O., Polevoi, A., Økland, B., and F. Götmark. 2007. Annotated checklist of fungus gnats from Sweden (Diptera: Bolitophilidae, Diadocidiidae, Ditomyiidae, Keroplatidae and Mycetophilidae). *Insect Systematics & Evolution Supplement* 65:1–128.

Liljeblad, J., Ronquist, F., Nieves-Aldrey, J. L., Fontal-Cazalla, F., Ros-Farré, P., Gaitros, D., and J. Pujade-Villar. 2008. A fully web-illustrated morphological phylogenetic study of relationships among oak gall wasps and their closest relatives. *Zootaxa* 1796:1–73.

Linné, C. von. 1749. *Oeconomia Naturae*. [Specimen academicum de Oeconomia Naturae, quod, ...] Uppsala.

Linné, C. von. 1761. *Fauna Suecica*. [Sistens Animalia Sueciae regni ...] 2nd ed. Stockholm: Laurentius Salvius.

Pyle, R., Earle, J. L., and B. D. Greene. 2008. Five new species of the damselfish genus *Chromis* (Perciformes: Labroidei: Pomacentridae) from deep coral reefs in the tropical western Pacific. *Zootaxa* 1671:3–31.

Tullgren, A. and E. Wahlgren. 1922. *Svenska insekter: En orienterande handbok vid studiet av vårt lands insektfauna*. Stockholm: Norstedt.

Appendix 1: Concordance of Linnaeus's Names for Rolander's Insects

This appendix lists Linnaeus's original names (*Systema Naturae*, 1758) for the insects collected by Daniel Rolander in Surinam and St. Eustatius (1755–1756). These names are cross-referenced with the original Rolandrian specimens, which Linnaeus used, preserved in the Charles De Geer collection in the Swedish Museum of Natural History (Naturhistoriska Riksmuseet). Moreover, the Linnaeus names have been cross-referenced with De Geer's own names (*Mémoires pour servir à l'histoire des Insectes*, vols 1–8, 1752–1778) as well as the page numbers of his descriptions and illustrations of these same insects. Finally, reference numbers are given for these insects in Retzius (*Caroli De Geer Genera et Species Insectorum*, 1783), which provides a full index, including descriptions, of De Geer's insects in Linnaeus's 1764 edition of *Systema Naturae*. See Appendix 2 for some information on the original drawings that served as models for De Geer's copperplates.

The entries read as follows:

Linnaeus's insect name in *Systema Naturae* (1758) = De Geer's name
Specific reference to location in *Sys. Nat.*
Box number at Swedish Museum of Natural History (NRM), De Geer Collection
Reference number in Retzius
Specific reference to location in De Geer's *Mémoires ... des Insectes*.

Acarus batatas L.
Sys. Nat. (1758), p. 617, #235
NRM box _____
Retzius _____
De Geer _____

Acarus sanguisugus L. = *Acarus ricinus* L. = *Acarus ricinoides* De Geer
Sys. Nat. (1758), p. 615, #234
NRM box
Retzius 1335
De Geer _____

Acarus scorpioides L. = *Phalangium acaroides* L. (1764) = *Chelifer americanus* De Geer
Sys. Nat. (1758), p. 616, #234
NRM box _____
Retzius 1413
De Geer T. 7, p. 353, t. 42, f. 1–5

Apis surinamnsis L. = *Apis abdomen flavum* De Geer
Sys. Nat. (1758), p. 575, #218
NRM box 5
Retzius 216

De Geer T. 3, p. 574, t. 28, f. 9–10
(De Geer has his own *Apis surinamnsis*, T. 3, p. 569, t. 28, f.1–2; Retzius 211)

Apis ichneumonea L. = *Nomada surinamnsis*
Sys. Nat. (1758), p. 578, #218
NRM box 5
Retzius 228
De Geer T. 2, p. 759, t. 32, f. 13–16

Apis surinamnsis L. (the same as on page 575?)
Sys. Nat. (1758), p. 579, #218
NRM box _____
Retzius _____
De Geer _____

Attelabus betulae L.
Sys. Nat. (1758), p. 387, #178
NRM box _____
Retzius _____
De Geer _____

Attelabus surinamnsis L.
Sys. Nat. (1758), p. 387, #178
NRM box _____
Retzius _____
De Geer T. 4, p. 80, t. 17, f. 16

Blatta oblongata L.
Sys. Nat. (1758), p. 425, #193
NRM box 10
Retzius 544
De Geer T. 3, p. 541, t. 44, f. 11–12

Buprestis linearis
Sys. Nat. (1758), p. 410, #184
NRM box 11
Retzius 644
De Geer T. 4, p. 137, t. 77, f. 26

Cantharis ignita L. = *Lampyris ignita* De Geer
Sys. Nat. (1758), p. 400, #181
NRM box 10
Retzius 567
De Geer T. 4, p. 49, t. 17, f. 2

Cantharis lampyris L. =
Sys. Nat. (1758), p. 400, #181
NRM box 10
Retzius _____
De Geer _____

Cantharis pectinata L. = *Lampyris pectinata* De Geer
Sys. Nat. (1758), p. 403, #181
NRM box 10
De Geer T. 4, p. 57, t. 17, f. 13–14

Cantharis phosphorea L. = *Lampyris phosphorea* De Geer
Sys. Nat. (1758), p. 400, #181
NRM box 10
Retzius 571
De Geer T. 4, p. 51, t. 17, f. 7

Cantharis serrata
Sys. Nat. (1758), p. 403, #181
NRM box 10
Retzius 575
De Geer T. 4, p. 55, t. 17, f. 12

Carabus americanus
Sys. Nat. (1758), p. 415, #186
NRM box 10
Retzius 618
De Geer T. 4, p. 107, t. 17, f. 21

Cassida bifasciata L. = *Cassida nigromaculata* De Geer
Sys. Nat. (1758), p. 363, #174
NRM box 15 (insect intact)
Retzius _____
De Geer _____

Cassida cruciata L.
Sys. Nat. (1758), p. 363, #174
NRM box _____
Retzius 950
De Geer T. 5, p. 187, t. 15, f. 15

Cassida discoides L. = *C. quadrimaculata*
Sys. Nat. (1758), p. 364, #174
NRM box 15 (label present; specimen missing)
Retzius
De Geer T. 5, p. 183, t. 15, f. 13

Cassida flava L.
Sys. Nat. (1758), p. 363, #174
NRM box 15 (insect intact)
Retzius _____
De Geer T. 5, p. 184, t. 15, f. 13

Cassida inaequalis L. = *C. bimaculata* De Geer
Sys. Nat. (1758), p. 364, #174
NRM box
Retzius
De Geer T. 5, p. 182, t. 15, f. 10

Cassida lateralis L.
Sys. Nat. (1758), p. 364, #174
NRM box 15 (label present; specimen missing)
Retzius
De Geer T. 5, p. 184, t. 15, f. 12

Cassida marginata L.
Sys. Nat. (1758), p. 363, #174,
NRM box 15 (insect intact)
Retzius _____
De Geer T. 5, p. 185

Cassida purpurea L.
Sys. Nat. (1758), p. 363, #174,
NRM box 15 (insect intact)
Retzius _____
De Geer T. 5, p. 190, t. 15, f. 19

Cassida reticularis L.
Sys. Nat. (1758), p. 363, #174
NRM box 15 (label present; specimen missing)
Retzius _____
De Geer T. 5, p. 188, t. 15, f. 17

Cassida variegata L.
Sys. Nat. (1758), p. 363, #174
NRM box 15
Retzius
De Geer T. 5, p. 178, t. 15, f. 6

Cerambyx auratus L.
Sys. Nat. (1758), p. 395, #179
NRM box 14 (label present; specimen missing)
Retzius
De Geer T. 5, p. 101, t. 13, f. 15

Cerambyx festivus L. = *C. spinosus* De Geer
Sys. Nat. (1758), p. 389, #179
NRM box 14
Retzius
De Geer T. 5, p. 100, t. 13, f. 14
Rolander originally named this *Leptura domestica*, which he deleted in order to add *Cerambux domesticus*, seu festivus (Sept. 11, 1755, ms p. 299)

Cerambyx glaucus L. = *C. tuberculatus*
Sys. Nat. (1758), p. 390, #179
NRM box 14 (label present; specimen missing)
Retzius
De Geer T. 5, p. 112, t. 14, f. 4

Cerambyx stigma L.
Sys. Nat. (1758), p. 395, #179
NRM box 14
Retzius _____
De Geer T. 5, p. 119, t. 14, f. 13

Chrysomela aequinoctialis L.
Sys. Nat. (1758), p. 374, #176,
NRM box _____
Retzius 1102
De Geer T. 5, p. 356, t. 16, f. 19

Chrysomela aestuans L. = *Ch. octoguttae* De Geer
Sys. Nat. (1758), p. 371, #176
NRM box _____
Retzius 1064
De Geer T. 5, p. 352, t. 16, f. 12

Chrysomela clavicornis L.
Sys. Nat. (1758), p. 370, #176
NRM box _____
Retzius 1063
De Geer T. 5, p. 351, t. 16, f. 11

Chrysomela occidentalis L.
Sys. Nat. (1758), p. 369, #176
NRM box _____
Retzius 1066
De Geer T. 5, p. 353, t. 16, f. 14

Chrysomela S. litera L.
Sys. Nat. (1758), p. 373, #176
NRM box _____
Retzius 1104
De Geer T. 5, p. 357, t. 16, f. 21

Chrysomela Surinamnsis L.
Sys. Nat. (1758), p. 373, #176
NRM box _____
Retzius 1101
De Geer T. 5, p. 355, t. 16, f. 17–18

Cicada 4-fasciata L. = *C. quadrifasciata* De Geer
Sys. Nat. (1758), p. 436, #195
NRM box 7
Retzius 395
De Geer T. 3, p. 225, t. 33, f. 11

Cicada fronditia L. = *C. foliata-sinuosa* De Geer
Sys. Nat. (1758), p. 435, #195
NRM box _____
Retzius 375
De Geer T. 3, p. 208, t. 32, f. 15–16

Cicada lucernaria L. = *Fulgora lucernaria* L. (1764) = *Cicada brevirostris* De Geer
Sys. Nat. (1758), p. 434, #195
NRM box _____
Retzius 370
De Geer T. 3, p. 203, t. 32, f. 6

Cicada noctivida L. = *Fulgora phosphorea* L. (1764) = *Cicada conirostris* De Geer
Sys. Nat. (1758), p. 434, #195
NRM box _____
Retzius 369
De Geer T. 3, p. 202, t. 32, f. 4–5

Cicada phosphorea L. = *Fulgora phosphorea* L. (1764) = *Cicada filirostris* De Geer
Sys. Nat. (1758), p. 434, #195
NRM box _____
Retzius 368
De Geer T. 3, p. 201, t. 3, f. 2–3

Cicada reticulata L.
Sys. Nat. (1758), p. 436, #195
NRM box 7
Retzius 398
De Geer T. 3, p. 227, t. 33, f. 15–16

Cicada rubra L.
Sys. Nat. (1758), p. 438, #195
NRM box 7
Retzius 394
De Geer T. 3, p. 224, t. 33, f. 8–10

Cicada squamigera L. = C. hastata
Sys. Nat. (1758), p. 435, #195
NRM box _____
Retzius 376
De Geer T. 3, p. 209, t. 32, f. 17–18

Cimex erosus L. = *C. scorpio* De Geer
Sys. Nat. (1758), p. 443, #198
NRM box _____
Retzius 446
De Geer T. 3, p. 350, t. 35, f. 13–15

Cimex lineola L.
Sys. Nat. (1758), p. 445, #198
NRM box _____
Retzius _____
De Geer _____

Cimex nigripes L. = *C. hirtipes* De Geer
Sys. Nat. (1758), p. 449, #198
NRM box ____ _
Retzius 430
De Geer T. 3, p. 344, t. 35, f. 1–3

Cimex pustulatus L.
Sys. Nat. (1758), p. 443, #198
NRM box 7
Retzius (not listed under *Cimex*)
De Geer T. 3, p. 270, t. 34, f. 2 (not confirmed)

Cimex Rolandri[1] L. (Not a Surinam insect) = *C. fulvo-maculatus*
Sys. Nat. (1758), *p. 448, #198
NRM box _____
Retzius 440
De Geer T. 3, p. 294

Cimex spinosus
Sys. Nat. (1758), p. 444, #198
NRM box _____
Retzius _____
De Geer _____

Cimex variolosus L.
Sys. Nat. (1758), p. 445, #198
NRM box _____
Retzius _____
De Geer _____

Cimex ypsilon L.
Sys. Nat. (1758), p. 443–4, #198
NRM box 7
Retzius (not listed under Cimex)
De Geer T. 3, p. 332, f. 34, f. 7–8 (not confirmed)

Coccus cacti coccinelliferi[2] L. (probably collected on St. Eustatius)
Sys. Nat. (1758), p. 457, #201
NRM box _____
Retzius 1309
De Geer T. 6, p. 447, t. 30, f. 12–18

Dermestes eustatius L. = Ips Eustatius De Geer
Sys. Nat. (175), p. 357, #171
NRM box _____
Retzius ____
De Geer T. 5, p. 197, t. 15, f. 20–21

*Dermestes hemipteru*s L.
Sys. Nat. (1758), p. 358, #171
NRM box 11 (insect intact)
Retzius 705
De Geer T. 4, p. 224

Dermestes surinamnsis L. = *Tenebrio surinamnsis* De Geer
Sys. Nat. (1758), p. 357, #171
NRM box 13 (insect intact)
Retzius _____
De Geer T. 5, p. 54, t. 13, f. 12

Elater phosphoreus
Sys. Nat. (1758), p. 404, #182
NRM box 11
Retzius 672
De Geer T. 4, p. 161, t. 18, f. 2

Formica atrata L. = *F. quadriden*s De Geer
Sys. Nat. (1758), p. 581, #218
NRM box _____
Retzius 338
De Geer T. 3, p. 609, t. 31, f. 17–20

Formica bidens L.
Sys. Nat. (1758), p. 581, #218
NRM box 6
Retzius 329
De Geer T. 3, p. 600, t. 31, f. 1–2

Formica foetida L. = *F. lobata* De Geer
Sys. Nat. (1758), p. 582, #218
NRM box 6
Retzius 331
De Geer T. 3, p. 602, t. 31, f. 6–8

Formica haematoda = *F. maxillosa* De Geer
Sys. Nat. (1758), p. 582, #218
NRM box 6
Retzius 330
De Geer T. 3, p. 601, t. 31, f. 3–5

Gryllus myrtifolius L. = *Locusta myrtifolius* De Geer
Sys. Nat. (1758), p. 429, #194
NRM box _____
Retzius 479
De Geer T. 3, p. 447, f. 38, f. 4

Hemerobius marginalis L. = *Perla nasuta* De Geer
Sys. Nat. (1758), p. 550, #210
NRM box _____
Retzius 206
De Geer T. 3, p. 568, t. 27, f. 6–7

Hemerobius testaceus L. = *Perla fusca* De Geer
Sys. Nat. (1758), p. 550, #210
NRM box _____
Retzius 205
De Geer T. 3, p. 567, t. 27, f. 4–5

Ichneumon manifestator L.
Sys. Nat. (1758), p. 563, #215
NRM box _____
Retzius _____
De Geer _____

Leptura necydalea L. = *Necydalia nitida* De Geer
Sys. Nat. (1758), p. 399, #180
NRM box _____
Retzius _____
De Geer _____

Libellula dimidiata L. = *L. marginata* De Geer
Sys. Nat. (1758), p. 545, #207
NRM box _____
Retzius 192
De Geer T. 3, p. 558, t. 26, f. 6

Libellula umbrata L. = *L. unifasciata* De Geer
Sys. Nat. (1758), p. 545, #207
NRM box _____
Retzius 190
De Geer T. 3, p. 557, t. 26, f. 4

Musca leprae L.
Sys. Nat. (1758), p. 598, #222
NRM box _____
Retzius _____
De Geer _____

Musca Radicum L.
Sys. Nat. (1758), p. 596, #222
NRM box _____
Retzius _____
De Geer _____

Pediculus ricinoides L.
Sys. Nat. (1758), p. 610, #233
NRM box _____
Retzius _____
De Geer _____

Phalaena noctua secalis[3] L. (not from Surinam)
Sys. Nat. (1758), p. 519, #205
NRM box _____
Retzius _____
De Geer _____

Phalaena phyralis pinguinalis[4] L. (not from Surinam)
Sys. Nat. (1758), p. 533, #206,
NRM box _____
Retzius _____
De Geer _____

Pulex penetrans L.
Sys. Nat. (1758), p. 615, #234
NRM box _____
Retzius _____
De Geer _____

Scarabaeus festivus
Sys. Nat. (1758), p. 350, #170
NRM box 12 (label present; specimen missing)
Retzius 741
De Geer T. 4, p. 315, t. 18, f. 15

Scarabaeus lineola
Sys. Nat. (1758), p. 350, #170
NRM box 12 (insect intact)
Retzius 752
De Geer T. 4, p. 320, t. 19, f. 5

Silpha seminulum L.
Sys. Nat. (1758), p. 360, #173
NRM box (appears not to be in box 11)
Retzius _____
De Geer _____

Sphex argillacea L.
Sys. Nat. (1758), p. 569, #216
NRM box _____
Retzius _____
De Geer T. 3, p. 569, t. 28, f. 1–2 (not confirmed)

Staphylinus boleti L.
Sys. Nat. (1758), p. 423, #191
NRM box _____
Retzius 560 _____
De Geer T. 4, p. 26, t. 1, f. 15–17

Tabanus antarticus L.
Sys. Nat. (1758), p. 602, #223
NRM box _____
Retzius _____
De Geer _____

Tabanus exaestuans L.
Sys. Nat. (1758), p. 601, #223
NRM box _____
Retzius 1239
De Geer T. 6, p. 229, t. 30, f. 5

Tabanus mexicanus L. = *T. olivaceus* De Geer
Sys. Nat. (1758), p. 602, #223
NRM box _____
Retzius 1240
De Geer T. 6, p. 230, t. 30, f. 6

Tenebrio fossor L. = *Attelabus fossor* De Geer
Sys. Nat. (1758), p. 417, #187
NRM box _____
Retzius 787
De Geer T. 4, p. 350, t. 13, f. 1–3

Tenthredo americana L.
Sys. Nat. (1758), p. 555, #214
NRM box _____
Retzius 301
De Geer T. 3, p. 598, t. 30, f. 21

Termes fatale L. = *T. destructor* De Geer
Sys. Nat. (1758), p. 609, #232
NRM box _____
Retzius 1322
De Geer T. 7, p. 50, t. 37, f. 1–8

ENDNOTES

1. Named after Rolander by Linnaeus. Here described, "Rolandri, 66. C. oblongus ater, elytris membrana-ceis macula flava. *Habitat in* Europa."
2. "*Habitat in* Cactis Opuntiis *variis* Americes. *Vivus transmissus a D. Rolandro in Hort. Upsaliensem* 1756."
3. Not from Surinam, "*Roland. Act. Stockh.* 1752. *p. 62.*" The *Act. Stockh.* are *Kungl. Svenska vetenskapsakademiens handlingar.*
4. "*Roland. Act. Stockh.* 1755. *p. 51. t. 2.*"

Appendix 2: A Partial List of Rolander's Insects in De Geer's Original Drawings

The room for rare books and manuscripts at Uppsala University Library provides access to two codices containing the original drawings made by Charles De Geer that served as models for the copperplates in his eight-volume *Mémoires … des Insectes*, 1752–1778. The final section of the larger volume contains the drawings of Rolander's Surinam insects. Unfortunately, only a few of the indices remain for that section. This appendix contains references to those images that mentioned Surinam. It is possible that some of these do not depict Rolander's specimens. An entomologist will have to compare the drawings to the illustrations in De Geer's published work to compose a definitive list. It is hoped that this appendix will aid in that work. The following is a verbatim transcription of De Geer's manuscript; the French has not been corrected.

Explication des Figures des Insectes exotiques, desineés par moi-même

---

TABLE 1.

Figure 6. Le Scarabé (élegant) … Il est de Surinam.

F. 7. Le Cerf-volant (interrompu) … Il est trowe à Surinam.

F. 8. Le Scarabé (d'ébene) .. Il se trowe à Surinam.

F. 9. Le Scarabé (Chrysis) … Il est encore de Surinam.

F. 12. Le Scarabé (à ligne jaune) … On le trowe à Surinam.

F. 13 Le Scarabé (à tête noire) … Il est encore de Surinam.

F. 16. Le Scarabé (nain) … Il se trowe à Surinam.

TABLE 2.

Figure 1. Le Dermeste (noir) … Il est de Surinam.

F. 2. Le Clairon (à pattes rousses) … Il est aussi de Surinam.

F. 3 Le Dermeste (à étuis courts) … Il est dessiné en grand et se trowe encore à Surinam.

F. 4 & 5. représentent en grand e' Ips (d'Eustassie) THIS IS *Dermestes eustatius* L. (p. 357, 171), corresponding to De Geer's Ips Eustatius (T. 5, p. 197, t. 15, f. 20 - 21)

F. 8. La Coccinelle (mouchetée) … Elle se trowe à Surinam.

F. 9. La Chrysomele (gigantesque) … Elle est de Surinam.

F. 10. La Chrysomele (ondée) … Elle est de Surinam.

F. 11. La Chrysomele (à points rouges) … Elle est de Surinam.

F. 12. La Chrysomele (à antennes à bouton) … Elle se trowe aussi à Surinam.

F. 13. La Chrysomele (à huit taches jaunes) … Elle est encore de Surinam.

F. 15. La Chrysomele (sauteuse de Surinam)

F. 16. Unde des pattes postérieures de cette Chrysomele, grossie. C, la grosse cuisse.

F. 17. La Chrysomele (sanguine) ... Elle est de Surinam.

F. 18 La Chrysomele (équinoctiale) ... elle est encore de Surinam.

F. 19. La Chrysomele (occidentale) ... Elle se trowe à Surinam.

F. 20. La Chrysomele (sombre) ... Elle est de Surinam.

F. 21. La Chrysomele (à deux couleurs) ... Elle se trowe à Surinam.

F. 22. La Chrysomele (à la lettre S) ... Elle est de Surinam.

F. 27. Le Charanson (du Palmier) ... Il se trowe à Surinam.

F. 29. Le Charanson (à étui) ... Il se trowe à Surinam.

F. 30. Le Charanson (poudré) ... Il est de Surinam.

F. 31. Le Charanson (rouse rayé) ... Il se trowe à Surinam.

F. 32. Le Charanson (à tête et trompe noires) ... Il est aussi de Surinam.

F. 33. Le Charanson (à long col) ... Il est encore de Surinam.

F. 34. La tête en trompe de ce Charanson, grossie. c b, la longue tête. b + b, la trompe. y y, les yeux. a a, les antennes.

F. 35. La Charanson (à broses) ... Il se trowe à Surinam.

F. 36. La tête de la Broche Fig. 24 et 25, grossie et ....

F. 37. La Bruche (rousse) (Not clear if this is from Surinam.)

Index

A

Actinobacillus actimycetemcomitans, 154
aggregation technology, 130
AGORA program, 87
Ailuropoda melanoleuca, 41
All Genera Index, 160. *See also* BIG index
All Taxa Biodiversity Inventory (ATBI), 249
alpha-taxonomy, 137
American Journal of Tropical Hygiene and Medicine, 159
American Ornithologists' Union, code of, 212
Amphibian Ark, 45
ancestor–descendant relationships, 118
AnimalBase (University of Göttingen), 91
AOU Code, 212
apostles, 2, 8
Approved Lists of Bacterial Names, 71, 72
Aristotelean essence, 34
ark of knowledge, engineering of, 53–61
    biodiversity crisis, 54, 55
    capacity, 53
    change in working practice, 56
    costly mistake, 55
    cybertaxonomy, 54, 57
    databases, 58
    deluge of species, 55–58
    descent with modification, 53
    engineering, 58–59
    funding, 56
    international collaboration, 59
    inventory of known species, 54
    macro-evolution, 56
    mission of taxonomy, 55
    moleculo-phenogram, 54
    National Aeronautics and Space Administration, 54
    species exploration, 59
    species treatments, numbers of, 56
    Systema Naturae 250 symposium, 60
    taxonomic moon shot, 56
    technological advances, 56
    trans-disciplinary literacy, 58
    unique identifiers, 53
ASC information model, 140
ATBI. *See* All Taxa Biodiversity Inventory
auto-immune disease syndrome, 41

B

barbarous names, 196
Barcode of Life Database (BOLD), 71
BDWD. *See* BioSystematic Database of World Diptera
BHL. *See* Biodiversity Heritage Library
Big Deals, 87
BIG index, 149–162
    accumulated information of a species, 149
    biodiversity informatics symposium, 149
    book of living things, 160

Catalogue of Fishes, 150
Catalogue of Life, 152, 156
    comprehensive index of genera, 157–160
    disparity among codes, 153
    endnotes, 160–161
    Global Biodiversity Information Facility, 150
    Globally Unique Identifiers, 151
    Global Species Datasets, 153
    information discovery, 157
    International Plant Names Index, 151
    library of biology, 150
    National Center for Biotechnology Information, 150
    occurrence records, 150
    Oenanthe, 158
    Pubmed Citation Index, 155
    qualities, 151–157
        GBIF, 156–157
        scientific names as labels for taxa, 152
        scientific names as strings of characters, 154–155
        scientific names as units of nomenclature, 152–154
    taxon, variation in formatting conventions, 154
    uBio project, 157
    validation of nomenclatural origins, 157
    Zoonomen, 156
biodiversity. *See also* biodiversity studies, major historical
        trends
    bioinformatics, founder of, 167
    crisis, 54, 55
    data, information and communication technology, 169
    indexing group. *See* BIG index
    informatics, 169
        father of, 167
        standards, 177
        symposium, 149
    information
        linking through names, 175
        tripletization of, 124
    inventorying, 250
Biodiversity Heritage Library (BHL), 119, 130, 131, 155, 181
Biodiversity Informatics Group (EOL), 128–130
Biodiversity Information Standards, 113, 120
biodiversity studies, major historical trends, 1–3
    apostles, 2
    diagnosis of entirety of biodiversity, 1
    Encyclopedia of Life, 3
    numbers of species, 2
    On the Origin of Species, 2
    Philosophie zoologique, 2
    Species Plantarum, 1
    Systema Naturae, 1
    systematic exploration of biosphere, 2
    systematics, evolution of, 1
Biodiversity Synthesis Group (EOL), 132–133
Biological Records Centre, 91
biophilia, 32
Biosystematic Database of World Diptera (BDWD), 79, 80–81